Solved Problems in Lagrangian and Hamiltonian Mechanics

Grenoble Sciences

Grenoble Sciences pursues a triple aim:
- ▶ to publish works responding to a clearly defined project, with no curriculum or vogue constraints,
- ▶ to guarantee the selected titles' scientific and pedagogical qualities,
- ▶ to propose books at an affordable price to the widest scope of readers.

Each project is selected with the help of anonymous referees, followed by an average one-year interaction between the authors and a Readership Committee whose members' names figure in the front pages of the book. Grenoble Sciences then signs a co-publishing agreement with the most adequate publisher.

Contact: Tel.: (33) (0) 4 76 51 46 95 - Fax: (33) (0) 4 76 51 45 79
E-mail: Grenoble.Sciences@ujf-grenoble.fr
Website: *http://grenoble-sciences.ujf-grenoble.fr*

Scientific Director of Grenoble Sciences: Jean BORNAREL,
Professor at the Joseph Fourier University, Grenoble, France

Grenoble Sciences is supported by the French Ministry of Higher Education and Research and the "Région Rhône-Alpes".

The *Solved Problems in Lagrangian and Hamiltonian Mechanics*
Reading Committee included the following members:

- ▶ **Robert ARVIEU**, Professor at the Joseph Fourier University, Grenoble, France
- ▶ **Jacques MEYER**, Professor at the Nuclear Physics Institute, Claude Bernard University, Lyon, France

with the contribution of:
- ▶ **Myriam REFFAY** and **Bertrand RUPH**

The translation of "Problèmes corrigés de Mécanique et résumés de cours. De Lagrange à Hamilton", published by Grenoble Sciences in partnership with EDP Sciences, was performed by:
- ▶ **Bernard SILVESTRE-BRAC** and **Anthony John COLE**, Senior researcher at the CNRS, Grenoble, France

Front cover illustration: composed by **Alice GIRAUD**

Claude Gignoux · Bernard Silvestre-Brac

Solved Problems in Lagrangian and Hamiltonian Mechanics

Dr. Claude Gignoux
Université Joseph Fourier
Lab. Physique Subatomique et
 Cosmologie
CNRS-IN2P3
53 avenue des Martyrs
38026 Grenoble CX
France
claude.gignoux@club-internet.fr

Dr. Bernard Silvestre-Brac
Université Joseph Fourier
Lab. Physique Subatomique et
 Cosmologie
CNRS-IN2P3
53 avenue des Martyrs
38026 Grenoble CX
France
silvestre@lpsc.in2p3.fr

ISBN 978-90-481-2392-6   e-ISBN 978-90-481-2393-3
DOI 10.1007/978-90-481-2393-3
Springer Dordrecht Heidelberg London New York

Library of Congress Control Number: 2009926952

© Springer Science+Business Media B.V. 2009
No part of this work may be reproduced, stored in a retrieval system, or transmitted in any form or by any means, electronic, mechanical, photocopying, microfilming, recording or otherwise, without written permission from the Publisher, with the exception of any material supplied specifically for the purpose of being entered and executed on a computer system, for exclusive use by the purchaser of the work.

Printed on acid-free paper

Springer is part of Springer Science+Business Media (www.springer.com)

Foreword

Mechanics is an old science, but it acquired its great reputation at the end of the 17th century, due to Newton's works. A century later, Euler and, above all, Lagrange renewed it and led it towards a formulation not only aesthetically elegant but also capable of applications to other fields of physics. Fifty years later, Hamilton and Jacobi gave their names to very important further contributions. Lastly, at the end of the XIXth century, Poincaré took a new step with the introduction of geometry in the analysis of physical problems. During the XXth century, physicists produced new developments from the works of their famous predecessors.

This book is addressed to readers already familiar with the Newtonian approach for mechanics. Several training textbooks – some of them excellent – rely on this approach. On the other hand, it seems that exercises based on Lagrangian and Hamiltonian formulations are rather scarce in the literature; we hope that the present work may help to fill this gap.

In a previous book published in French by EDP, Grenoble-Sciences collection, under the name *La mécanique: de la formulation lagrangienne au chaos hamiltonien*, we proposed, for undergraduate students, a comprehensible synthesis of all the modern facets of mechanics and their relationships to other domains of physics. This textbook contains a great number of exercises and problems, many of them original, dealing with the theories of Lagrange, Hamilton and Poincaré. We gave only the results or brief hints for solving these problems. Some problems can be considered as difficult, or even disconcerting, and readers encouraged us to provide the solution of those exercises which illustrate all the topics presented in the book. This is the aim of the present work. We retained from the foregoing book most of the problems presented here, very often trying to make them clearer, sometimes trying to find interesting extensions. We also proposed new ones better suited to our pedagogic goal. In the same spirit, others have been withdrawn because we judged them less instructive for physics, even if the mathematical points they dealt with were logical consequences of features treated in the textbook.

Of course, the present work is a natural complement of the course-book; nevertheless, we have tried to make it self-contained and, with this in mind, we add in each chapter a succinct, although clear and complete, summary of each topic. These summaries are adequate to tackle and solve the problems presented afterwards; every concept or notion necessary for obtaining the solution is presented and developed therein. Our aim is to make the reader familiar with the Lagrangian and Hamiltonian approaches, which may be difficult to grasp, to demonstrate the power of this formalism and help to develop skills for managing the techniques essential for this kind of study. The problems are selected with this purpose and they illustrate very often practical physical situations and sometimes aspects of everyday life.

This book is built around eight chapters entitled:

1. The Lagrangian formulation

2. Lagrangian systems

3. Hamilton's principle (also called the least action principle)

4. The Hamiltonian formalism

5. The Hamilton-Jacobi formalism

6. Integrable systems

7. Quasi-integrable systems

8. From order to chaos

In each chapter, the reader will find:

- A clear, succinct and rather deep *summary* of all the notions that must be understood, the important points that must be memorized and the notations and symbols used in the problems.

- The statements of the problems which are presented consecutively. These statements are sufficiently detailed so that, with the help of the lesson summaries, it is unnecessary for the reader to search for other sources of information. Whenever a figure turns out to be essential for a good understanding of the text, it appears in the statement. The progressive difficulty of the problems is symbolized with an increasing number of stars (from 1 to 3) added to the title.

- Detailed answers to the problems which are grouped together at the end of the chapter. A number of additional figures are inserted in the corresponding text in order to exhibit essential points or to avoid lengthy circumlocutions.

At the beginning of the book, a synoptic table gathers, chapter by chapter, the set of all the proposed problems, giving, for each of them, its reference (number, title, page number), its difficulty (1 to 3 stars), as well as the important features or the peculiar aspects treated in this problem.

Foreword

Two purposes are pursued

- To concretize, through simple and often academic examples, notions that seem apparently very abstract. The methods to find the solution are not necessarily the most elegant or the quickest, but it is important to check, via these simple examples, one's understanding of these new tools for mechanics. Sometimes the same problem, or the same physical situation, is studied once more in a subsequent chapter with new tools in order to emphasize some novel feature of the method.

- To emphasize the power of these new tools in physics applied to fields as miscellaneous as traditional mechanics, optics, electromagnetism, waves in general, and quantum mechanics. Concerning these fascinating and up to date subjects in physics, we will focus only on the mechanical aspect. However, the reader could satisfy his curiosity, with help of keywords, by looking for further information firstly in a good and complete encyclopedia, then on the web using a engine such as Google (or, for more exotic subjects, Yahoo). He will find exhaustive lectures as well as recent articles.

We strongly recommend that the reader carries out some applications and draws the figures proposed in the detailed solutions. For the drawing or plotting of curves, the authors have used the freewares *xfig* or *xmgrace* that can be downloaded from the web. The differential equations have been solved with the help of the *Mathematica* software package.

Acknowledgment

We are very grateful to the reading committee for all remarks and suggestions which were very helpful in achieving a greater degree of consistency and an improved pedagogical quality in this work. We also very much appreciated contributions from our students whose enthusiasm and various questions allowed us to make the text more understandable.

Finally, we wish to thank members of the editorial board, Nicole Sauval and Jean Bornarel, for their competent advice, Sylvie Bordage, Julie Ridard and Thierry Morturier for their role in improving our figures and Konstantin Protasov for his helpful work concerning the layout of the manuscript.

A work such as ours is never perfect neither in the form nor in the essence of the matter. If, despite the care brought by the authors in the construction of this work, you find mistakes, shortcomings or incoherences, we will be grateful to you for mentioning them via one of the following electronic addresses:

claude.gignoux@club-internet.fr
silvestre@lpsc.in2p3.fr

Contents

Foreword v

Contents xi

Synoptic Tables of the Problems 1

Chapter 1. The Lagrangian Formulation 9
Summary 9
   1.1. Generalized Coordinates 9
   1.2. Lagrange's Equations 10
   1.3. Generalized Forces 12
   1.4. Lagrange Multipliers 13
Problem Statements 14
   1.1. The Wheel Jack 14
   1.2. The Sling 15
   1.3. Rope Slipping on a Table 16
   1.4. Reaction Force for a Bead on a Hoop 16
   1.5. Huygens Pendulum 17
   1.6. Cylinder Rolling on a Moving Tray 18
   1.7. Motion of a Badly Balanced Cylinder 18
   1.8. Free Axle on a Inclined Plane 19
   1.9. The Turn Indicator 21
   1.10. An Experiment to Measure the Rotational Velocity
       of the Earth 22
   1.11. Generalized Inertial Forces 23
Problem Solutions 24
   1.1. The Wheel Jack 24
   1.2. The Sling 26
   1.3. Rope Slipping on a Table 27
   1.4. Reaction Force for a Bead on a Hoop 28
   1.5. The Huygens Pendulum 31
   1.6. Cylinder Rolling on a Moving Tray 33
   1.7. Motion of a Badly Balanced Cylinder 35
   1.8. Free Axle on a Inclined Plane 39
   1.9. The Turn Indicator 43
   1.10. An Experiment to Measure the Rotational Velocity
       of the Earth 46
   1.11. Generalized Inertial Forces 48

Chapter 2. Lagrangian Systems 51
Summary 51
   2.1. Generalized Potential 51
   2.2. Lagrangian System 52
   2.3. Constants of the Motion 53
   2.4. Two-body System with Central Force 55
   2.5. Small Oscillations 56
Problem Statements 57
   2.1. Disc on a Movable Inclined Plane 57
   2.2. Painlevé's Integral 58
   2.3. Application of Noether's Theorem 58
   2.4. Foucault's Pendulum 59
   2.5. Three-particle System 61
   2.6. Vibration of a Linear Triatomic Molecule:
        The "Soft" Mode 63
   2.7. Elastic Transversal Waves in a Solid 64
   2.8. Lagrangian in a Rotating Frame 65
   2.9. Particle Drift in a Constant Electromagnetic Field 66
   2.10. The Penning Trap 67
   2.11. Equinox Precession 68
   2.12. Flexion Vibration of a Blade 71
   2.13. Solitary Waves 73
   2.14. Vibrational Modes of an Atomic Chain 75
Problem Solutions 76
   2.1. Disc on a Movable Inclined Plane 76
   2.2. Painlevé's Integral 77
   2.3. Application of Noether's Theorem 78
   2.4. Foucault's Pendulum 79
   2.5. Three-particle System 82
   2.6. Vibration of a Linear Triatomic Molecule:
        The "Soft" Mode 86
   2.7. Elastic Transversal Waves in a Solid 88
   2.8. Lagrangian in a Rotating Frame 89
   2.9. Particle Drift in a Constant Electromagnetic Field 91
   2.10. The Penning Trap 94
   2.11. Equinox Precession 97
   2.12. Flexion Vibration of a Blade 102
   2.13. Solitary Waves 105
   2.14. Vibrational Modes of an Atomic Chain 107

Chapter 3. Hamilton's Principle 111
Summary 111
   3.1. Statement of the Principle 111
   3.2. Action Functional 112

3.3.	Action and Field Theory	112
3.4.	Some Well Known Actions	113
3.5.	The Calculus of Variations	114

Problem Statements 116

3.1.	The Lorentz Force	116
3.2.	Relativistic Particle in a Central Force Field	117
3.3.	Principle of Least Action?	118
3.4.	Minimum or Maximum Action?	119
3.5.	Is There Only One Solution Which Makes the Action Stationary?	120
3.6.	The Principle of Maupertuis	121
3.7.	Fermat's Principle	122
3.8.	The Skier Strategy	122
3.9.	Free Motion on an Ellipsoid	123
3.10.	Minimum Area for a Fixed Volume	124
3.11.	The Form of Soap Films	125
3.12.	Laplace's Law for Surface Tension	127
3.13.	Chain of Pendulums	128
3.14.	Wave Equation for a Flexible Blade	128
3.15.	Precession of Mercury's Orbit	128

Problem Solutions 131

3.1.	The Lorentz Force	131
3.2.	Relativistic Particle in a Central Force Field	132
3.3.	Principle of Least Action?	135
3.4.	Minimum or Maximum Action?	137
3.5.	Is There Only One Solution Which Makes the Action Stationary?	138
3.6.	The Principle of Maupertuis	141
3.7.	Fermat's Principle	144
3.8.	The Skier Strategy	146
3.9.	Free Motion on an Ellipsoid	150
3.10.	Minimum Area for a Fixed Volume	152
3.11.	The Form of Soap Films	154
3.12.	Laplace's Law for Surface Tension	158
3.13.	Chain of Pendulums	160
3.14.	Wave Equation for a Flexible Blade	161
3.15.	Precession of Mercury's Orbit	162

Chapter 4. Hamiltonian Formalism 165
Summary 165

4.1.	Generalized Momentum	165
4.2.	Hamilton's Function	166
4.3.	Hamilton's Equations	167
4.4.	Liouville's Theorem	167

| 4.5. Autonomous One-dimensional Systems 168
| 4.6. Periodic One-dimensional Hamiltonian Systems 169
| **Problem Statements** 171
| 4.1. Electric Charges Trapped in Conductors 171
| 4.2. Symmetry of the Trajectory 171
| 4.3. Hamiltonian in a Rotating Frame 172
| 4.4. Identical Hamiltonian Flows 173
| 4.5. The Runge-Lenz Vector 173
| 4.6. Quicker and More Ecologic than a Plane 174
| 4.7. Hamiltonian of a Charged Particle 176
| 4.8. The First Integral Invariant 177
| 4.9. What About Non-Autonomous Systems? 178
| 4.10. The Reverse Pendulum 178
| 4.11. The Paul Trap 180
| 4.12. Optical Hamilton's Equations 181
| 4.13. Application to Billiard Balls 183
| 4.14. Parabolic Double Well 184
| 4.15. Stability of Circular Trajectories in a Central Potential 185
| 4.16. The Bead on the Hoop 186
| 4.17. Trajectories in a Central Force Field 188
| **Problem Solutions** 188
| 4.1. Electric Charges Trapped in Conductors 188
| 4.2. Symmetry of the Trajectory 190
| 4.3. Hamiltonian in a Rotating Frame 192
| 4.4. Identical Hamiltonian Flows 194
| 4.5. The Runge-Lenz Vector 195
| 4.6. Quicker and More Ecologic than a Plane 198
| 4.7. Hamiltonian of a Charged Particle 200
| 4.8. The First Integral Invariant 204
| 4.9. What About Non-Autonomous Systems? 206
| 4.10. The Reverse Pendulum 207
| 4.11. The Paul Trap 211
| 4.12. Optical Hamilton's Equations 214
| 4.13. Application to Billiard Balls 216
| 4.14. Parabolic Double Well 219
| 4.15. Stability of Circular Trajectories in a Central Potential 222
| 4.16. The Bead on the Hoop 224
| 4.17. Stability of Circular Trajectories in a Central Potential 228

Chapter 5. Hamilton-Jacobi Formalism 233
Summary 233
      5.1. The Action Function 233
      5.2. Reduced Action 234
      5.3. Maupertuis' Principle 235

5.4.	Jacobi's Theorem	236
5.5.	Separation of Variables	236
5.6.	Huygens' Construction	238

Problem Statements 239

5.1.	How to Manipulate the Action and the Reduced Action	239
5.2.	Action for a One-dimensional Harmonic Oscillator	241
5.3.	Motion on a Surface and Geodesic	241
5.4.	Wave Surface for Free Fall	242
5.5.	Peculiar Wave Fronts	243
5.6.	Electrostatic Lens	243
5.7.	Maupertuis' Principle with an Electromagnetic Field	245
5.8.	Separable Hamiltonian, Separable Action	246
5.9.	Stark Effect	247
5.10.	Orbits of Earth's Satellites	248
5.11.	Phase and Group Velocities	251

Problem Solutions 252

5.1.	How to Manipulate the Action and the Reduced Action	252
5.2.	Action for a One-Dimensional Harmonic Oscillator	258
5.3.	Motion on a Surface and Geodesic	260
5.4.	Wave Surface for Free Fall	261
5.5.	Peculiar Wave Fronts	264
5.6.	Electrostatic Lens	265
5.7.	Maupertuis' Principle with an Electromagnetic Field	268
5.8.	Separable Hamiltonian, Separable Action	270
5.9.	Stark Effect	271
5.10.	Orbits of Earth's Satellites	275
5.11.	Phase and Group Velocities	279

Chapter 6. Integrable Systems 281

Summary 281

6.1.	Basic Notions	281
	6.1.1. Some Definitions	281
	6.1.2. Good Coordinates: The Angle–Action Variables	283
6.2.	Complements	286
	6.2.1. Building the Angle Variables	286
	6.2.2. Flow/Poisson Bracket/Involution	287
	6.2.3. Criterion to Obtain a Canonical Transformation	288

Problem Statements 289

6.1.	Expression of the Period for a One-Dimensional Motion	289
6.2.	One-dimensional Particle in a Box	290
6.3.	Ball Bouncing on the Ground	290
6.4.	Particle in a Constant Magnetic Field	291
6.5.	Actions for the Kepler Problem	292
6.6.	The Sommerfeld Atom	293
6.7.	Energy as a Function of Actions	294

6.8. Invariance of the Circulation Under a Continuous
      Deformation 296
6.9. Ball Bouncing on a Moving Tray 297
6.10. Harmonic Oscillator with a Variable Frequency 298
6.11. Choice of the Momentum 298
6.12. Invariance of the Poisson Bracket
      Under a Canonical Transformation 299
6.13. Canonicity for a Contact Transformation 299
6.14. One-Dimensional Free Fall 300
6.15. One-Dimensional Free Fall Again 301
6.16. Scale Dilation as a Function of Time 301
6.17. From the Harmonic Oscillator to Coulomb's Problem .. 302
6.18. Generators for Fundamental Transformations 303

Problem Solutions 305
6.1. Expression of the Period for a One-Dimensional Motion 305
6.2. One-Dimensional Particle in a Box 306
6.3. Ball Bouncing on the Ground 308
6.4. Particle in a Constant Magnetic Field 310
6.5. Actions for the Kepler Problem 314
6.6. The Sommerfeld Atom 316
6.7. Energy as a Function of Actions 318
6.8. Invariance of the Circulation Under a Continuous
     Deformation 322
6.9. Ball Bouncing on a Moving Tray 324
6.10. Harmonic Oscillator with a Variable Frequency 324
6.11. Choice of the Momentum 325
6.12. Invariance of the Poisson Bracket
      Under a Canonical Transformation 326
6.13. Canonicity for a Contact Transformation 327
6.14. One-dimensional Free Fall 329
6.15. One-dimensional Free Fall Again 330
6.16. Scale Dilation as a Function of Time 332
6.17. From the Harmonic Oscillator to Coulomb's Problem .. 333
6.18. Generators for Fundamental Transformations 336

Chapter 7. Quasi-Integrable Systems 341
Summary 341
7.1. Introduction 341
7.2. Perturbation Theory 342
7.3. Canonical Perturbation Theory 342
7.4. Adiabatic Invariants 345
Problem Statements 347
7.1. Limits of the Perturbative Expansion 347
7.2. Non-canonical Versus Canonical Perturbative Expansion 347

7.3.	First Canonical Correction for the Pendulum	348
7.4.	Beyond the First Order Correction	349
7.5.	Adiabatic Invariant in an Elevator	350
7.6.	Adiabatic Invariant and Adiabatic Relaxation	351
7.7.	Charge in a Slowly Varying Magnetic Field	352
7.8.	Illuminations Concerning the Aurora Borealis	354
7.9.	Bead on a Rigid Wire: Hannay's Phase	356

Problem Solutions 358

7.1.	Limits of the Perturbative Expansion	358
7.2.	Non-canonical Versus Canonical Perturbative Expansion	361
7.3.	First Canonical Correction for the Pendulum	363
7.4.	Beyond the First Order Correction	367
7.5.	Adiabatic Invariant in an Elevator	370
7.6.	Adiabatic Invariant and Adiabatic Relaxation	372
7.7.	Charge in a Slowly Varying Magnetic Field	375
7.8.	Illuminations Concerning the Aurora Borealis	379
7.9.	Bead on a Rigid Wire: Hannay's Phase	382

Chapter 8. From Order to Chaos 385

Summary 385

8.1.	Introduction	385
8.2.	The Model of the Kicked Rotor	386
8.3.	Poincaré's Sections	388
8.4.	The Rotor for a Null Perturbation	388
8.5.	Poincaré's Sections for the Kicked Rotor	390
8.6.	How to Recognize Fixed Points	393
8.7.	Separatrices/Homocline Points/Chaos	394
8.8.	Complements	395

Problem Statements 396

8.1.	Disappearance of Resonant Tori	396
8.2.	Continuous Fractions or How to Play with Irrational Numbers	396
8.3.	Properties of the Phase Space of the Standard Mapping	398
8.4.	Bifurcation of the Periodic Trajectory 1:1 for the Standard Mapping	398
8.5.	Chaos–Ergodicity : A Slight Difference	399
8.6.	Acceleration Modes: A Curiosity of the Standard Mapping	401
8.7.	Demonstration of a Kicked Rotor?	401
8.8.	Anosov's Mapping (or Arnold's Cat)	403
8.9.	Fermi's Accelerator	405
8.10.	Damped Pendulum and Standard Mapping	407
8.11.	Stability of Periodic Orbits on a Billiard Table	409
8.12.	Lagrangian Points: Jupiter's Greeks and Trojans	412

Problem Solutions 415
    8.1. Disappearance of Resonant Tori 415
    8.2. Continuous Fractions or How to Play with Irrational
        Numbers 417
    8.3. Properties of the Phase Space of the Standard Mapping 418
    8.4. Bifurcation of the Periodic Trajectory 1:1
        for the Standard Mapping 419
    8.5. Chaos–Ergodicity: A Slight Difference 423
    8.6. Acceleration Modes: A Curiosity of the Standard
        Mapping 425
    8.7. Demonstration of a Kicked Rotor? 427
    8.8. Anosov's Mapping (or Arnold's Cat) 432
    8.9. Fermi's Accelerator 438
    8.10. Damped Pendulum and Standard Mapping 443
    8.11. Stability of Periodic Orbits on a Billiard Table 447
    8.12. Lagrangian Points: Jupiter's Greeks and Trojans 450

Bibliography 457

Index 461

Synoptic Tables of the Problems

Chapter 1. Lagrangian Formulation

No	Title	Level	P.	Features
1.1	The wheel jack	★	14	Lagrangian mechanics. D'Alembert's principle
1.2	The sling	★	15	Lagrange equations for a very simple system
1.3	The rope slipping on a table	★	16	Lagrange equations in the presence of friction
1.4	Reaction force for a bead on a hoop	★★	16	Reaction force calculated by adding one generalized coordinate
1.5	Huygens pendulum	★★★	17	Work of contact forces
1.6	Cylinder rolling on a movable tray	★★	18	Lagrange equations with two different coordinates
1.7	Motion of a badly balanced cylinder	★★★	18	Koenig's theorem. Holonomic constraint
1.8	Free axle on a inclined plane	★★★	19	Constrained Lagrange equations. Lagrange multipliers
1.9	The turning indicator	★★★	21	The gyroscope studied with the Lagrangian formalism
1.10	An experiment to measure the rotational velocity of the Earth	★★★	22	A well controlled gyroscope as an alternative experiment to Foucault's pendulum
1.11	Generalized inertial forces	★★★	23	Lagrange equations in a non Galilean frame

Chapter 2. Lagrangian Systems

No	Title	Level	P.	Features
2.1	Disc on a movable inclined plane	★	57	Example of a time-dependent Lagrangian
2.2	Painlevé's integral	★★	58	Search for a first integral for a time-dependent Lagrangian
2.3	Application of Noether's theorem	★	58	Very simple application of Noether's theorem
2.4	Foucault's pendulum	★★	59	A famous experiment explained with Lagrangian formalism
2.5	Three-particle system	★★	61	Appearance of symmetries by a change of variables
2.6	Vibration of a linear triatomic molecule: the "soft" mode	★★	63	Eigenmodes for an oscillating system
2.7	Elastic transversal waves in a solid	★★	64	Passing from a discrete model to a continuous model for the study of a solid bar
2.8	Lagrangian in a rotating frame	★★	65	Modification of the Lagrangian in the passage to a rotating frame
2.9	Particle drift in a constant electromagnetic field	★★	66	Drift motion for a particle in an electromagnetic field exhibited in the Lagrangian formalism
2.10	The Penning trap	★★★	67	An astute electromagnetic system to trap a particle
2.11	Equinox precession	★★★	68	Lagrangian formalism applied to an astronomical phenomenon
2.12	Flexion vibration of a blade	★★★	71	Passing from a discrete model to a continuous model for the study of the vibration of an embedded blade
2.13	Solitary waves	★★	73	Non-linear equation for a wave propagating without deformation
2.14	Vibrational modes of an atomic chain	★★★	75	System composed of an infinite number of coupled oscillators

Chapter 3. Hamilton's Principle

No	Title	Level	P.	Features
3.1	The Lorentz force	★★	116	Hamilton's principle applied to an electromagnetic problem
3.2	Relativistic particle in a central force field	★★★	117	Relativistic Binet's equation
3.3	Principle of least action?	★★★	118	Justification of the concept of "least action"
3.4	Minimum or maximum action?	★★	119	Why the action is not always minimal
3.5	Is there only one solution which makes the action stationary?	★★	120	Hamilton's principle. Through two points may pass several trajectories
3.6	The principle of Maupertuis	★★	121	Alternative to the Hamilton principle for the determination of the trajectories
3.7	Fermat's principle	★★	122	Hamilton's principle in the domain of optics
3.8	The skier strategy	★★★	122	Calculus of variations for the brachistochrone
3.9	Free motion on an ellipsoid	★★	123	Calculus of variations with a holonomic constraint. Lagrange multipliers
3.10	Minimum area for a fixed volume	★★	124	Calculus of variations with an integral constraint. Lagrange multipliers
3.11	The form of soap films	★★★	125	Amusing application of Hamilton's principle. Calculus of variations
3.12	Laplace's law for surface tension	★★★	127	Hamilton's principle applied to hydrostatics
3.13	Chain of pendulums	★★	128	Hamilton's principle for a continuous system
3.14	Wave equation for a flexible blade	★★	128	Building a Lagrangian density
3.15	Precession of Mercury's orbit	★★★	128	Hamilton's principle in the context general relativity

Chapter 4. Hamiltonian Formalism

No	Title	level	P.	Features
4.1	Electric charges trapped in conductors	★★	171	Electrostatic image. Hamilton's equations. First integral
4.2	Symmetry of the trajectory	★	171	Binet's equation. Its use for treating symmetries
4.3	Hamiltonian in a rotating frame	★★	172	Change of frame. Legendre transform
4.4	Identical Hamiltonian flows	★	173	Hamilton's equations and their flow
4.5	The Runge–Lenz vector	★★	173	Building a constant vector. Relationship with other constants of motion
4.6	Quicker and more ecologic than a plane	★★	174	Hamilton's equations in a gravitational field
4.7	Hamiltonian of a charged particle	★★★	176	Hamilton's equations. Covariant relativistic formalism
4.8	The first integral invariant	★★	177	Integral invariant in the Hamiltonian formalism
4.9	What about non-autonomous systems?	★	178	To render autonomous a non-autonomous system. Corresponding flow
4.10	The reverse pendulum	★★★	178	Hamilton's equations. Propagator. Stability conditions. Arnold's tongue
4.11	The Paul trap	★★★	180	Electromagnetic system. Propagator. Stability
4.12	Optical Hamilton's equations	★★★	181	Snell–Descartes law obtained from Hamilton's equations
4.13	Application to billiard balls	★★	183	Choice of variables for trajectories on a billiard table. Conservation of the area
4.14	Parabolic double well	★★	184	Simple motion and phase portrait
4.15	Stability of circular trajectories in a central potential	★★	185	Motion for a power-law potential. Stability conditions

Synoptic Tables of the Problems

No	Title	level	P.	Features
4.16	The bead on the hoop	★★	186	Phase portrait. Stability. Bifurcation
4.17	Trajectories in a central force field	★★	188	Relativistic Binet's equation. Phase portrait

Chapter 5. Hamilton–Jacobi Formalism

No	Title	level	P.	Features
5.1	How to manipulate the action and the reduced action	★★	239	Relation between the total action and the reduced action. Hamilton–Jacobi equation
5.2	Action for a one-dimensional harmonic oscillator	★★	241	Reduced action. Hamilton–Jacobi equation
5.3	Motion on a surface and geodesic	★★	241	Action with a general metric. Principle of Maupertuis
5.4	Wave surface for free fall	★★	242	Hamilton–Jacobi equation. Wave fronts in a gravitational field
5.5	Peculiar wave fronts	★★	243	Wave fronts and trajectories in a gravitational field. Hamilton–Jacobi equation
5.6	Electrostatic lens	★★★	243	Electromagnetism and the principle of Maupertuis for a system with cylindrical symmetry
5.7	Maupertuis' principle with an electromagnetic field	★★★	245	Electromagnetic field and the principle of Maupertuis. Cyclotron motion
5.8	Separable Hamiltonian, separable action	★	246	Separation of the variables in the Hamilton–Jacobi equations
5.9	Stark effect	★★★	247	Parabolic coordinates which separate the variables
5.10	Orbits of Earth's satellites	★★★	248	Elliptic coordinates which separate the variables
5.11	Phase and group velocities	★	251	A notion used in optics which is also valid in mechanics

Chapter 6. Integrable Systems

No	Title	level	P.	Features
6.1	Expression of the period for a one-dimensional motion	★	289	Reduced action and angular frequency
6.2	One-dimensional particle in a box	★★	290	Angle-action variables. Quantization
6.3	Ball bouncing on the ground	★★	290	Angle-action variables. Quantization
6.4	Particle in a constant magnetic field	★★★	291	Action variable. Phase portrait. Landau's levels
6.5	Actions for the Kepler problem	★★★	292	Energy as a function of actions. Quantization
6.6	The Sommerfeld atom	★★★	293	Energy as a function of actions. Relativistic systems. Quantization
6.7	Energy as a function of actions	★★★	294	Form of the Hamiltonian as a function of actions
6.8	Invariance of the circulation under a continuous deformation	★★★	296	Functions in involution. Circulation on a torus
6.9	Ball bouncing on a moving tray	★	297	Time-dependent canonical transformation for the free fall
6.10	Harmonic oscillator with a variable frequencys	★★	298	Time-dependent canonical transformation for the harmonic oscillator
6.11	Choice of the momentum	★★	298	A particular canonical transformation
6.12	Invariance of the Poisson bracket under a canonical transformation	★	299	Poisson brackets. Canonical transformation
6.13	Canonicity for a contact transformation	★	299	Poisson brackets. Canonical transformation
6.14	One-dimensional free fall	★★	300	Canonical transformation. Trajectory
6.15	One-dimensional free fall again	★★	301	Angle-action variables. Generating function
6.16	Scale dilation as a function of time	★★★	301	Time dependent canonical transformation. Generating function

No	Title	level	P.	Features
6.17	From the harmonic oscillator to Coulomb's problem	★★	302	Angle-action variables. Canonical transformation. Energy as a function of actions
6.18	Generators for fundamental transformations	★★	303	Flow parameters. Generators. Relativity

Chapter 7. Quasi-integrable Systems

No	Title	level	P.	Features
7.1	Limits of the perturbative expansion	★	347	Differential equation. Classical perturbation theory
7.2	Non-canonical versus canonical perturbative expansion	★★	347	Anharmonic oscillator. Canonical and non-canonical perturbation theories
7.3	First canonical correction for the pendulum	★★★	348	Simple pendulum. Quartic perturbation. Exact and perturbative treatments
7.4	Beyond the first order correction	★★★	349	Canonical perturbation theory. Second order
7.5	Adiabatic invariant in an elevator	★★	350	Action variable. Adiabatic invariant
7.6	Adiabatic invariant and adiabatic relaxation	★★★	351	Action variable. Adiabatic invariant. Monatomic ideal gaz
7.7	Charge in a slowly varying magnetic field	★★★	352	Generating function. Canonical transformation. Adiabatic invariant
7.8	Illuminations concerning the aurora borealis	★★	354	Electromagnetism. Adiabatic invariant. Cyclotron motion and drift
7.9	Bead on a rigid wire: Hannay's phase	★★★	356	Angle-action variables. Adiabatic invariance. Hannay's phase

Chapter 8. From Order to Chaos

No	Title	level	P.	Features
8.1	Disappearance of resonant tori	★★	396	Resonant tori. KAM theorem
8.2	Continuous fractions or how to play with irrational numbers	★★	396	Continuous fractions. Rational and irrational numbers. Convergence
8.3	Properties of the phase space of the standard mapping	★	398	Poincaré's section. Symmetries. Parameters
8.4	Bifurcation of the periodic trajectory 1:1 for the standard mapping	★★	398	Standard mapping. Fixed points. Bifurcation
8.5	Chaos–ergodicity: a slight difference	★★	399	Condition for ergodicity
8.6	Acceleration modes: a curiosity of the standard mapping	★★	401	Poincaré's section. Standard mapping. Acceleration of the momentum
8.7	Demonstration of a kicked rotor?	★★★	401	Jerky pendulum. Sawtooth mapping
8.8	Anosov's mapping (or Arnold's cat)	★★★	403	Fibonacci sequence. Anosov's mapping. Fixed points
8.9	Fermi's accelerator	★★★	405	Moving walls. Conservation of the area. Ulam's mapping
8.10	Damped pendulum and standard mapping	★★	407	Non-Hamiltonian system. Friction. Spiral point
8.11	Stability of periodic orbits on billiard table	★★★	409	Mapping on a billiard table. Derivative matrix of the mapping. Stability of periodic trajectories. Examples of billiard tables
8.12	Lagrangian points: Jupiter's Greeks and Trojans	★★★	412	Restricted three-body problem. Lagrangian points and their stability

Chapter 1
The Lagrangian Formulation

Summary

1.1. Generalized Coordinates

A mechanical system is composed, *in fine*, of a given number N of elements α, with a mass m_α, which can be considered as pointlike and located at position $\boldsymbol{r}_\alpha$. The configuration of this system is specified by the set of the constituent coordinates. However, in most situations, internal constraints (for example in a rigid body the distance between the constituents is independent of the configuration) or external constraints (for example a point subjected to remain on a given surface) impose a number of relationships between the coordinates; in such cases, a smaller set of specifications allows us to characterize the configuration of the system.

The n variables ($n \leq 3N$), which unambiguously define the configuration of the system are called **generalized coordinates**; they are denoted generically as q, for the set $(q_1, q_2, \ldots, q_n)$ of the n generalized coordinates q_i. In any practical case, generalized coordinates are either lengths, or angles. Generalized coordinates being sufficient to completely describe the configuration, there exist N mathematical relations $\boldsymbol{r}_\alpha(q, t)$ ($\alpha = 1, \ldots, N$), each coordinate position depending only on n variables q_i. Sometimes one encounters an explicit time dependence of the constraints, for example when a point moves on a surface which moves itself.

1.2. Lagrange's Equations

The Lagrangian formulation of mechanics consists in writing Newton's equations, which depend on N vectorial quantities $\boldsymbol{r}_\alpha$, in terms of n scalar quantities q_i ($q_1, q_2, \ldots, q_n$). To begin with, let us consider the case for which the n generalized coordinates are independent; in this case, n is called the number of **degrees of freedom for the system**. The Lagrangian formalism relies on the kinetic energy T, which is a kinematic quantity defined in terms of velocities[1] $\boldsymbol{v}_\alpha = \dot{\boldsymbol{r}}_\alpha = d\boldsymbol{r}_\alpha/dt$ of each element by

$$T = \frac{1}{2}\sum_{\alpha=1}^{N} \boldsymbol{v}_\alpha^2.$$

If the particle positions are given in terms of generalized coordinates, the kinetic energy is expressed not only in terms of n generalized coordinates q_i, but also in terms of n **generalized velocities** $\dot{q}_i = dq_i/dt$ and, possibly, in terms of time: $T(q, \dot{q}, t)$. From the kinetic energy, one builds n kinematical quantities A_i, called **generalized accelerations**, defined by the following relation:[2]

$$A_i(q, \dot{q}, \ddot{q}, t) = \frac{d}{dt}\partial_{\dot{q}_i}T(q, \dot{q}, t) - \partial_{q_i}T(q, \dot{q}, t). \tag{1.1}$$

Then, Newton's equations are translated into the Lagrangian formalism through a set of n dynamical equations, called **Lagrange's equations**, which are written

$$A_i(q, \dot{q}, \ddot{q}, t) = Q_i(q, \dot{q}, t), \tag{1.2}$$

where $Q_i(q, \dot{q}, t)$ are dynamical quantities, called generalized forces, which will be defined later.

The Lagrange equations are a set of n **coupled differential equations of second order**.

Directions for use and precisions

The first task is to obtain the expression of the kinetic energy as a function[3] of the generalized velocities, possibly of the generalized coordinates, and

---

[1] As usual in mechanics, a dot above a quantity means its first derivative with respect to time, two dots its second derivative, ...: $\dot{f} = df/dt$, $\ddot{f} = d^2f/dt^2$, ...

[2] With typographical simplicity in view, we will use a simplified notation to define partial derivatives for a function of several variables

$$\partial_x f(x,y) = \frac{\partial f(x,y)}{\partial x}, \quad \partial_{x^2}^2 f(x,y) = \frac{\partial^2 f(x,y)}{\partial x^2}, \quad \partial_{xy}^2 f(x,y) = \frac{\partial^2 f(x,y)}{\partial x \partial y}.$$

[3] The choice for generalized coordinates is, a priori, arbitrary. The best starting choice is that which gives the most simple form to the kinetic energy.

Summary

(although rarely) of time[4] (see Exercise 1.4). To obtain the kinetic energy, one supposes first that the generalized coordinates depend on time $q(t)$; naturally the derivatives of these functions with respect to time $\dot{q}(t)$ appear in the expression of the velocities. Thus the kinetic energy is expressed in terms of q and $\dot{q}$. **Subsequently, these functions are considered as independent.** Sometimes, the kinetic energy exhibits only generalized velocities, and sometimes both generalized velocities and coordinates.

Just as an example, let us consider a particle with mass m, moving on a plane: if one locates the particle by the Cartesian coordinates (x, y) the kinetic energy is expressed only as function of generalized velocities since $T(\dot{x}, \dot{y}) = \frac{1}{2}m(\dot{x}^2 + \dot{y}^2)$, whereas if one chooses polar coordinates (ρ, ϕ) the same kinetic energy contains, in addition to the generalized velocities, the coordinate ρ since $T(\rho, \dot{\rho}, \dot{\phi}) = \frac{1}{2}m(\dot{\rho}^2 + \rho^2\dot{\phi}^2)$.

Once the expression for the kinetic energy is obtained the rest of the treatment is as follows:

- One derives the function $T(q, \dot{q}, t)$ with respect to the generalized coordinates q_i to get $\partial_{q_i} T(q, \dot{q}, t)$.

- One derives the function $T(q, \dot{q}, t)$ with respect to the generalized velocities $\dot{q}_i$ to get $\partial_{\dot{q}_i} T(q, \dot{q}, t)$.

- One derives with respect to time the function $\partial_{\dot{q}_i} T(q, \dot{q}, t)$, considering that one handles a function $q(t)$, for which $\dot{q} = dq(t)/dt$ and $\ddot{q} = d\dot{q}(t)/dt$. The generalized acceleration $A_i(q, \dot{q}, \ddot{q}, t)$ follows from (1.1).

Proceeding with the previous example and polar coordinates, this series of operations leads to the generalized acceleration:

$$A_\rho = m(\ddot{\rho} - \rho\dot{\phi}^2); \quad A_\phi = m(\rho^2\ddot{\phi} + 2\rho\dot{\rho}\dot{\phi}).$$

From Newton's equations, the product of mass with acceleration is determined by the forces acting on the system; similarly the link between the generalized accelerations and generalized forces through Lagrange's equations matches Newton's equations.

As long as the forces are not specified, the functions $q(t)$ entering the generalized accelerations are arbitrary. Equating generalized accelerations to generalized forces leads to a system of differential equations which are fulfilled only for special functions $q(t)$, which are precisely the solutions of the true physical motion and which are called **trajectories**. To determine them unambiguously, it is necessary to set the initial values $q(0)$ and $\dot{q}(0)$.

---

[4] In the kinetic energy, the generalized coordinates and the generalized velocities must be considered as independent variables. Only once the forces are given is the kinetic energy a function of $q(t)$ and of its derivatives.

1.3. Generalized Forces

To define generalized forces, one must first introduce the notion of virtual displacement. Let us imagine that, **at a given time**, two configurations of the system are described by the coordinates q et $q + \delta q$, compatible with the constraints imposed on the system. The quantity δq is called a **virtual displacement**.

In this displacement, the constituents α are displaced by a quantity δr_α and the forces f_α acting on them produce a total work

$$\delta W = \sum_{\alpha=1}^{N} f_\alpha \cdot \delta r_\alpha.$$

This last quantity is said to be a **virtual work** and it can be put under a form expressed in terms of the virtual displacements δq:

$$\delta W = \sum_{i=1}^{n} Q_i \, \delta q_i. \tag{1.3}$$

This expression defines the **generalized forces**[5] $Q_i(q, \dot{q}, t)$. Let us note that a virtual displacement is only compatible with the constraints and can be entirely different from a real displacement of the system which results from the temporal evolution given by Lagrange's equations (1.2).

Let us emphasize a point. In the Lagrangian formalism, the forces responsible for the constraints are inaccessible, since the generalized coordinates were chosen precisely to get rid of them. Since they are generally uninteresting quantities, this is of little consequence and, in fact, lies at the origin of the elegance of Lagrange's equations. If, after all, we insist on obtaining the expression of these constraint forces, we have to introduce supplementary generalized coordinates (to get rid of cumbersome constraints) in order to obtain a non vanishing virtual work concerning this type of force (see Problems 1.4 and 1.7).

When the system is at rest, generalized velocities and accelerations vanish; Lagrange's equations (1.2) then imply a vanishing value for the generalized force. The relation:

$$Q_i = 0 \quad \text{at rest} \tag{1.4}$$

represents **d'Alembert's principle.**

---

[5] The generalized force depends on generalized coordinates, on time and sometimes also on generalized velocities. This is in particular the case for friction forces, magnetic forces and Coriolis forces. Concerning this point, let us note that if the real work of these latter forces vanishes because they are perpendicular to the displacement, this is no longer the case for the virtual work, the displacement being arbitrary.

1.4. Lagrange Multipliers

Let us consider now the case where the n generalized coordinates are not independent. It is useful to remind ourselves that these coordinates were introduced with the purpose of taking into account a number of constraints.

The present case thus corresponds to a situation for which the system is subject to additional constraints. Practically, this happens when the constraints are not able to reduce the number of generalized coordinates or when the search for new generalized coordinates turns out to be a too painful procedure.

All the virtual displacements are no longer possible, but compelled to obey new conditions taking into account the supplementary constraints. For simplicity, let us consider only one condition written under a differential form:

$$\sum_{i=1}^{n} \Lambda_i \, \delta q_i = 0. \tag{1.5}$$

This equation defines the quantity Λ_i, an important ingredient in constrained Lagrange equations. There exists a special very simple kind of constraint, known as **holonomic**, for which this quantity is the differential of a single function Φ:

$$\sum_{i=1}^{n} \Lambda_i \delta q_i = d\Phi(q).$$

The constraint is thus equivalent to the fact that $\Phi(q)$ is a constant. This allows us, in principle, to express one generalized coordinate as a function of the $n-1$ others; it is enough then to proceed like this in the expressions of the kinetic energy and generalized forces[6] in order to work now with $n-1$ generalized coordinates. Indeed the system depends on $n-1$ rather than n degrees of freedom.

If the constraint is not holonomic, or if elimination is not an easy task, then we keep the original generalized coordinates and introduce **Lagrange multipliers**. It is not our intention, in this brief summary, to develop

---

[6] In particular, this is the case for rolling without slipping motion in two dimensions. This is no more the case in three dimensions.

Let us stress the fact that a rolling without slipping motion necessarily implies a non vanishing tangential reaction force acting on the rolling surface. Nevertheless in a virtual displacement, this force does not perform work. The deep reason for this is a consequence of the fact that, in this virtual displacement $\delta\phi$, the application point of the force follows a cycloid and there is a displacement only of second order in $\delta\phi$ in the perpendicular direction and of third order in the tangential direction (see Exercise 1.5).

the general theory of Lagrange multipliers. We simply give the form of constrained Lagrange equations when the system is subject to l differential constraints of type (1.5):

$$A_i(q,\dot{q},\ddot{q},t) = Q_i(q,\dot{q},t) + \sum_{k=1}^{l} \lambda_k \Lambda_i^k(q,t). \qquad (1.6)$$

The quantities λ_k are the Lagrange multipliers. Their values must be determined at the same time as the trajectories $q(t)$ by solving the n differential equations **and** the l constraints equations imposed on the coordinates. The Lagrange multipliers can be interpreted in terms of reaction forces associated with the constraints (see Exercise 1.8).

Problem Statements

1.1. The Wheel Jack [Solution and Figure p. 24] ★

This exercise is simply an application of d'Alembert principle

A wheel jack is an articulated machine which is designed to lift up heavy burdens (a coach for example); it is represented in Fig. 1.1. It is composed of two rigid bases (one resting on the ground, the other sustaining the burden with weight P); they form an articulated lozenge with side l, which may be deformed by mean of a threaded stem with a step h, (the axis of the jack changes by a length h for each revolution of the crank) driven by a force F applied on the crank of arm length a.

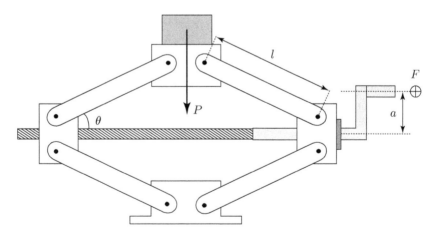

Fig. 1.1 Principle of a wheel jack

We assume no friction on the mechanical parts of the jack (fortunately frictional forces do exist and allow the mass to be supported without any effort!).

1. Examine the constraints imposed on the system and show that it has only one degree of freedom. Which generalized coordinate seems to you the most appropriate?
2. Using d'Alembert's principle, deduce the relationship between the weight to be lifted up and the exerted force, as a function of the characteristics of the jack and of the angle θ between the lozenge side and the threaded stem.

Numerical application: Calculate the ratio of the weight and the force for a jack with crank arm length $a = 20$ cm, with $h = 2$ mm, at the beginning of the lifting process when $\theta = 30°$.

1.2. The Sling [Solution p. 26] ★

Very simple application of Lagrange's equations

A rigid stem is maintained fixed at one of its ends O. It turns around O in the horizontal plane with a constant angular speed $\omega = \dot{\phi}$ (see Fig. 1.2). A pointlike mass m slips without friction on this stem. It is placed at rest at a point A such that $OA = a$.

1. Find the most natural generalized coordinate and assess the real and generalized forces.
2. Write and solve the Lagrange equation.

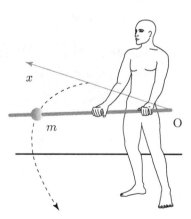

Fig. 1.2 The sling

1.3. Rope Slipping on a Table [Solution p. 27] ★

Classical problem for which the Lagrangian formalism is well suited

A part $L - l$ of a rope with length L and with constant linear mass μ is originally at rest on a horizontal table. The rest of the rope, with length l, hangs vertically in a constant gravitational field g.

The rope is placed without any initial velocity. One assumes that the part which hangs over the edge of the table remains always vertical[7] (Fig. 1.3).

Fig. 1.3 Rope slipping on a table

1. In a first study, assume that friction on the table is absent. Find a generalized coordinate and assess the real and generalized forces. Write and solve the Lagrange equation.

2. Assume a solid friction, with a constant friction coefficient f (the dynamical friction coefficient is assumed equal to the static coefficient). What is the minimum length l_0 necessary to induce sliding of the rope. If $l > l_0$ write and solve the Lagrange equation.

1.4. Reaction Force for a Bead on a Hoop
[Solution and Figures p. 28] ★ ★

Calculation of a reaction force by adding a generalized coordinate

Let us consider a system composed of a pierced bead M, with mass m, which slides without friction on a massless hoop with center O, radius R, which itself rotates around one fixed diameter Oz, parallel to the vertical.

---

[7] Indeed the linear momentum acquired by the horizontal part of the rope when it falls has the consequence that it keeps falling and the rope does not turn at right angles. Moreover a rope is a flexible system which can exhibit transverse deformations and the fall can cause undulations. Of course, all these complications are neglected.

The angle ϕ, between the plane of the hoop and the fixed vertical plane xOz, varies in time according to a known law imposed by the operator: $\phi(t)$. A generalized coordinate is chosen as the angle θ between the direction OM and the vertical direction Oz.

This system is embedded in a constant gravitational field g, acting along the vertical axis.

1. Give the expression of the kinetic energy in terms of the generalized coordinate.

2. Give the expression of the generalized force.

3. Write down the corresponding Lagrange equation.

4. This simple question illustrates the difference between virtual work and real work. We are interested in the reaction force of the hoop on the bead. Introducing a new generalized coordinate which allows virtual work for the component of the reaction force normal to the plane of the hoop, determine this force. Check your result with the Coriolis inertial force.

1.5. Huygens Pendulum
[Solution and Figures p. 31] ⋆ ⋆ ⋆

Work done by contact forces responsible for a motion without slipping

In a vertical plane xOz, a point M, with mass m, is fixed to a massless hoop, with radius R, which can roll without slipping on a horizontal stem Ox, placed above it. It is well known that the curve followed by M is a cycloid. We choose as generalized coordinate the angle ϕ, such that $R\phi$ is the abscissa of the center C of the hoop. The origin O is taken when M is in its lowest position, CM being then parallel to Oz. The system is subject to a constant gravitational field g directed along the downward vertical.

1. Write the Lagrange equation relative to the coordinate ϕ.

2. Make a change of variable and take instead $x = \sin(\phi/2)$. Show that x varies in time following a harmonic motion with angular frequency $\omega = \sqrt{g/(4R)}$. Deduce that ϕ evolves periodically with the same angular frequency, independently of its amplitude. This pendulum is said to be isochronous and is known as Huygens pendulum.

1.6. Cylinder Rolling on a Moving Tray
[Solution and Figure p. 33] ⋆ ⋆

Work performed by contact forces responsible for a motion without slipping

A homogeneous cylinder, with radius R, mass M and moment of inertia I around its axis, rolls without slipping on a horizontal tray. We impose a translational motion on the tray, perpendicular to the axis of the cylinder, with a given time law $a(t)$. This situation represents for instance the motion of a bottle in the boot of a car.

1. As generalized coordinate, one can choose the angle θ that specifies an arbitrary point of the cylinder along the horizontal direction. Show that the position X of the center of the cylinder in the Galilean frame is linked to θ by an holonomic constraint which is to be determined. What is the corresponding generalized acceleration? Solve the corresponding Lagrange equation and give the real acceleration of the cylinder.

2. Repeat this question choosing now as the generalized coordinate the position X of the center of the cylinder.

1.7. Motion of a Badly Balanced Cylinder
[Solution and Figure p. 35] ⋆ ⋆ ⋆

Application of Koenig's theorem; holonomic forces

An **inhomogeneous** cylinder (center C), with radius R and mass M, has its center of mass G at a distance a from its axis. The mass density is constant along a straight line parallel to the axis. This property implies that one of the principal axes for the cylinder is also parallel to its axis. We denote by I the moment of inertia of the cylinder with respect to the straight line parallel to the axis which passes through G. We study the motion without slipping of the cylinder subject to a constant vertical gravitational field g; the cylinder rolls on a fixed horizontal plane, the plane of its circular section being always fixed (the instantaneous rotation vector $\boldsymbol{\omega}$ is always parallel to the axis).

Equation of motion

1. To define the cylinder configuration, let us take as the single generalized coordinate the angle θ between the downward vertical and the direction CG. Taking into account the constraint for rolling without slipping, write the corresponding Lagrange equation.

2. After multiplying both sides of this equation by the angular velocity $\dot{\theta}$, express the conservation of energy E (we speak of a constant of the motion). To get an idea of this type of motion, plot the angular velocity as a function of the angle, for several different values of the energy: $\dot{\theta}(\theta, E)$.

Vertical reaction force

Explain why the cylinder can exhibit singular behaviour if it rolls too quickly. Using a second generalized coordinate, which breaks the contact with the plane, determine the vertical component $F_v(\theta, E)$ of the reaction force to the plane. It is naturally supposed that this force is weakest when the center of mass is in its highest position. Deduce the maximum energy of the system.

Horizontal reaction force

Rolling without slipping is possible only because the plane exerts a horizontal reaction force to the cylinder. But we know that the ratio between the horizontal and vertical components of the reaction force cannot exceed the friction coefficient f. To obtain this horizontal reaction force $F_h(\theta, E)$, we must consider two generalized coordinates in order to break the constraint of rolling without slipping. Study graphically, as a function of the energy, the conditions that must be fulfilled to achieve rolling without slipping.

1.8. Free Axle on a Inclined Plane
[Solution and Figures p. 39] ★ ★ ★

To understand how to use Lagrange multipliers

A massless axle CC' maintains two identical wheels, of centers C and C' and radius R, in planes normal to it and separated by a distance $L = CC'$. These wheels, for which the axle is a symmetry axis, have a mass m, and the three moments of inertia are $I_1 = I_2 = I$ (in the plane of the wheel) and I_3 (along CC').

The mechanical system consists of the set of the axle and the two wheels (see Fig. 1.4). We study the rolling without slipping of this system on a inclined plane making an angle α with the horizontal plane. For rigidly locked wheels, the motion is identical to that of a cylinder, that is a uniformly accelerated motion.

The aim of this problem is to study the motion when the wheels roll independently of each other.

One chooses a system of perpendicular axes in the inclined plane: horizontal XX', and YY' along the direction of steepest upward slope. The center O of the axle is characterized by its coordinates X et Y in this frame with an arbitrary origin A. The direction $C'C$ makes an angle θ with the horizontal line XX'. We denote by ϕ and ϕ' the angles which mark the positions of reference points on the circumference of the wheels with respect to the line normal to the inclined plane. Thus, the system is described in terms of 5 generalized coordinates $(X, Y, \theta, \phi, \phi')$.

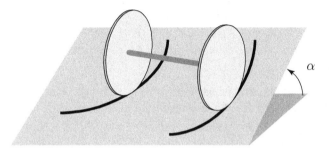

Fig. 1.4 Axle with independent wheels rolling without slipping on a inclined plane

1. There exist four scalar relationships concerning the constraints of rolling without slipping for each of the wheels (two per wheel). In fact, two of them are identical. Give the three independent constraint relationships and show that one of them is holonomic whereas the other two are not.

2. Introducing three Lagrange multipliers λ_1, λ_2, λ_3, write the five constrained Lagrange equations.

3. Interpret the three Lagrange multipliers in terms of contact forces.

4. To solve the eight equations (five Lagrange equations plus three constraint equations), it is judicious to change variables by defining $\sigma = (\phi + \phi')/2$ and $\delta = (\phi - \phi')$.

   Rewrite the Lagrange equations in terms of these new variables. According to the initial conditions, study the various types of behavior for the axle. In particular, give the equations of the motion if, initially, the axle center is located at A and sets off down the slope with a speed V_0, the axle itself being horizontal and having an initial angular velocity $\dot{\theta}(0) = \omega$.

5. In this framework, calculate the Lagrange multipliers λ_i which represent the reaction forces.

1.9. The Turn Indicator
[Solution and Figure p. 43] ★ ★ ★

Mechanics in the "clouds"

In the absence of any visual reference, the pilot of an aircraft would ignore whether he is turning or not, without a small gyroscope (10 cm or so), refereed to as "turn indicator" or "needle". Such a gyroscope of center O is presented in Fig. 1.5. An axis $X'OX$, parallel and firmly attached to the fuselage of the aircraft, is assumed to remain horizontal during the turn. A frame, with normal OZ, is free to oscillate around $X'X$. The mechanical system under study is the inertia flywheel of the gyroscope which is a cylinder with symmetry axis $Y'OY$. The axes of inertia are OX, OY, OZ, which form a direct orthogonal trihedron $OXYZ$, and the corresponding moments of inertia are respectively $I_X = I_Z$ and $I_Y = I$.

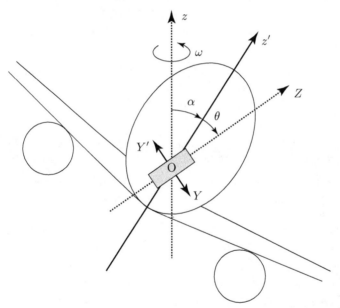

Fig. 1.5 Gyroscope inside a plane. Only the axis $Y'Y$ of the gyroscope, the true vertical Oz and the apparent vertical Oz' are represented

A small electric motor maintains the flywheel rotation around the axis $Y'Y$ and imposes a constant angular velocity Ω on it. The apparent vertical for the aircraft – namely the normal to the wing plane– is denoted Oz'. A small spiral spring acts to force the axes Oz' and OZ to coincide with a restoring torque $C = -k\theta$, where θ is the angle between the Oz' and OZ axes. A needle measures the angle θ.

The aircraft turns, with constant angular velocity ω (see Fig. 1.5; the effective rotation axis is located out of the plane, but this does not matter for the reasoning), around the true vertical Oz (parallel to gravitational acceleration $\boldsymbol{g}$). The angle between Oz and Oz' is denoted α and it is assumed to be constant throughout the turn.

One chooses as generalized coordinate the angle θ which is the only freedom left to the gyroscope.

In a first study, the apparent vertical is supposed to coincide with the true vertical: $\alpha = 0$.

1. Write down the kinetic energy of the flywheel.
2. In the following, we assume the condition $\omega \ll \Omega$, which is always satisfied in practical circumstances.
3. We are interested in the equilibrium solution (θ = const) (obtained in practice with a small pneumatic shock absorber). Give ω as a function of θ; the second order terms ω^2 are neglected. The pilot reads the angle θ and deduces the value ω.

As a matter of fact, the real situation is a little more complicated. Exactly as does a cyclist, the aircraft banks during the turn, and this corresponds to an angle $\alpha \neq 0$. The pilot maintains the inclination and the velocity V of the plane during the turn.

1. Give α as a function of V, ω and g. The simplest method is to consider a static problem in the frame of the plane.
2. How is the relationship between ω and θ modified if the (obligatory) inclination of the plane is correctly taken into account. What relation should exist between V, g and the characteristics of the instrument in order to achieve maximum sensitivity (to give the biggest value of θ for a given ω). It is legitimate to employ the approximation $\alpha \cong \tan(\alpha)$.

1.10. An Experiment to Measure the Rotational Velocity of the Earth
[Solution p. 46]   ★ ★ ★

An alternative to the Foucault pendulum, realized by A.H. Compton

Imagine yourself sitting on a seat of a carousel turning with constant angular velocity ω. You now take hold of the axis of a disc which can rotate without friction with an angular velocity $\dot{\phi}$. Initially, the axis is maintained in the vertical direction and the disc is motionless in the frame of the carousel. Now pivot the axis into the horizontal plane. In so doing, you feel a reaction due to the axis and, to your great surprise, the disc starts turning spontaneously around its axis. Pursue the change of orientation of the axis until its complete reversal along the vertical. You will notice that the rotation velocity increases.

It is easy to do the experiment, sitting on a turning stool, with a bike wheel grasped in your hands.

Being located at the pole and considering the Earth as the carousel, this simple experiment directly establishes the earth's rotation, using a much less cumbersome set up than Foucault's pendulum. It was proposed and realized by A.H. Compton (*Phys. Rev.* 5, February 1915, 109).

1. Determine the angular velocity of the disc $\dot\phi$ around its axis for each angle $\theta(t)$ between its axis and the rotation axis of the carousel.

2. What is the prediction of the calculation if the experiment is realized not at the pole but at a place located at latitude λ ?

1.11. Generalized Inertial Forces
[Solution p. 48] ★ ★ ★

How to use the Lagrangian formalism in a non Galilean frame?

In establishing formula (1.1), there is no hypothesis concerning the choice of the physical frame. The kinetic energy and the acceleration that come out are those relative to this peculiar frame. If this frame is not Galilean, one has to take into account inertial forces and equate $m_\alpha a_\alpha$ to $f_\alpha^{(v)} + f_\alpha^{(i)}$, the sum of the true force acting on the particle α and the corresponding inertial force.

It is important to recall that the inertial force is itself the sum of a driving force due to the acceleration of the origin, of a Coriolis force (depending on the velocity v_α in the given frame) and of a centrifugal force:

$$f_\alpha^{(i)} = -m_\alpha\, a^{(e)} - 2m_\alpha \boldsymbol{\omega} \times v_\alpha - m_\alpha (d\boldsymbol{\omega}/dt) \times r_\alpha - m_\alpha \boldsymbol{\omega} \times (\boldsymbol{\omega} \times r_\alpha).$$

In this case, the formalism leads to Lagrange equations containing additional generalized forces $Q_i \to Q_i^{(v)} + Q_i^{(i)}$.

1. If the given frame moves translationally with an acceleration $a^{(e)}$ (which can depend on time) with respect to the Galilean frame, show that the generalized inertial force is simply:

$$Q_i^{(e)} = -M a^{(e)} \partial_{q_i} R_{cm},$$

where M is the total mass of the system and R_{cm} is the center of mass coordinate. As an application, write Lagrange's equations for a pendulum of length l in a constant gravitational field, whose point of suspension is subject to an imposed arbitrary vertical motion $h(t)$.

2. If the given frame rotates uniformly with a **constant** instantaneous rotation vector $\boldsymbol{\omega}$ with respect to the Galilean frame, show that there exists a generalized Coriolis force

$$Q_i^{(\text{cor})} = \partial_{q_i}(\boldsymbol{\omega} \cdot \boldsymbol{L}) - \frac{d}{dt}[\partial_{\dot{q}_i}(\boldsymbol{\omega} \cdot \boldsymbol{L})]$$

and a generalized centrifugal force

$$Q_i^{(\text{cent})} = \partial_{q_i} T,$$

where
$$\boldsymbol{L} = \sum_\alpha m_\alpha \boldsymbol{r}_\alpha \times \boldsymbol{v}_\alpha$$

is the angular momentum of the system about a point of the axis in the chosen frame and

$$T = \frac{1}{2} \sum_\alpha m_\alpha (\boldsymbol{\omega} \times \boldsymbol{r}_\alpha)^2$$

is the driving kinetic energy of the system (energy of the coincident points).

Hints: it is expedient to introduce the mixed product $[\boldsymbol{a}, \boldsymbol{b}, \boldsymbol{c}] = \boldsymbol{a} \cdot (\boldsymbol{b} \times \boldsymbol{c})$ and its invariance properties under even permutations and change of sign under odd permutations. The following vectorial calculus formulae may also be useful.

$$\begin{aligned} \boldsymbol{a} \times (\boldsymbol{b} \times \boldsymbol{c}) &= (\boldsymbol{a} \cdot \boldsymbol{c})\,\boldsymbol{b} - (\boldsymbol{a} \cdot \boldsymbol{b})\,\boldsymbol{c}; \\ (\boldsymbol{a} \times \boldsymbol{b}) \cdot (\boldsymbol{c} \times \boldsymbol{d}) &= (\boldsymbol{a} \cdot \boldsymbol{c})(\boldsymbol{b} \cdot \boldsymbol{d}) - (\boldsymbol{a} \cdot \boldsymbol{d})(\boldsymbol{b} \cdot \boldsymbol{c}). \end{aligned}$$

Problem Solutions

1.1. The Wheel Jack [Statement and Figure p. 14]

1. Let $ABCD$ denote the lozenge of the jack, the apex A lying beneath the weight, B at the crank and C on the ground; let O be the center of the lozenge, in the middle of the threaded stem BD (see Fig. 1.6).

A priori the configuration of the system is given by α, the angle between the crank and the vertical, and by the form of the lozenge, that is by the values of DB and AC. A first holonomic constraint is due to the invariance of the length, l, of the side of the lozenge: $OA^2 + OB^2 = l^2$.

One has a second holonomic constraint due to the threaded stem, which gives a relation between α and DB (when α varies by 2π, DB varies by h). Finally only the angle α is needed to describe the configuration of the system; it has one degree of freedom.

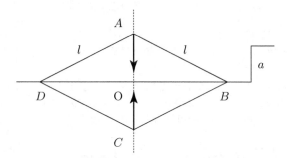

Fig. 1.6 Lozenge $ABCD$ representing schematically the wheel-jack

Let us make a virtual displacement $\delta\alpha$ such that the stem length DB increases by an amount $\delta x = h\,\delta\alpha/(2\pi)$ and the length $OB = OD$ by $\delta OB = \delta x/2$. The length OA decreases, in order to fulfill the relation $OA^2 + OB^2 = l^2$.

Using these conditions, it is easy to see that $\delta CA = 2\delta OA = -\delta x/\tan\theta$. Thus the variation in the altitude of the weight is given as a function of the virtual displacement by: $\delta z = \delta CA = -h\,\delta\alpha/(2\pi\tan\theta)$.

2. The forces concerned are
   - the weight P acting at A, which produces an amount of work:
   $\delta W_P = -P\,\delta z = Ph\,\delta\alpha/(2\pi\tan\theta)$;

   - the reaction force of the ground acting at C which remains at rest; thus the work due to this force vanishes;

   - the force F acting on the crank, the virtual work of which is given by $\delta W_F = -Fa\,\delta\alpha$ (if one wishes to maintain equilibrium, the force must be opposed to the direct rotation considered previously).

The total virtual work is the sum of all these contributions, namely

$$\delta W = \left[\frac{Ph}{2\pi\tan\theta} - Fa\right]\delta\alpha = Q_\alpha\,\delta\alpha.$$

One deduces the generalized force $Q_\alpha = Ph/(2\pi\tan\theta) - Fa$.

At equilibrium, d'Alembert's principle imposes a null generalized force and leads to the required expression:

$$\frac{P}{F} = \frac{2\pi a \tan\theta}{h}.$$

To insure a ratio as large as possible (this is precisely the justification of the jack principle), one must thus choose a large crank arm and/or a small step h for the screw.

Numerical application: With $a = 20$ cm, $h = 0.2$ cm and $\tan\theta = 0.577$, one finds $P/F \cong 363$; these values allow us to maintain a 16,000 N coach with a force of only $F = 11$ N (remember that $P = 16,000/4$ in this special case).

1.2. The Sling [Statement and Figure p. 15]

The system $Oxyz$ is Galilean; the axis Oz is vertical and its unit vector $\mathbf{k}$ is directed upwards. In the plane xOy, it is natural to specify the position of the mass by its distance to O: $OM = \rho$. The angle ϕ between Ox and OM is proportional to time, $\phi = \omega t$, since the angular velocity is kept constant $\dot\phi = \omega$. Note that ϕ is not a coordinate, since it is an externally imposed function.

1. The forces are the weight, along Oz, and the reaction force of the mass on the stem, perpendicular to the stem since we have a frictionless contact. None of these forces performs work during the virtual displacement $\delta\rho$ along OM. The virtual work thus vanishes and the resulting generalized force is null:

$$Q_\rho = 0.$$

2. The expression for the kinetic energy is easy to obtain; it is the usual expression in polar coordinates

$$T = \frac{1}{2}m\left(\dot\rho^2 + \omega^2\rho^2\right).$$

From this, one obtains, with (1.1), the generalized acceleration $A_\rho = m\left(\ddot\rho - \omega^2\rho\right)$. Lastly, the Lagrange equation $A_\rho = Q_\rho = 0$ provides the differential equation $\ddot\rho - \omega^2\rho = 0$. The general solution is well known: $\rho(t) = A\cosh(\omega t) + B\sinh(\omega t)$. The integration constants are determined from the initial conditions $\dot\rho(0) = 0$ and $\rho(0) = a$. One finds $A = a$, $B = 0$. The solution is thus given by:

$$\rho(t) = a\cosh(\omega t).$$

Problem Solutions

Note: The equation $\ddot{\rho} - \omega^2 \rho = 0$ represents the fundamental principle of dynamics in a rotating frame (acceleration = centrifugal force). In this case, a classical treatment is even simpler.

1.3. Rope Slipping on a Table
[Statement and Figure p. 16]

1. Let M be the total mass of the rope and $\mu = L/M$ its linear mass. One can choose as generalized coordinate the length x which hangs vertically. Since the rope is not elastic, all the points α of the rope have the same velocity $v_\alpha = \dot{x}$, $\forall \alpha$. The kinetic energy of the rope is deduced:

$$T = \frac{1}{2} \sum_\alpha m_\alpha v_\alpha^2 = \frac{1}{2} M \dot{x}^2 = \frac{1}{2} \mu L \dot{x}^2.$$

The acceleration follows from (1.1): $A = \mu L \ddot{x}$.

Concerning the real external forces, one distinguishes the weight of the rope and the reaction force of the table (perpendicular to the table since there is no friction); both are vertical. Let us make a virtual displacement δx. The work produced by the weight and by the reaction force on the horizontal part of the table vanishes because the displacement is perpendicular to them. There remains the work of the hanging portion $\mu g x$ of the weight. This work is equal to $\delta W = \mu g x \, \delta x = Q \, \delta x$; the expression of the generalized force follows: $Q = \mu g x$.

The Lagrange equation $A = Q$ leads to $\mu L \ddot{x} = \mu g x$, or $\ddot{x} - \omega^2 x = 0$ with $\omega = \sqrt{g/L}$. The solution of this differential equation is $x(t) = A \cosh(\omega t) + B \sinh(\omega t)$. The integration constants are determined from the initial conditions $\dot{x}(0) = 0$, $x(0) = l$. They imply $A = l$, $B = 0$, hence the solution:

$$x(t) = l \, \cosh\left(t\sqrt{g/L}\right).$$

2. In this case, the reaction force $\boldsymbol{R}$ has both a vertical component R_v and a horizontal one R_h. At equilibrium, the part on the table is subject to the weight $\boldsymbol{P}$, to the reaction force $\boldsymbol{R}$ and to the rope tension $\boldsymbol{T}$ due to the hanging part. One must have $\boldsymbol{P} + \boldsymbol{R} + \boldsymbol{T} = \boldsymbol{0}$. Projection of this equality on the vertical axis gives $P = R_v = \mu(L - l)g$. Projection on the horizontal axis gives $R_h = T$. On the other hand, the tension is also equal to the weight of the hanging part (in order to insure equilibrium): $T = \mu l g = R_h$.

This reasoning "à la Newton" is simpler to understand. The condition of static solid friction imposes $R_h \leq f R_v$ or $l \leq f(L-l)$. It follows that there exists a critical length l_0 for equilibrium: $l \leq l_0$. This minimum length necessary for the motion of the rope is thus:

$$l_0 = \frac{f}{1+f} L.$$

The kinetic energy takes the same form as before and hence $A = \mu L \ddot{x}$. In contrast to the previous case, the horizontal reaction produces work (the tension is an internal force that does no work) and its virtual work is $-R_h \delta x$ (the force acts against the motion). The total virtual work is thus $\delta W = (\mu g x - R_h) \delta x$. On the other hand, for a dynamical friction action, one has: $R_h = f R_v = f P = f \mu g (L - x)$. The generalized force is derived as: $Q = \mu g \left[(1+f)x - fL \right]$.

The Lagrange equation $A = Q$ implies

$$\ddot{x} = \frac{g}{L} \left[(1+f)x - fL \right] \quad \text{or} \quad \ddot{x} = \frac{g}{L}(1+f)(x - l_0) = \omega_d^2 (x - l_0),$$

where we introduced a new dynamical angular frequency in the presence of friction $\omega_d = \sqrt{g(1+f)/L}$. The solution of the resulting differential equation, with the correct initial conditions, is given by an expression of the form:

$$x(t) = l_0 + (l - l_0) \cosh\left(t \sqrt{\frac{g(1+f)}{L}} \right).$$

which is valid for a time less than the time required for the rope to fall.

One should think about the fact, which may seem paradoxical, that, in the presence of friction, the variation in time for the hanging length is greater than the corresponding rate without friction: $\omega_d > \omega$.

1.4. Reaction Force for a Bead on a Hoop
[Statement p. 16]

1. It is possible to begin with Cartesian coordinates expressed in terms of R, θ, ϕ, but it is as simple to deal directly with spherical coordinates since the proposed variables are precisely this type of coordinate. Let us denote as usual the unit vector $\boldsymbol{u}_r$ (along OM), $\boldsymbol{u}_\theta$ (along the motion on the circle) and $\boldsymbol{u}_\phi$ (along the normal to the hoop plane). The expression for the bead velocity is given by $\boldsymbol{v} = R(\dot{\theta} \boldsymbol{u}_\theta + \dot{\phi} \sin\theta \boldsymbol{u}_\phi)$. This is simply the velocity expressed with spherical coordinates when the bead is

constrained to move on a circle ($\dot{R} = 0$). From the velocity, the kinetic energy is expressed as

$$T = \frac{1}{2}mR^2 \left(\dot{\theta}^2 + \dot{\phi}(t)^2 \sin^2 \theta \right).$$

In this particular case, it would be incorrect to consider ϕ as a generalized coordinate; it is simply an externally imposed function. We are faced with a constraint (the hoop) which varies with time. As a consequence, the kinetic energy depends explicitly on time through the function $\dot{\phi}(t)$ (see Fig. 1.7).

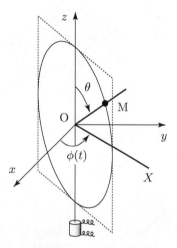

Fig. 1.7 Bead M slipping without rubbing on a hoop with an imposed external rotation

2. Let us give the bead a virtual displacement $\delta\theta$; it moves physically with $\delta\boldsymbol{r} = R\,\delta\theta\,\boldsymbol{u}_\theta$. Since we have a contact without friction, the reaction force is always perpendicular to the hoop and does no work. The only force which produces work is the weight

$$\boldsymbol{P} = mg(\sin\theta\,\boldsymbol{u}_\theta - \cos\theta\,\boldsymbol{u}_r).$$

The corresponding virtual work is

$$\delta W = \boldsymbol{P} \cdot \delta\boldsymbol{r} = mgR\sin\theta\,\delta\theta.$$

Identifying this expression to $Q_\theta\,\delta\theta$, one obtains the generalized force:

$$Q_\theta = mgR\sin\theta.$$

3. From the kinetic energy and using (1.1), one deduces the acceleration:

$$A_\theta = mR^2(\ddot{\theta} - \dot{\phi}^2\cos\theta\sin\theta).$$

The corresponding Lagrange equation $A_\theta = Q_\theta$ leads, after simplification, to the differential equation:[8]

$$\ddot{\theta} = \frac{g}{R}\sin\theta + \dot{\phi}(t)^2 \cos\theta \sin\theta.$$

It is easy to check this result with the help of the fundamental principle of dynamics using the momentum of the weight and of the centrifugal force; this gives the time derivative of the angular momentum (see Fig. 1.8).

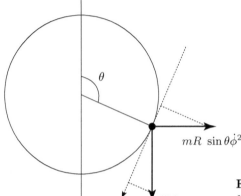

Fig. 1.8 Weight and centrifugal force acting on the bead

The system has only one degree of freedom and it is meaningless to calculate an acceleration A_ϕ, because ϕ is not a generalized coordinate.

4. We are concerned with the component f of the reaction force along the normal $\boldsymbol{u}_\phi$ to the plane of the hoop.[9] If one wishes to calculate it, one must introduce generalized coordinates which give a non null work for this force during the virtual displacement. In order to do this, let us introduce, in addition to θ, the angle ϕ between the plane of the hoop and the plane xOz. It coincides with the angle corresponding to the imposed rotation but, now, instead of considering it as a given function, it must be considered as a full generalized coordinate.

Let $\delta\phi$ be a virtual displacement of the bead. It moves with $\delta\boldsymbol{r} = R\sin\theta \cdot \delta\phi\, \boldsymbol{u}_\phi$. The virtual work is $\delta W = \boldsymbol{f} \cdot \delta\boldsymbol{r} = f\, R\sin\theta\, \delta\phi = Q_\phi\, \delta\phi$. Hence the expression for the generalized force is $Q_\phi = f\, R\sin\theta$.

---

[8] The sign in front of the gravitational restoring term may seem strange. It follows from our choice concerning the definition of the angle θ (from the vertical axis directed upwards).

[9] We could be interested as well by the component in the hoop plane, but along the radial direction.

Problem Solutions   31

Using the kinetic energy and (1.1), the Lagrange equation $A_\phi = Q_\phi$ explicitly gives

$$mR^2\ddot\phi \sin^2\theta + 2mR^2\dot\phi\,\dot\theta \sin\theta \cos\theta = fR\sin\theta.$$

After simplification, one arrives at the required expression:

$$f(t) = 2m\,R\,\dot\phi(t)\,\dot\theta\cos\theta + m\,R\,\ddot\phi(t)\,\sin\theta.$$

In contrast to its virtual work which vanishes, this force produces work during a real displacement of the bead.

Once more, one can check this result classically. In the rotating frame, there exist two terms in the component of the inertial force perpendicular to the hoop plane: the usual Coriolis force and a contribution due to the variation of the angular velocity.

1.5. The Huygens Pendulum
[Statement and Figure p. 17]

Remarks concerning the cycloid

Let us consider an arbitrary point P on a circle (not depicted) such that when the center C lies on the vertical through O, the angle CP with the upward vertical is α. When the circle has rolled by an angle ϕ, the contact point I is horizontally displaced by $R\phi$ (rolling without slipping). The coordinates for P are easily obtained:

$$(R(\phi - \sin(\phi + \alpha)), R(\cos(\phi + \alpha) - 1)).$$

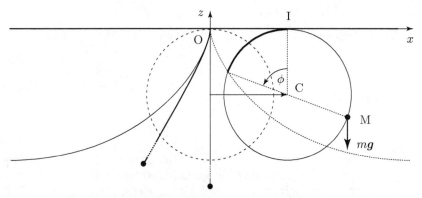

Fig. 1.9 The rope of the Huygens pendulum oscillates between two cycloids. Thus, its length decreases with the deviation from the vertical

The trajectory followed by the point P is a cycloid which exhibits cusps when $\phi = -\alpha$ modulo 2π.

1. For the point M under consideration, $\alpha = \pi$ and its coordinates are $(R(\phi + \sin\phi), -R(1 + \cos\phi))$, see Fig. 1.9.

   The kinetic energy of this point can be calculated at once:
   $$T = \frac{1}{2}m\boldsymbol{v}^2 = mR^2\dot\phi^2(1 + \cos\phi)$$
   and the generalized acceleration follows from (1.1):
   $$A_\phi = mR^2\left[2\ddot\phi\,(1 + \cos\phi) - \dot\phi^2\sin\phi\right].$$

   Let us make a virtual displacement $\delta\phi$ and study the various forces.
   – First, the weight: the corresponding work is expressed as
   $$\delta W = \boldsymbol{P} \cdot \delta\boldsymbol{OM} = -mgR\sin\phi\,\delta\phi,$$
   which provides us with the generalized force $Q_\phi = -mgR\sin\phi$.
   – Secondly, the contact force exerted at the point of contact I between the circle and the axis Ox. To first order in $\delta\phi$ the arbitrary point P is displaced by
   $$(R\delta\phi(1 - \cos(\phi + \alpha)), -R\delta\phi\sin(\phi + \alpha)).$$
   For the given point I, $\alpha = -\phi$, and, to first order, this displacement vanishes. It is a cusp for which both velocity components are null. The contact force does not furnish virtual work and the corresponding generalized force is null.

   The Lagrange equation (1.2) $A_\phi = Q_\phi$ leads, with the definition
   $$\omega = \frac{1}{2}\sqrt{g/R},$$
   to the following differential equation:
   $$2\ddot\phi\,(1 + \cos\phi) - \dot\phi^2\sin\phi = -4\omega^2\sin\phi.$$

2. Let us transform first the expression for the Lagrange equation with the help of well known trigonometric formulae in terms of $\phi/2$; after simplification, we are left with the equation: $2\ddot\phi\cos(\phi/2) - \dot\phi^2\sin(\phi/2) = -4\omega^2\sin(\phi/2)$. Now, let us switch to the variable $x = \sin(\phi/2)$. This last equation is then transformed into the much simpler differential equation:
   $$\ddot x + \omega^2 x = 0.$$

The solution is $x(t) = x_0 \sin \omega t = \sin(\phi(t,x_0)/2)$. Let $T = 2\pi/\omega$; then, it is easily seen that $\phi(t+T, x_0) = \phi(t, x_0)$, independently of the amplitude x_0. In other words, we have to deal with a synchronous pendulum with a period T given by:

$$T = 4\pi \sqrt{\frac{R}{g}}.$$

Now let us consider a simple pendulum whose string is attached at one end to a fixed point on the Oz axis, with a length $4R$ and which is constrained by two cycloids symmetric with respect to Oz, in such a way that the free string length decreases with the oscillation amplitude. It is possible to show that the pendulum bob follows a cycloid similar to that studied in this problem. Indeed, a very good isochronism can be obtained by attaching the string to a flexible blade.

1.6. Cylinder Rolling on a Moving Tray
[Statement p. 18]

Let $Oxyz$ represent a Galilean frame, C the center of the cylinder with abscissa X, H the contact point of the cylinder on the tray (see Fig. 1.10). Considering this point as belonging to the tray, its velocity is $\dot{a}$ (imposed by the operator); considering this point as belonging to the cylinder, its velocity is $\dot{X} + R\dot{\theta}$. The non slipping rolling condition imposes equality for both velocities $\dot{a} = \dot{X} + R\dot{\theta}$. This expression gives a link between the generalized coordinate X and the generalized coordinate θ. The constraint is holonomic.

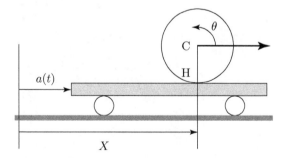

Fig. 1.10 Cylinder rolling without slipping on a tray driven with a motion $a(t)$. X denotes the coordinate of the center of the cylinder and θ its rotation angle

Now consider the cylinder; its kinetic energy is given by $T = \frac{1}{2}M\dot{X}^2 + \frac{1}{2}I\dot{\theta}^2$. The system is described by one degree of freedom.

1. Let us choose first the θ coordinate. Taking into account the previous relation, the kinetic energy can be recast as

$$T = \frac{1}{2}M\left(\dot{a} - R\dot{\theta}\right)^2 + \frac{1}{2}I\dot{\theta}^2.$$

Using (1.1), the value for the acceleration is easily obtained:

$$A_\theta = \left(I + MR^2\right)\ddot{\theta} - MR\ddot{a}(t).$$

The only force to be considered for the virtual work is the weight (the force necessary for the rolling has already be taken into account through the relation between the X and θ coordinates (see Problem 1.5)). For a virtual displacement $\delta\theta$, the center of mass altitude does not vary and the work performed by the weight is null. One deduces a vanishing generalized force: $Q_\theta = 0$. The Lagrange equation leads to:

$$\ddot{\theta} = \frac{MR}{I + MR^2}\ddot{a},$$

which, coupled with the already quoted relation $\ddot{\theta} = (\ddot{a} - \ddot{X})/R$, allows us to find the acceleration of the center of the cylinder:

$$\ddot{X} = \frac{I}{I + MR^2}\ddot{a}(t).$$

2. Let us now choose X as the coordinate. The expression for the kinetic energy is at present:

$$T = \frac{1}{2}M\dot{X}^2 + \frac{1}{2}\frac{I}{R^2}(\dot{a} - \dot{X})^2.$$

One obtains the corresponding acceleration as:

$$A_X = \left(M + \frac{I}{R^2}\right)\ddot{X} - \frac{I}{R^2}\ddot{a}.$$

For a virtual displacement δX, the virtual work furnished by the weight is still null, with the consequence of a vanishing generalized force $Q_X = 0$. In this case, the Lagrange equation leads to

$$\left(M + \frac{I}{R^2}\right)\ddot{X} - \frac{I}{R^2}\ddot{a} = 0,$$

or, in other words:

$$\ddot{X} = \frac{I}{I + MR^2}\ddot{a}(t).$$

One finds the same result, as required.

1.7. Motion of a Badly Balanced Cylinder
[Statement p. 18]

1. The Galilean frame $Oxyz$ is depicted in Fig. 1.11 and the angle θ is defined positively in the trigonometric sense.

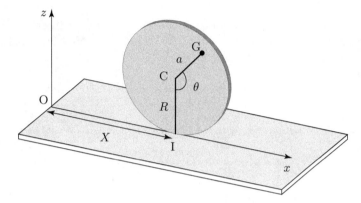

Fig. 1.11 Cylinder rolling without slipping on a horizontal plane. The center of gravity G is out of true by a distance a

The velocity component for point G can be calculated very easily:
$$\boldsymbol{v}_G = (\dot{X} + a\dot{\theta}\cos\theta,\ a\dot{\theta}\sin\theta).$$

The non slipping rolling condition imposes the constraint
$$\dot{X} + R\dot{\theta} = 0.$$

This holonomic constraint allows us to retain the angle θ as the unique coordinate (because $X = -R\theta$). Applying Koenig's theorem, one obtains the total kinetic energy of the cylinder as the sum of the translational energy for the center of mass $\frac{1}{2}Mv_G^2$ and the rotational energy in the center of mass frame which is simply $\frac{1}{2}I\dot{\theta}^2$. Explicitly:
$$T(\theta, \dot{\theta}) = \frac{1}{2}\dot{\theta}^2 \left[I + M(R^2 + a^2 - 2aR\cos\theta)\right].$$

With the help of formula (1.1), the acceleration is derived
$$A_\theta = \ddot{\theta}\left[I + M(R^2 + a^2 - 2aR\cos\theta)\right] + MaR\dot{\theta}^2\sin\theta.$$

The generalized force must now be calculated; the weight $\boldsymbol{P}$ is the only force that performs work during a virtual displacement $\delta\theta$:
$$\delta W = \boldsymbol{P} \cdot \delta z_G = -Mga\sin\theta\ \delta\theta,$$

which is identified with the expression $\delta W = Q_\theta\, \delta\theta$, in order to give the generalized force $Q_\theta = -Mga\sin\theta$. The Lagrange equation follows from (1.2):

$$\ddot{\theta}\left[I + M(R^2 + a^2 - 2aR\cos\theta)\right] + MaR\dot{\theta}^2 \sin\theta = -Mga\sin\theta. \quad (1.7)$$

Multiplying by $\dot\theta$, the Lagrange equation, $A_\theta - Q_\theta = 0$, can be recast in the form $dE(\theta,\dot\theta)/dt = 0$ where $E(\theta,\dot\theta)$ is the energy function, which remains constant at the value E.

$$E = \frac{1}{2}\dot{\theta}^2\left[I + M(R^2 + a^2 - 2aR\cos\theta)\right] - Mga\cos\theta.$$

From this last equation, it is easy to deduce the velocity in terms of the coordinate

$$\dot{\theta} = \pm\sqrt{\frac{2(E + Mga\cos\theta)}{I + M(R^2 + a^2 - 2aR\cos\theta)}}. \quad (1.8)$$

It is useful to discuss the problem of sign and distinguish several regimes which depend on the sign of the quantity $E + Mga\cos\theta$.

- If $E < -Mga$, the sign of the numerator under the square root is always negative and Equation (1.8) cannot be satisfied. No motion is possible.

- If $E > Mga$, the sign is always positive and $\dot\theta$ maintains a constant sign, which depends on the initial conditions. The cylinder always rolls in the same direction, the angular velocity being comprised between two extreme values.

- If $-Mga < E < Mga$, the numerator of $\dot\theta$ in (1.8) vanishes for two values of the angle: $\theta = \pm\theta_0$, with $\cos\theta_0 = |E|/(Mga)$. The cylinder moves by oscillating between these two values where the velocity vanishes and then changes sign.

The curve which corresponds to the value $E = Mga$ discriminating the last two regimes is called a separatrix.

All these regimes are illustrated in the upper part of Fig. 1.12.

2. To obtain the vertical component F_v of the reaction force, it is necessary that it does work to which end one must introduce another coordinate which allows such work. Thus the ordinate of the center C is no longer considered to be a constant R but rather a new coordinate q, subject to a virtual variation. In contrast, the cylinder radius is still R and the constraint relation remains unchanged. The Cartesian coordinates for point G are changed to $x_G = X + a\sin\theta$, $z_G = q - a\cos\theta$.

We recalculate the total kinetic energy as:

$$T(\dot{q}, \theta, \dot{\theta}) = \frac{1}{2} M \left[\dot{q}^2 + 2 a \dot{q} \dot{\theta} \sin\theta \right] + \frac{1}{2} \dot{\theta}^2 \left[I + M(R^2 + a^2 - 2aR\cos\theta) \right].$$

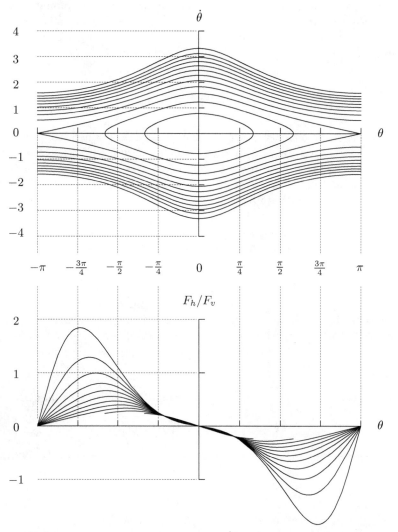

Fig. 1.12 Upper part: phase portrait $\dot{\theta}(\theta, E)$ for several values of the energy E. The outermost curve corresponds to maximum energy just before the cylinder takes off

Lower part: ratio between the tangential and normal components of the reaction force. This ratio must be less than the f coefficient.

The parameters are: $M = R = g = 1$, $a = 0,3$, $I = 0,4$.

The acceleration relative to q can now be obtained:

$$A_q = M\left[\ddot{q} + a\ddot{\theta}\sin\theta + a\dot{\theta}^2\cos\theta\right].$$

For a virtual displacement δq, the weight produces an amount of work $-Mg\delta q$ and work corresponding to the reaction force an amount $F_v\,\delta q$ (it acts so as to oppose the weight). Hence, we obtain the generalized force $Q_q = (F_v - Mg)$. The Lagrange equation, $A_q = Q_q$, in which we substitute $q = R$ ($\ddot{q} = 0$) finally provides the reaction force:

$$F_v(\theta, E) = M\left[g + a(\ddot{\theta}\sin\theta + \dot{\theta}^2\cos\theta)\right] = M\left[g - a\frac{d^2\cos\theta}{dt^2}\right].$$

The energy dependence for the F_v component is obtained using the relation $\dot{\theta}(\theta, E)$ given by (1.8) and the expression $\ddot{\theta}(\theta, E)$ from (1.7). Moreover the condition $F_v > 0$ must be satisfied in order to keep contact with the ground; this is an effect of a centrifugal force which is too strong. The outer curve of the upper part of the Fig. 1.12 corresponds to an energy $\tilde{E}$ responsible for the critical situation $F_v(\pi, \tilde{E}) = 0$. For a greater value of the energy, the cylinder no longer stays on the ground and all the previous equations are meaningless.

3. We now investigate the horizontal component F_h of the reaction force. In order to make it perform work, we have to consider the X coordinate as an independent coordinate no longer connected to θ by a constraint relationship. The Cartesian coordinates of point G are, in this case, $x_G = X + a\sin\theta$, $z_G = R - a\cos\theta$. The total kinetic energy becomes:

$$T(\dot{X}, \theta, \dot{\theta}) = \frac{1}{2}M\left[\dot{X}^2 + 2a\dot{X}\dot{\theta}\cos\theta\right] + \frac{1}{2}\dot{\theta}^2\left[I + Ma^2\right].$$

The acceleration relative to the X coordinate is deduced from (1.1)

$$A_X = M\left[\ddot{X} + a\,d(\dot{\theta}\cos\theta)/dt\right].$$

During a virtual displacement δX, the only work comes from the F_h force: $\delta W = F_h\,\delta X$ (F_h as given here includes its sign which can be positive or negative); the value of the generalized force is derived at once: $Q_X = F_h$. The Lagrange equation $A_X = Q_X$, in which one inserts $\dot{X} = -R\dot{\theta}$, gives the expression for the component of the reaction force:

$$F_h(\theta, E) = M\frac{d}{dt}\left[\dot{\theta}(a\cos\theta - R)\right].$$

The ratio between the tangential and normal components of the force is plotted in the lower part of Fig. 1.12. For a given friction coefficient f, characteristic of the materials, the energy must be such as to allow this ratio to be less than $< f$.

1.8. Free Axle on a Inclined Plane
[Statement p. 19]

Let us use the convention for axes proposed in the statement. The natural frame is defined by the inclined plane, with origin A, with horizontal axis AX, axis AY in the direction of steepest slope, and axis AZ normal to the plane. The coordinates of the center of the axle, O, are denoted X and Y. Let $\boldsymbol{K}$ be the unit vector normal to the plane, $\boldsymbol{u}$ the unit vector along $\boldsymbol{OC}$ and $\boldsymbol{v}$ the unit vector of the plane perpendicular to $\boldsymbol{OC}$. Of course, one has $\boldsymbol{OC} = (L/2)\boldsymbol{u}$. Starting with $\boldsymbol{AC} = \boldsymbol{AO} + \boldsymbol{OC}$ the velocity for point C follows: $\boldsymbol{V}_C = \boldsymbol{V}_O + (L/2)\dot{\theta}\boldsymbol{v}$. In the following discussion, to obtain a quantity relative to wheel C', it is enough to change $L \to -L$ and $\phi \to \phi'$ in the corresponding quantity relative to the wheel C (see the configuration in Fig. 1.13).

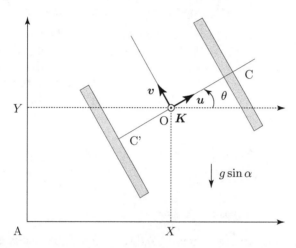

Fig. 1.13 Position of the axle and the two wheels on the inclined plane. The axle center O is specified by its two coordinates X, Y and the axle direction by the angle θ made with the horizontal

1. The axle being massless, only the wheels contribute to the kinetic energy. The instantaneous rotation vector for wheel C is $\boldsymbol{\omega} = \dot{\theta}\boldsymbol{K} + \dot{\phi}\boldsymbol{u}$. The kinetic energy for this wheel is $T_C = \frac{1}{2}m\boldsymbol{V}_C^2 + T_C^{(r)}$. The rotational energy $T_C^{(r)}$ is calculated using $\boldsymbol{\omega}$, the moments of inertia of the wheel and the translational kinetic energy using the expression for the velocity given previously. Thus, one obtains:

$$T_C = \frac{1}{2}m\left[\boldsymbol{V}_O^2 + (L^2/4)\dot{\theta}^2 + L\dot{\theta}\boldsymbol{v}\cdot\boldsymbol{V}_O\right] + \frac{1}{2}\left[I\dot{\theta}^2 + I_3\dot{\phi}^2\right]$$

with a corresponding expression for the other wheel. With $V_O^2 = \dot{X}^2 + \dot{Y}^2$, the kinetic energy of the system, which is the sum of the kinetic energy of the two wheels, is calculated as:

$$T = m\left(\dot{X}^2 + \dot{Y}^2\right) + \left(I + \frac{1}{4}mL^2\right)\dot{\theta}^2 + \frac{1}{2}I_3\left(\dot{\phi}^2 + \dot{\phi}'^2\right).$$

From this expression and from definition (1.1), one deduces the accelerations

$$A_X = 2m\ddot{X}, \qquad A_Y = 2m\ddot{Y},$$
$$A_\theta = 2\left(I + \frac{1}{4}mL^2\right)\ddot{\theta},$$
$$A_\phi = I_3\ddot{\phi}, \qquad A_{\phi'} = I_3\ddot{\phi}'.$$

We may now express the conditions for rolling without slipping.

Let H be the contact point of the wheel C with the plane; the required conditions impose $\mathbf{V}_H = \mathbf{0}$. This last velocity is calculated from that of C and from the instantaneous rotation vector: $\mathbf{V}_H = \mathbf{V}_C + \boldsymbol{\omega} \times \mathbf{CH}$. The non slipping rolling condition provides two scalar conditions. The same thing is applied to wheel C'. Among these four conditions, two of them are identical (those referred to (1.9)). Finally, one has three constraint equations:

$$\dot{X}\cos\theta + \dot{Y}\sin\theta = 0; \tag{1.9}$$

$$\dot{Y}\cos\theta - \dot{X}\sin\theta + \frac{1}{2}L\dot{\theta} + R\dot{\phi} = 0; \tag{1.10}$$

$$\dot{Y}\cos\theta - \dot{X}\sin\theta - \frac{1}{2}L\dot{\theta} + R\dot{\phi}' = 0 \tag{1.11}$$

After simple elimination, these conditions can be recast in the simpler form:

$$2\dot{X} - R\left(\dot{\phi} + \dot{\phi}'\right)\sin\theta = 0; \tag{1.12}$$

$$2\dot{Y} + R\left(\dot{\phi} + \dot{\phi}'\right)\cos\theta = 0; \tag{1.13}$$

$$L\dot{\theta} + R\left(\dot{\phi} - \dot{\phi}'\right) = 0. \tag{1.14}$$

The conditions (1.14) (holonomic) and (1.12), (1.13) (non holonomic) are the relations required for a non slipping rolling motion. We are faced with 5 generalized coordinates $X, Y, \theta, \phi, \phi'$ and 3 differential conditions of type (1.5):

$$\sum_{i=1}^{5} \Lambda_i^{(k)} \delta q_i = 0.$$

The vectors $\mathbf{\Lambda}^{(k)}$ possess the following components:

$$\begin{aligned}
\mathbf{\Lambda}^{(1)} &= (2, 0, 0, -R\sin\theta, -R\sin\theta); \\
\mathbf{\Lambda}^{(2)} &= (0, 2, 0, R\cos\theta, R\cos\theta); \\
\mathbf{\Lambda}^{(3)} &= (0, 0, L, R, -R)
\end{aligned}$$

2. The application points for the reaction forces due to the ground are not displaced during a virtual displacement, because of the non slipping rolling condition (see the reasoning of Problem 1.5); these reaction forces do not imply generalized forces. The only non vanishing virtual work comes from the weight; it is calculated from the displacement of the center of mass O. We easily get $\delta W = -2mg\,\delta z_O = -2mg\,\sin\alpha\,\delta Y$. The only non vanishing generalized force is thus $Q_Y = -2mg\sin\alpha$.

Introducing three Lagrange multipliers $\lambda_1, \lambda_2, \lambda_3$, the constrained Lagrange equations (1.6) are written

$$m\ddot{X} = \lambda_1; \tag{1.15}$$
$$m\ddot{Y} = -mg\sin\alpha + \lambda_2; \tag{1.16}$$
$$2\left(I + \frac{1}{4}mL^2\right)\ddot{\theta} = L\lambda_3; \tag{1.17}$$
$$I_3\ddot{\phi} = -\lambda_1 R\sin\theta + \lambda_2 R\cos\theta + R\lambda_3; \tag{1.18}$$
$$I_3\ddot{\phi}' = -\lambda_1 R\sin\theta + \lambda_2 R\cos\theta - R\lambda_3. \tag{1.19}$$

3. *The interpretation of the Lagrange multipliers*

The right hand side of Equation (1.6), multiplied by δq_i and summed over i, gives, in the case of only one constraint:

$$\sum Q_i \delta q_i + \sum \lambda \Lambda_i \delta q_i.$$

The last term, which is null because of the constraint, takes the form of a virtual work, product of the force responsible for the constraint by the displacement of the application point. In our problem, the expression corresponding to the multiplier λ_1 is $\lambda_1 (2\delta X - R(\delta\phi + \delta\phi')\sin\theta)$. It produces the work performed by the sum of the horizontal components of the reaction force λ_1 to cancel the horizontal displacement $\delta X - R\delta\phi\sin\theta$ of the first wheel and $\delta X - R\delta\phi'\sin\theta$ of the second wheel with respect to the plane.

λ_2 is interpreted as the sum of the components of reaction forces along OY, and, lastly, λ_3 is interpreted as the difference of the components along $\boldsymbol{v}$ which acts against the axle rotation.[10]

4. Let us make the proposed change of variables. After some rearrangements based on (1.14), one obtains $L\dot\theta + R\dot\delta = 0$; and from (1.18–1.19) $\lambda_3 = (I_3/2R)\ddot\delta$. Using (1.17), one arrives at

$$\left[2I + \frac{1}{2}mL^2 + (I_3 L^2/2R^2)\right]\ddot\theta = 0,$$

or $\ddot\theta = 0$, then $\ddot\delta = 0$. The axle spins with a constant angular velocity ω and, with a convenient choice of the time origin, $\theta = \omega t$. It follows that: $\delta - \delta_0 = -(L\omega/R)\,t$ and $\lambda_3 = 0$.

Let us derive the two non-holonomic constraints (1.10), (1.11). Using the proposed variables, after some algebra, one obtains

$$\left(I_3 + mR^2\right)\ddot\sigma = mgR\cos\alpha\,\cos(\omega t),$$

which can be integrated to give

$$\sigma - \sigma_0 = -\frac{4\Gamma}{R\omega^2}\left(\cos(\omega t) - \frac{V_0}{R}t\right),$$

where $\Gamma = \frac{1}{4}\left(mgR^2\sin\alpha\right)/\left(I_3 + mR^2\right)$.

Other results are obtained with no particular difficulty. Let us summarize the solution of the problem

$$\theta(t) = \omega t;$$
$$\delta(t) - \delta_0 = -\frac{L\omega}{R}t;$$
$$\sigma(t) - \sigma_0 = -\frac{4\Gamma}{R\omega^2}\cos(\omega t) - \frac{V_0}{R}t \quad \text{with} \quad \Gamma = \frac{mgR^2\sin\alpha}{4(I_3+mR^2)},$$
$$X(t) = \frac{V_0}{\omega}\left[\cos(\omega t) - 1 + \frac{\Gamma}{\omega V_0}(2\omega t - \sin(2\omega t))\right];$$
$$Y(t) = \frac{V_0}{\omega}\left[\sin(\omega t) + \frac{\Gamma}{\omega V_0}(\cos(2\omega t) - 1)\right],$$

δ_0 and σ_0 are two integration constants which fix the initial values of the angles ϕ and ϕ'.

---

[10] In order to find each reaction force separately, a relation is missing. In fact, very much as in a hyperstatic system (a table with four legs or more on the ground) it is impossible, without further information, to obtain the distribution of the reaction forces (reaction force on each leg).

If we move in a frame which drifts horizontally with constant speed $2\Gamma/\omega$, we recognize a periodic trajectory which passes through the four particular points $(0,0)$, $(-2,0)$, $(-1, 1 - 2\Gamma/(\omega V_0))$, $(-1, -1 - 2\Gamma/(\omega V_0))$.

In the inclined plane, the trajectory of the axle center is plotted in Fig. 1.14 in units of V_0/ω and for several values of the Γ parameter. The line of steepest slope is directed downwards while the horizontal is from left to right.

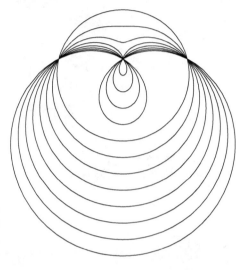

Fig. 1.14 Trajectories of the axle center for several values of parameter Γ in the inclined plane

5. Finally, the reaction forces are obtained quite easily

$$\lambda_1(t) = m\left(-\omega V_0 \cos(\omega t) + 4\Gamma \sin(2\omega t)\right);$$
$$\lambda_2(t) = m\left(-\omega V_0 \sin(\omega t) - 4\Gamma \cos(2\omega t) + g\sin\alpha\right);$$
$$\lambda_3(t) = 0.$$

1.9. The Turn Indicator [Statement p. 21]

We will refer to Fig. 1.15 for the axis and frame conventions. The flywheel rotates around $\hat{Y}$ with a constant angular velocity Ω. The frame rotates with respect to the plane XOz with the instantaneous rotation vector $\dot{\theta}\hat{X}$. Finally, the plane XOz itself rotates with respect to the Earth's frame of reference (assumed to be Galilean) with the instantaneous rotation vector $\omega\hat{z}$. The instantaneous rotation vector of the flywheel with respect to the Galilean frame is thus $\boldsymbol{\omega} = \omega\hat{z} + \dot{\theta}\hat{X} + \Omega\hat{Y}$. This vector is projected onto the axes (XYZ), which are the principle axes of the flywheel, to obtain the components: $\omega_X, \omega_Y, \omega_Z$.

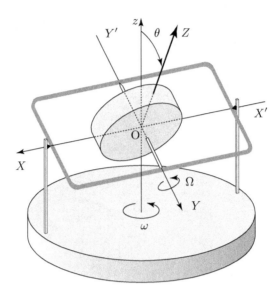

Fig. 1.15 Gyroscope in rotation around the axis $Y'Y$ of its frame with an imposed constant velocity Ω. The frame can oscillate also around an axis $X'X$ locked on a tray rotating at constant velocity ω. The only degree of freedom is the angle θ between the normal to the frame $Z'Z$ and the rotation axis Ox of the tray

1. The rotational kinetic energy is equal to

$$T_{\rm rot} = \frac{1}{2}\left(I_X\omega_X^2 + I_Y\omega_Y^2 + I_Z\omega_Z^2\right).$$

To this energy one must, in principle, add the center of mass kinetic energy of the flywheel. Since this energy is independent of θ, it is of no consequence for our study. Performing the calculations with the previously obtained components of $\boldsymbol{\omega}$ one obtains:

$$T = \frac{1}{2}I_X\left(\dot{\theta}^2 + \omega^2\cos^2\theta\right) + \frac{1}{2}I\left(\Omega - \omega\sin\theta\right)^2.$$

2. The only dynamical variable is θ and the system has only one degree of freedom, since ω and Ω are imposed variables. The acceleration is calculated from (1.1):

$$A_\theta = I_X\ddot{\theta} + I\omega\Omega\cos\theta + \frac{1}{2}\omega^2\left(I_X - I\right)\sin(2\theta).$$

The forces acting on the flywheel are the weight and the efforts exerted on the axis by the frame and the restoring force of the spring. For a

Problem Solutions

virtual displacement $\delta\theta$, the center of mass altitude does not vary and the work of the weight vanishes. The axis $X'X$ does not change its direction and the forces that maintain it perform no work. Lastly, the restoring force performs an amount of work $\delta W = -C\,\delta\theta = -k\theta\,\delta\theta$; this implies a generalized force $Q_\theta = -k\theta$. The Lagrange equation $A_\theta = Q_\theta$ leads to the differential equation:

$$I_X\ddot\theta + I\omega\Omega\cos\theta + \frac{1}{2}\omega^2\,(I_X - I)\sin(2\theta) = -k\theta.$$

3. At equilibrium, one has $\theta = const \Rightarrow \ddot\theta = 0$. If the terms in ω^2 are neglected with respect to $\omega\Omega$, the previous equation gives the required relation:

$$\omega = -\frac{k\theta}{I\Omega\cos\theta}.$$

4. The lift (component of the air reaction force perpendicular to the relative velocity) is perpendicular to the wings and is thus directed along the apparent vertical. In the frame of the aeroplane, this lift balances the weight (vertical) and the centrifugal force (horizontal). A simple drawing shows immediately that $\tan(\alpha) = $ centrifugal force/weight, or (R being the radius of the circle corresponding to the turn):

$$\tan\alpha = \frac{V^2}{Rg} = \frac{\omega V}{g}.$$

5. The restoring torque is exerted between the apparent vertical and the normal to the frame; the angle between these two directions is θ. However the angle between the instantaneous rotation vector $\boldsymbol\omega$ and the true vertical is now $\theta + \alpha$; this is precisely the angle which appears in the expression of the kinetic energy whose value is presently

$$T = \frac{1}{2}I_X\left(\dot\theta^2 + \omega^2\cos^2(\theta + \alpha)\right) + \frac{1}{2}I\left(\Omega - \omega\sin(\theta + \alpha)\right)^2.$$

The rest of the calculation is similar to that quoted in questions 2 and 3. We arrive at the following result:

$$\omega = -\frac{k\theta}{I\Omega\cos(\theta + \alpha)}.$$

To obtain a maximum sensitivity, one requires that a small speed variation leads to a large variation of the reading; this happens when $\cos(\theta+\alpha)$ is maximum. In the vicinity of $\theta \cong -\alpha$, the relation $\omega(\theta)$ becomes linear and

$$|\omega| \cong \frac{k\alpha}{I\Omega} \cong \frac{k\tan\alpha}{I\Omega},$$

or, with the result of the last question, $\omega = k\omega V/(I\Omega g)$, that is:

$$\frac{\Omega I}{k} = \frac{V}{g}.$$

Consequently, the rotational direction and the speed of the engine must be correctly chosen in order that the flywheel plane is as close as possible to the true vertical; this is obviously not the situation depicted in the drawing of the statement!

1.10. An Experiment to Measure the Rotational Velocity of the Earth
[Statement p. 22]

We still refer to the Fig. 1.15 of Problem 1.9 since we deal with the same system, a gyroscope, but employed in a different context.

With respect to its frame $OXYZ$, the disc rotates around the axis OY with angular velocity Ω which, following the statement notation, is simply $\dot{\phi}$. The frame itself rotates with angular velocity $\dot{\theta}$ around the axis OX. Lastly, the carousel rotates around axis Oz with angular velocity ω with respect to a Galilean frame.

1. As a consequence, the instantaneous rotation vector is $\mathbf{\Omega} = \omega\hat{z} + \dot{\theta}\hat{X} + \dot{\phi}\hat{Y}$, which can be rewritten in terms of its components in the frame of the inertial axes of the disc

$$\mathbf{\Omega} = \Omega_X \hat{X} + \Omega_Y \hat{Y} + \Omega_Z \hat{Z}.$$

Simple projection leads to the expression $\mathbf{\Omega} = \dot{\theta}\hat{X} + (\omega \sin\theta + \dot{\phi})\hat{Y} + \omega \cos\theta\,\hat{Z}$.

The kinetic energy, written first as $T = \frac{1}{2}\left[I\Omega_X^2 + I\Omega_Y^2 + I_Z\Omega_Z^2\right]$, is recast as

$$T = \frac{1}{2}\left[\left(\dot{\theta}^2 + \omega^2 \sin^2\theta\right) I + \left(\dot{\phi} + \omega \cos\theta\right)^2 I_Y\right].$$

The rotation ω and the pivoting motion $\theta(t)$ are imposed externally. The only coordinate describing the system is thus ϕ. As a result, the acceleration is

$$A_\phi = I_Y \frac{d}{dt}\left(\omega \cos\theta + \dot{\phi}\right).$$

Let us perform a virtual displacement which consists, **at a given time**, of a small rotation $\delta\phi$ for the disc while maintaining the frame fixed.

Problem Solutions

The weight performs no work since the center of mass is kept fixed; the forces that lock the frame do not furnish work since the frame does not move. Finally, if we suppose a frictionless rotation of the disc around the axis OY, the reaction force on the rotational axis also does not produce work. In summary, the virtual work of external forces vanishes and the generalized force Q_ϕ is null.

Be careful: In a **real** displacement, the momentum of the forces exerted on the frame is not null.

The Lagrange equation $A_\phi = Q_\phi = 0$ leads to the interesting conclusion $\dot\phi + \omega\cos\theta = const$. At the initial time $t = 0$, one has $\theta = 0$ and $\dot\phi = 0$ (the disc is at rest with its axis in the vertical position), whence $\omega = const$. We are led to the desired expression:

$$\dot\phi(t) = \omega\left(1 - \cos\theta(t)\right).$$

When the axis is horizontal ($\theta = \pi/2$) one obtains $\dot\phi = \omega$ and after a complete turn ($\theta = \pi$) $\dot\phi = 2\omega$.

2. It is always possible to choose the carousel axes $OXYz$ with the axis Oz along the true vertical, the axis OX in the southerly direction and axis OY in the easterly direction. The instantaneous rotation vector of the carousel is now given by $(\omega\sin\lambda)\,\hat{z} - (\omega\cos\lambda)\,\hat{X}$. The rest of the treatment is completely similar to that of the previous question. The instantaneous rotation vector for the disc is therefore written:

$$\mathbf{\Omega} = (\omega\sin\lambda)\hat{z} + (\dot\theta - \omega\cos\lambda)\hat{X} + \dot\phi\hat{Y}.$$

With respect to the previous study, it is sufficient to replace ω by $\omega\sin\lambda$ and $\dot\theta$ by $\dot\theta - \omega\cos\lambda$. We then arrive at the equation $\dot\phi + \omega\sin\lambda\cos\theta = const$, which, using the initial conditions, leads to the final expression:

$$\dot\phi(t) = \omega\sin\lambda\left(1 - \cos\theta(t)\right).$$

After a complete turn, the angular velocity of the disc is: $\dot\phi = 2\omega\sin\lambda$. There is no effect at all at the equator, whereas the effect is maximum at the pole.

Remarks:
- In contrast to some other problems, the Lagrangian formalism is much more convenient here than a treatment "à la Newton".
- Mistakes are not reserved to beginners. A.H Compton, Nobel prize, who imagined this experiment, found a result which was half the exact value.

1.11. Generalized Inertial Forces
[Statement p. 23]

The previously derived formula

$$A_i = \frac{d}{dt}(\partial_{\dot{q}_i} T) - \partial_{q_i} T = \sum_\alpha m_\alpha \boldsymbol{a}_\alpha \cdot (\partial_{q_i} \boldsymbol{r}_\alpha)$$

is valid in any frame. On the other hand, Newton's formula $m_\alpha \boldsymbol{a}_\alpha = \boldsymbol{f}_\alpha^{(v)}$, in terms of the true force, is valid only in a Galilean frame. If we want to work in the frame under consideration, we must replace this last equation by $m_\alpha \boldsymbol{a}_\alpha = \boldsymbol{f}_\alpha^{(v)} + \boldsymbol{f}_\alpha^{(e)} + \boldsymbol{f}_\alpha^{(cor)} + \boldsymbol{f}_\alpha^{(cent)}$ where $\boldsymbol{f}_\alpha^{(e)} = -m_\alpha \boldsymbol{a}_\alpha^{(e)}$ is the driven inertial force of the origin, $\boldsymbol{f}_\alpha^{(cor)} = -m_\alpha \boldsymbol{a}_\alpha^{(cor)}$ is the inertial Coriolis force and $\boldsymbol{f}_\alpha^{(cent)} = -m_\alpha \boldsymbol{a}_\alpha^{(cent)}$ is the inertial centrifugal force. Substituting these values in the expression A_i and performing virtual displacements, the virtual work can be expressed in the form

$$\sum_i A_i\, \delta q_i = \delta W^{(v)} + \delta W^{(e)} + \delta W^{(cor)} + \delta W^{(cent)}.$$

The work $\qquad \delta W^{(u)} = \sum_i Q_i^{(u)} \delta q_i$

is that due to the generalized force of type u:

$$Q_i^{(u)} = \sum_\alpha \boldsymbol{f}_\alpha^{(u)} \cdot (\partial_{q_i} \boldsymbol{r}_\alpha).$$

1. **Translation case**

   In case of pure translation, the instantaneous rotation vector is null $\boldsymbol{\omega} = 0$. This means that the Coriolis and centrifugal forces are also null; there remains only the driven inertial force. Moreover, the corresponding acceleration is independent of the point mass: $\boldsymbol{a}_\alpha^{(e)} = \boldsymbol{a}^{(e)}$; $\forall \alpha$. The generalized force is derived as:

$$Q_i^{(e)} = -\boldsymbol{a}^{(e)} \cdot \sum_\alpha m_\alpha (\partial_{q_i} \boldsymbol{r}_\alpha), \quad \text{or} \quad Q_i^{(e)} = -\boldsymbol{a}^{(e)} \cdot \left[\partial_{q_i}\left(\sum_\alpha m_\alpha \boldsymbol{r}_\alpha\right)\right].$$

   Introducing the center of mass coordinate

$$\boldsymbol{R}_{cm} = \frac{1}{M} \sum_\alpha m_\alpha \boldsymbol{r}_\alpha,$$

   one easily arrives at the desired formula:

$$Q_i^{(e)} = -M \boldsymbol{a}^{(e)} \cdot \frac{\partial \boldsymbol{R}_{cm}}{\partial q_i}.$$

Problem Solutions

Application to the pendulum:

Let us direct the vertical downwards and let us specify the direction of the pendulum, with mass m and length l, by the angle θ. The point of suspension, A, is subject to a variation $\overline{OA} = h(t)$. In the system of reference of the pendulum, the kinetic energy is simply $T = \frac{1}{2}ml^2\dot{\theta}^2$ and the generalized force due to the weight is $Q^{(v)} = -mgl\sin\theta$. However, one must add to this force the driven inertial force. Since $\boldsymbol{a}^{(e)} = (0, \ddot{h})$ and $(\partial \boldsymbol{R}_{cm}/\partial\theta) = (l\cos\theta, -l\sin\theta)$, the inertial force is easily derived as $Q^{(e)} = ml\ddot{h}\sin\theta$. The Lagrange equation in the pendulum frame is written, after simplification:

$$\ddot{\theta} + \frac{g - \ddot{h}(t)}{l}\sin\theta = 0.$$

2. Uniform rotation case

In the case of a uniform rotation around a fixed point taken as origin, the driven acceleration of the origin vanishes and the instantaneous rotation vector $\boldsymbol{\omega}$ is constant, with the consequence $d\boldsymbol{\omega}/dt = 0$. Students often forget the term $d\boldsymbol{\omega}/dt$ in generalized forces which vanishes only in the case of a uniform rotation; this is often a good approximation (the Earth's rotation around its axis, or the revolution of the Earth around the sun), but not a general situation. With our hypothesis, we are faced with two inertial forces:

– the Coriolis force: $\boldsymbol{f}_\alpha^{(\text{cor})} = -2m_\alpha \boldsymbol{\omega} \times \boldsymbol{v}_\alpha$ where $\boldsymbol{v}_\alpha$ is the velocity of the point α in the considered frame (relative velocity);

– the centrifugal force: $\boldsymbol{f}_\alpha^{(\text{cent})} = -m_\alpha \boldsymbol{\omega} \times (\boldsymbol{\omega} \times \boldsymbol{r}_\alpha)$.

They give rise to two generalized forces $Q_i^{(\text{cor})}$ and $Q_i^{(\text{cent})}$.

We consider first the Coriolis force

$$Q_i^{(\text{cor})} = -2\sum_\alpha m_\alpha (\boldsymbol{\omega} \times \boldsymbol{v}_\alpha) \cdot (\partial_{q_i}\boldsymbol{r}_\alpha) = -2\sum_\alpha m_\alpha [\boldsymbol{\omega}, \boldsymbol{v}_\alpha, \partial_{q_i}\boldsymbol{r}_\alpha],$$

using the notation [] for the mixed product.

Let us introduce, in our system of reference, the angular momentum with respect to an arbitrary point O chosen on the axis: $\boldsymbol{L} = \sum_\alpha \boldsymbol{r}_\alpha \times m_\alpha \boldsymbol{v}_\alpha$, which, with the known relation $\boldsymbol{v}_\alpha = d\boldsymbol{r}_\alpha/dt = \sum_i \dot{q}_i(\partial_{q_i}\boldsymbol{r}_\alpha) + \partial_t \boldsymbol{r}_\alpha$, allows us to write

$$\boldsymbol{\omega} \cdot \boldsymbol{L} = \sum_{\alpha,i} m_\alpha \dot{q}_i [\boldsymbol{\omega}, \boldsymbol{r}_\alpha, \partial_{q_i}\boldsymbol{r}_\alpha] + \sum_{\alpha,i} m_\alpha [\boldsymbol{\omega}, \boldsymbol{r}_\alpha, \partial_t \boldsymbol{r}_\alpha].$$

Then
$$\partial_{\dot{q}_i}(\boldsymbol{\omega} \cdot \boldsymbol{L}) = \sum_\alpha m_\alpha [\boldsymbol{\omega}, \boldsymbol{r}_\alpha, \partial_{q_i}\boldsymbol{r}_\alpha].$$

We take the total derivative with respect to time, in which we put $d\boldsymbol{r}_\alpha/dt = \boldsymbol{v}_\alpha$ and $d(\partial_{q_i}\boldsymbol{r}_\alpha)/dt = \partial_{q_i}\boldsymbol{v}_\alpha$, to find

$$\frac{d}{dt}[\partial_{\dot{q}_i}(\boldsymbol{\omega} \cdot \boldsymbol{L})] = \sum_\alpha m_\alpha [\boldsymbol{\omega}, \boldsymbol{v}_\alpha, \partial_{q_i}\boldsymbol{r}_\alpha] + \sum_\alpha m_\alpha [\boldsymbol{\omega}, \boldsymbol{r}_\alpha, \partial_{q_i}\boldsymbol{v}_\alpha].$$

Moreover, one has

$$\partial_{q_i}(\boldsymbol{\omega} \cdot \boldsymbol{L}) = \sum_\alpha m_\alpha [\boldsymbol{\omega}, \partial_{q_i}\boldsymbol{r}_\alpha, \boldsymbol{v}_\alpha] + \sum_\alpha m_\alpha [\boldsymbol{\omega}, \boldsymbol{r}_\alpha, \partial_{q_i}\boldsymbol{v}_\alpha].$$

It then suffices to take the difference between these two last equations to find:

$$Q_i^{(\text{cor})} = \frac{\partial(\boldsymbol{\omega} \cdot \boldsymbol{L})}{\partial q_i} - \frac{d}{dt}\left(\frac{\partial(\boldsymbol{\omega} \cdot \boldsymbol{L})}{\partial \dot{q}_i}\right).$$

Note that it is not necessary to suppose a uniform rotation; if a term $\dot{\boldsymbol{\omega}}$ is present in the Coriolis force, it appears in the term $d[\partial_{\dot{q}_i}(\boldsymbol{\omega} \cdot \boldsymbol{L})]/dt$ and the previous formula is still valid.

Let us consider now the centrifugal force

$$Q_i^{(\text{cent})} = -\sum_\alpha m_\alpha [\boldsymbol{\omega} \times (\boldsymbol{\omega} \times \boldsymbol{r}_\alpha)] \cdot (\partial_{q_i}\boldsymbol{r}_\alpha).$$

A well known formula in vector analysis gives $\boldsymbol{\omega} \times (\boldsymbol{\omega} \times \boldsymbol{r}_\alpha) = (\boldsymbol{\omega} \cdot \boldsymbol{r}_\alpha)\boldsymbol{\omega} - \omega^2 \boldsymbol{r}_\alpha$ and allows us to write:

$$Q_i^{(\text{cent})} = -\sum_\alpha m_\alpha [(\boldsymbol{\omega} \cdot \boldsymbol{r}_\alpha)(\boldsymbol{\omega} \cdot (\partial_{q_i}\boldsymbol{r}_\alpha)) - \omega^2 \boldsymbol{r}_\alpha \cdot (\partial_{q_i}\boldsymbol{r}_\alpha)].$$

On the other hand, the driving rotational kinetic energy (energy of the coincident points) is

$$T = \frac{1}{2}\sum_\alpha m_\alpha (\boldsymbol{\omega} \times \boldsymbol{r}_\alpha)^2.$$

After derivation, one obtains

$$\partial_{q_i} T = \sum_\alpha m_\alpha (\boldsymbol{\omega} \times \boldsymbol{r}_\alpha) \cdot (\boldsymbol{\omega} \times \partial_{q_i}\boldsymbol{r}_\alpha).$$

Finally, let us use the vectorial property

$$(\boldsymbol{\omega} \times \boldsymbol{r}_\alpha) \cdot (\boldsymbol{\omega} \times \partial_{q_i}\boldsymbol{r}_\alpha) = \omega^2 (\boldsymbol{r}_\alpha \cdot (\partial_{q_i}\boldsymbol{r}_\alpha)) - (\boldsymbol{\omega} \cdot \boldsymbol{r}_\alpha)(\boldsymbol{\omega} \cdot (\partial_{q_i}\boldsymbol{r}_\alpha)).$$

Then:
$$Q_i^{(\text{cent})} = \sum_\alpha m_\alpha (\boldsymbol{\omega} \times \boldsymbol{r}_\alpha) \cdot (\boldsymbol{\omega} \times \partial_{q_i}\boldsymbol{r}_\alpha) = \frac{\partial T}{\partial q_i}.$$

Chapter 2
Lagrangian Systems

Summary

2.1. Generalized Potential

A generalized force Q_i, associated with the generalized coordinate q_i, is said to arise from a generalized potential V, also called potential energy, if it can be expressed in the form:

$$Q_i(q, \dot{q}, t) = \frac{d}{dt}\left(\frac{\partial V(q,\dot{q},t)}{\partial \dot{q}_i}\right) - \left(\frac{\partial V(q,\dot{q},t)}{\partial q_i}\right). \tag{2.1}$$

If the potential does not depend on velocities, the corresponding force is simply the negative of the gradient of the potential. This is a situation that occurs frequently: constant gravitational field, Coulomb-type or Newton-type law, Hooke's law for springs.

The first term appears for particular forces that depend on velocities. The most important case concerns electric and magnetic forces acting on a particle with charge q_e, which arises from the electromagnetic potential (scalar potential U, vector potential $\boldsymbol{A}$):

$$V(\boldsymbol{r}, \dot{\boldsymbol{r}}, t) = q_e \left(U(\boldsymbol{r},t) - \dot{\boldsymbol{r}} \cdot \boldsymbol{A}(\boldsymbol{r},t) \right). \tag{2.2}$$

There exist also macroscopic forces, which do not arise from a potential, such as solid friction forces.

2.2. Lagrangian System

Lagrange's equations

If all generalized forces arise from a potential, we say that we deal with a **Lagrangian system**.

In this case, it is particularly useful to introduce the **Lagrange function** (or **Lagrangian**) $L(q, \dot{q}, t)$, which is the difference between the kinetic energy function and the potential,

$$L(q, \dot{q}, t) = T(q, \dot{q}, t) - V(q, \dot{q}, t). \tag{2.3}$$

As was the case for the kinetic energy, one considers, in this function, the generalized coordinates and velocities as independent variables. Newton's equations, equivalent to Lagrange's equations (1.2), are expressed in this case in a very elegant way, using the Lagrangian

$$\frac{d}{dt}\left(\frac{\partial L(q, \dot{q}, t)}{\partial \dot{q}_i}\right) = \left(\frac{\partial L(q, \dot{q}, t)}{\partial q_i}\right). \tag{2.4}$$

These equations are still known as Lagrange's equations. They form a system of n **coupled second order differential equations**. Their solution gives the real path $q(t)$ followed by the system; we speak of **trajectories** for the system.[1]

Generalized momentum

The quantity

$$p_i(q, \dot{q}, t) = \frac{\partial L(q, \dot{q}, t)}{\partial \dot{q}_i} \tag{2.5}$$

plays a basic role in Lagrangian theory. We say that p_i is the **generalized momentum** (or simply momentum) associated with the coordinate q_i; alternatively, p_i and q_i are said to be **conjugate variables**.

Lagrange's equation (2.4) can be also written:

$$\dot{p}_i = \left(\frac{\partial L(q, \dot{q}, t)}{\partial q_i}\right). \tag{2.6}$$

---

[1] "Much better" than the potential which is defined up to a constant, it is always possible to add to the Lagrange function, without changing the solution, an arbitrary function of the coordinates which has the form $\dot{q}\partial_q f(q, t) + \partial_t f(q, t)$ or, for a given trajectory, $df(q(t), t)/dt$.

Summary

2.3. Constants of the Motion

Solving the Lagrange equations, which are differential equations of second order in time, is made easier if it is possible to display **constants of the motion**. These are functions that depend only on generalized coordinates and velocities (but not on accelerations), which remain constant in time when **they are calculated along the trajectory**, hence their name.

In some sense, we have lowered the degree of the equation and transformed a second order equation into a first order one. This favorable situation is often possible for Lagrangian systems. There exists no general algorithm for the obtention of constants of the motion, but rather recipes that can be tested in each practical case.

Energy

Let us assume that, for a judicious choice[2] of generalized coordinates, the Lagrange function does not depend explicitly on time. In this case, there exists a constant of the motion, known as **energy**,[3] which is given by the following formula:

$$E(q, \dot{q}) = \sum_i p_i(q, \dot{q}) \dot{q}_i - L(q, \dot{q}) = \text{const.} \tag{2.7}$$

This case corresponds to a very frequent situation and one should think of it before any further procedure.

Cyclic or ignorable coordinate

If the Lagrange function does not depend on the generalized coordinate[4] q_i, this last coordinate is said to be **cyclic** or **ignorable**. In this case, as is easily seen from Equation (2.6), the momentum p_i associated with this generalized coordinate is a constant of the motion

$$p_i(q, \dot{q}, t) = \frac{\partial L(q, \dot{q}, t)}{\partial \dot{q}_i} = \text{const.} \tag{2.8}$$

Once more, one should try to find a set of generalized coordinates such that some of them are cyclic.

---

[2] For instance for systems subject to rotation, if one works in the rotating frame.

[3] Very often, but not always, this quantity can be identified with the sum of the kinetic and potential energies.

[4] On the other hand, it depends on the generalized velocity $\dot{q}_i$.

Let us consider for instance a particle in a plane subject to a central potential. Expressed in terms of Cartesian coordinates, the Lagrangian

$$L(x, y, \dot{x}, \dot{y}) = \frac{1}{2}m(\dot{x}^2 + \dot{y}^2) - V\left(\sqrt{x^2 + y^2}\right)$$

does not exhibit cyclic coordinates. However, the same Lagrangian expressed with polar coordinates, i.e.,

$$L(\rho, \phi, \dot{\rho}, \dot{\phi}) = \frac{1}{2}m(\dot{\rho}^2 + \rho^2\dot{\phi}^2) - V(\rho)$$

shows clearly that ϕ coordinate is cyclic.

Noether's theorem – constants of the motion associated with symmetries

Imagine a set of generalized coordinates $q(s)$ which depend on a continuous parameter s. If the complete Lagrangian does not depend on this parameter, then the following quantity, I, is a constant of the motion ($q = q(s = 0)$):

$$I(q, \dot{q}, t) = \sum_i \left(\frac{\partial L(q, \dot{q}, t)}{\partial \dot{q}_i}\right) \frac{dq_i(s)}{ds}\bigg|_{s=0} = \text{const.} \qquad (2.9)$$

This property is known as the **Noether theorem**.

Application – translational invariance

The total momentum[5] is a constant of the motion.

Let us consider a system with a set of generalized coordinates $\tilde{q}$ (which are lengths), such that if the frame origin is displaced by a along axis Oz, some coordinates (known as intrinsic coordinates) are unchanged $q = \tilde{q}$, whereas others, q, obey the relation $q(a) = \tilde{q} + a$.

If the system is invariant under translation along Oz, its Lagrangian cannot depend on the choice of origin on Oz, i.e., on the choice of a. From Noether's theorem, the quantity associated with this translational invariance

$$P_z = \sum_i \partial_{\dot{q}_i} L(q, \dot{q})$$

is a constant of the motion, called the component of the total momentum of the system along Oz (the sum over i is effective only for the non-intrinsic coordinates). If the invariance concerns the three orthogonal axes of the frame, the component along each of the axes is a constant of the motion and the total momentum $\boldsymbol{P} = (P_x, P_y, P_z)$ provides three constants of the motion.

---

[5] Generalized momentum and linear momentum are often identical quantities, but this is not always the case. For instance for a particle embedded in a uniform magnetic field, translational invariance holds: the generalized momentum along the field direction is a constant of the motion but not the linear momentum.

Application – rotational invariance

The total angular momentum[6] is a constant of the motion.

If the system is invariant under a rotation around the Oz axis, its Lagrangian is unchanged when we switch from the original coordinates $\tilde{q}$ (which are angles) to new ones $q(\phi) = \tilde{q} + \phi$, where ϕ corresponds to a change in the angle origin in a plane perpendicular to the Oz axis. Noether's theorem provides us with a constant of the motion

$$L_z = \sum_i \partial_{\dot{q}_i} L(q, \dot{q}),$$

which is referred to as the component of the total angular momentum of the system along Oz. If the system is isotropic, this invariance is fulfilled for the three orthogonal axes of the frame, and the total angular momentum $\boldsymbol{L} = (L_x, L_y, L_z)$ provides three constants of the motion.

2.4. Two-body System with Central Force

The system is composed of two particles with masses m_1 and m_2, subject to a force arising from a potential which depends only on the distance between the particles, $V(|\boldsymbol{r}_1 - \boldsymbol{r}_2|)$. Writing the corresponding Lagrangian, it is easily shown that there exist three constants of the motion identified with the total momentum. With a convenient choice of the frame – the center of mass frame – it is possible to cancel the total momentum.

The study of the system can then be applied to a fictitious particle with mass $m = m_1 m_2/(m_1 + m_2)$, the **reduced mass**, subject to a central force field acting from the center of mass.[7] There exist also constants of the motion due to angular momentum; it can be deduced that the motion takes place in a plane and that the modulus of the angular momentum[8] $\sigma = m\rho^2 \dot{\phi}$ is a constant of the motion ($\rho = |\boldsymbol{r}_1 - \boldsymbol{r}_2|$ is the relative distance between the two bodies and ϕ is the angle that specifies the relative direction in the plane of motion with respect to an arbitrary axis).

---

[6] Just as we made a subtle distinction between generalized momentum and linear momentum, it is necessary to distinguish between the angular momentum and the kinetic momentum. Very often these two quantities refer to the same thing, but not necessarily. In case of rotational invariance, the angular momentum is a constant of the motion but not the kinetic momentum.

[7] Of course, one can treat with the same formalism the case of a single particle in a central potential.

[8] It is also the sum of kinetic momenta of the two particles with respect to the center of mass.

This property is equivalent to the famous law on "areal velocity": the areal velocity A remains constant in time $dA/dt = \sigma/(2m)$ (the particle sweeps out equal areas in equal times).

In polar coordinates, the trajectory[9] $\rho(\phi)$ is given by Binet's equation

$$\frac{d^2u}{d\phi^2} + u + \frac{m}{\sigma^2}\frac{dV(1/u)}{du} = 0 \tag{2.10}$$

with $u = 1/\rho$. The constant of the motion $\sigma = m\rho^2\dot\phi$ allows us to recover the temporal motion $\phi(t)$ from the trajectory with the help of an integral.

Kepler problem (or attractive Coulomb) for a confined motion

For a potential $V(\rho) = -K/\rho$, the trajectory of relative motion is an ellipse, with major axis a and with one focus at the origin. The energy E, which is a negative quantity, is a constant of the motion. The revolution period T_r and eccentricity e of this ellipse are given by:

$$T_r = 2\pi\sqrt{\frac{m}{K}}\,a^{3/2}; \qquad e = \sqrt{1 + \frac{2E\sigma^2}{mK^2}}. \tag{2.11}$$

The harmonic problem

In this case, the force varies linearly with the distance (example Hooke's law) and the corresponding potential is $V(\rho) = \frac{1}{2}k\rho^2$. The trajectory is again an ellipse but the origin is at the center of the ellipse and not at a focus. The revolution period and the values of axes of the ellipse are obtained through:

$$T_r = 2\pi\sqrt{\frac{m}{k}}; \qquad E = \frac{ka^2}{2} + \frac{\sigma^2}{2ma^2} = \frac{kb^2}{2} + \frac{\sigma^2}{2mb^2}. \tag{2.12}$$

2.5. Small Oscillations

The equilibrium configurations of a Lagrangian system correspond to the extrema of the potential function. Around the potential minima, the equilibrium is stable,[10] whereas around maxima the equilibrium is unstable. Weakly deviated from a stable equilibrium configuration, the potential is

---

[9] More precisely the trajectories of each body are similar to that of the "relative motion" since one has $r_{1,2} = m_{2,1}\rho/(m_1 + m_2)$.

[10] This means that if we displace the system from its equilibrium position, it will return if there are friction forces, or oscillate around this position in absence of friction.

identified with that of a harmonic oscillator. The motion is a superposition of the n proper modes, the typical combination being a function of the initial conditions. A proper mode is a solution for which all the generalized coordinates have a harmonic dependence in time (same angular frequency) but with distinct phases and amplitudes.

Practically, to find a proper mode, one begins by writing the Lagrangian in terms of new coordinates $Q = q - q^{(\text{equil})}$, which measure the deviation from the equilibrium position, and their corresponding velocities $\dot{Q}$. The linear terms are absent because of the equilibrium condition, and we retain only the quadratic terms.

We then seek the proper modes (k), that is the complex numbers $Q_i^{\max}$ and angular frequencies $\omega^{(k)}$ (proper frequencies) such that

$$Q_i^{(k)}(t) = \mathrm{Re}\left(Q_i^{\max} e^{i\omega^{(k)} t}\right)$$

are solutions of Lagrange's equations. Generally, one finds a homogeneous linear system whose n eigenvalues are the n angular frequencies $\omega^{(k)}$ of proper modes (in general the frequencies appear as squares in this system). The solution of the problem is a superposition of proper modes

$$Q_i(t) = \sum_{k=1}^{n} Q_i^{(k)}(t),$$

the amplitudes of which are determined from the initial conditions.

Problem Statements

2.1. Disc on a Movable Inclined Plane
[Solution p. 76] ★

Study of a very simple Lagrangian system

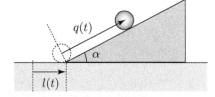

Fig. 2.1 Disc rolling without slipping on a movable inclined plane

In a two-dimensional space, a disc is able to roll without slipping on an inclined plane under the influence of a constant gravitational vertical field g.

We note by R, I, m, respectively the radius, the moment of inertia with respect to the axis and the mass of the disc, α, the slope of the inclined plane and $l(t)$ the temporal equation of the inclined plane imposed by an operator (see Fig. 2.1).

1. Describe the motion of the disc for a horizontal displacement imparted to the inclined plane? Is the disc able to climb the slope?
2. Check this result, in a more elegant way, working in the non Galilean frame of the inclined plane. You will first verify that the inertial translational force, studied in Problem 1.11, arises from the potential

$$V_{\text{trans}} = m\boldsymbol{a}^{(e)} \cdot \boldsymbol{R}_{\text{cm}}.$$

2.2. Painlevé's Integral [Solution p. 77] ★ ★

How to search for a constant of the motion when the Lagrangian depends on time

A simple pendulum, with length l and mass m, free to oscillate in a vertical plane, is subject to the Earth's gravity. One imposes on its point of suspension a motion described by the law $a(t)$ on a horizontal straight line in this plane. One may choose, as a generalized coordinate, q, the angle between the pendulum direction and the vertical oriented downwards.

1. Is this pendulum a Lagrangian system?
2. Give the Lagrange equation. Interpret your result.
3. We now consider the case of a constant acceleration $\ddot{a}$. Determine a function $W(q, \dot{q}, t)$ such that $\partial_t L(q, \dot{q}, t) = d(W(q, \dot{q}, t))/dt$.
4. Deduce from this that $I(q, \dot{q}, t) = E(q, \dot{q}, t) + W(q, \dot{q}, t)$ is a constant of the motion, called Painlevé's integral. Interpret your result.
5. Find again this result working in a non Galilean frame. In the case of a constant acceleration, show that there exists a constant of the motion, which is precisely Painlevé's integral.

2.3. Application of Noether's Theorem
[Solution p. 78] ★

Very simple application of a fundamental theorem

In a two-dimensional space, a particle of mass m, located by its Cartesian coordinates x, y, is subject to a potential of the form $V(x - 2y)$.

Problem Statements

1. Write down the Lagrangian of the system. If x increases by the arbitrary quantity s, what is the increase in y in order for the potential to be unchanged ? Deduce that the Lagrangian is invariant in a group of oblique translations.

2. Write down the new set of coordinates, which depend continuously on s, and which leave the Lagrangian invariant. With the help of Noether's theorem show that the quantity $\dot{x} + \frac{1}{2}\dot{y}$ is a constant of the motion.

   Check – This result can be checked writing the two Lagrange equations and eliminating the derivative of the potential.

2.4. Foucault's Pendulum
[Solution and Figure p. 79] ★ ★

Study of a famous experiment in the Lagragian formalism

This experiment was realized, in March 31st 1851, with a 67 m pendulum beneath the dome of the Pantheon; it was revived in 1902, after Foucault's death (1819–1868), by Camille Flammarion (1842–1925).

Let a simple pendulum of length l and with mass m be located at a latitude λ (complementary angle between the vertical at this point and the Earth's rotation axis $\boldsymbol{Z}$) on the Earth surface. The pendulum thus moves on a sphere (Fig. 2.2). One wishes to study the effect of the Earth's rotation on the motion of the pendulum, in a very elegant way, using the Lagrangian formalism. The effect due to the Earth's revolution around the Sun is neglected.

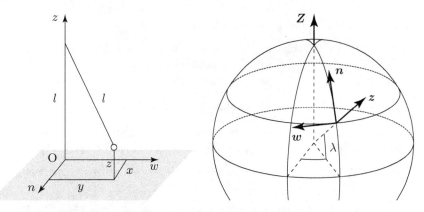

Fig. 2.2 Foucault's pendulum. On the left hand side, the trajectory of Foucault's pendulum on the ground. On the right hand side, the position of the pendulum on the Earth's surface

We will take as generalized coordinates x, y the deviations with respect to the vertical along the south-north direction n and along the east-west direction w.

1. In the limit of **small deviations** with respect to the vertical, show that the potential energy to within a constant, is

$$V(x,y) \cong mg\frac{x^2 + y^2}{2l}.$$

   Here g represents the acceleration due to gravity only, the direction of which is the true vertical.

2. Give, with respect to the northerly axis n, the westerly axis w and the true vertical (passing through the Earth's center) z at this position, the components of the pendulum velocity with respect to the Earth. One neglects $\dot{z}$. Why?

3. Give the components of the unit vector Z (rotation axis of the Earth, see Fig. 2.2) on n and z.

4. Give the driving velocity due to the Earth's rotation (radius R_E, angular velocity Ω). The term in z is neglected. Why?

5. Derive the expression of the kinetic energy and of the Lagrangian. One neglects all terms containing the square of the Earth's rotational speed, except those containing the Earth's radius. Justify.

6. Write down the Lagrange equations.

7. The equation for the x coordinate contains a constant term. Show that it can be eliminated by the substitution $\tilde{x} = x - x_e$. Give the value of the constant x_e and its interpretation.

8. To solve the two Lagrange equations on an equal footing, the complex function $u(t) = \tilde{x}(t) + i\,y(t)$ is introduced. Show that the equation to be solved is

$$\ddot{u} + 2i\Omega \sin\lambda\,\dot{u} + gu/l = 0.$$

   Rather than using a systematic method for finding the solution, it is convenient to make the change of function $u(t) = U(t)\exp(irt)$.

   Choose r and ω as real numbers in order to obtain the equation $\ddot{U} + \omega^2 U = 0$.

   What is the nature of the motion seen in terms of variables X, Y with $U = X + iY$?

   Describe the motion in terms of variables x, y?

   What happens at the pole and on the equator?

Hints:
- The velocity of a point M, which is part of a solid and which rotates with angular velocity Ω around the axis $\boldsymbol{u}$ is $\Omega \boldsymbol{u} \times \boldsymbol{OM}$ where O is any point on the rotation axis.
- In the complex plane, the multiplication by $\exp(irt)$ rotates a point by an angle rt. This method is known as the switch towards rotating axes. It is also employed for a charged particle in a magnetic field.
- The equilibrium position is not the vertical at the point under consideration (there exists also the centrifugal force which acts against gravity).

Alternative derivation

One works in the non Galilean frame $\boldsymbol{n}, \boldsymbol{w}, \boldsymbol{z}$.

Express the Coriolis force (the centrifugal force is neglected).

Find the potential corresponding to the Coriolis force?

Write down the Lagrangian and the Lagrange equations.

Compare with the first method.

The Lagrangian is time-independent. What is the constant of the motion?

2.5. Three-particle System
[Solution and Figure p. 82]   ★ ★

How astute changes of variables allow us to exhibit symmetries

A – Changing coordinates

Let us consider a system formed with three particles, of equal mass m, constrained to move on a straight line $x'Ox$. They interact via a potential that depends only on the relative distance between them. This system can represent the vibrations of a linear triatomic molecule.

Let q_1, q_2, q_3, the abscissae, be chosen as generalized coordinates. The corresponding Lagragian is then written:

$$L = \frac{1}{2}m(\dot{q}_1^2 + \dot{q}_2^2 + \dot{q}_3^2) - V_1(q_2 - q_3) - V_2(q_3 - q_1) - V_3(q_1 - q_2).$$

1. Give the constant of the motion, associated with spatial translations.

2. One works with the Jacobi coordinates[11] X, x, y defined by:

$$X = \frac{1}{3}(q_1 + q_2 + q_3); \quad x = q_1 - q_2; \quad y = \alpha\left(q_3 - \frac{1}{2}(q_1 + q_2)\right),$$

where α is a constant.

Adjust α so that the kinetic energy is written

$$T = \frac{3}{2}m\dot{X}^2 + \frac{1}{4}m(\dot{x}^2 + \dot{y}^2),$$

and give the expression of the new Lagrange function. Find again the previous constant of the motion. What is the number of intrinsic degrees of freedom for the system?

In this change of coordinates, particle 3 was privileged. One can as well favor particle 1 using new coordinates X, x', y', obtained by a circular permutation.

3. Show that, if one introduces the complex numbers $z = x + iy$; $z' = x' + iy'$, one obtains the very simple relation $z' = \exp(2i\pi/3)\, z$. Deduce that $x^2 + y^2 = x'^2 + y'^2$. Express the Lagrangian with these new coordinates.

B – Harmonic approximation case

One assumes now that the potentials are identical[12] and have the same expression $kr^2/2$ where r is the relative distance between the interacting particles (harmonic potential). It is possible to obtain this condition for an arbitrary potential, if one considers small variations with respect to the equilibrium position.

4. Give the expression of the Lagrangian and find three independent constants of the motion. Interpret the different proper modes of the system.

C – From the three-body system to the system studied by Hénon and Heiles

In the case of three identical potentials $V(r) = V_0 \exp(r/a)$, employed in the so-called Toda's net, it was numerically noticed that, in addition to the constants of the motion associated with the invariance with respect to space and time translations, there exists a third constant of the motion, a very infrequent situation. It was only in 1974 that two researchers – M. Hénon and H. Flachka – found mathematically and independently this constant of the motion. As we shall see in Chapter 6, the problem is considered as entirely solved; one says that the system is integrable.

---

[11] It is possible to generalize this transformation to a three dimensional space and, less easily, for three particles with different masses.

[12] It is quite easy to generalize to potentials with different strengths.

5. Let us consider the case where the distances between the particles are so small so as to justify truncating the expansion of the potential at third order. In this limit, give the expression of the Lagrangian. Plot schematically the equipotential curves.

6. Using the numbers z and z' of the third question, show that the previous potential is invariant under the permutation $(x, y) \to (x', y')$. Deduce that the equipotential curves exhibit a symmetry of order 3. Prove that one of these curves is an equilateral triangle.

It happens that this kind of potential can mimic the motion of a star in the galaxy. The study of this model, which was carried out by Hénon and Heiles, is connected to the stability of the solar system. Very surprisingly, this approximate potential does not exhibit a constant of the motion.

2.6. Vibration of a Linear Triatomic Molecule: The "Soft" Mode
[Solution p. 86] ★ ★

Academic case of the appearance of a mode with a null angular frequency

We are interested in the proper modes of longitudinal vibrations for a triatomic molecule, linear and symmetric. We employ a simple model with respective point-like masses m M m, whose attractive forces are represented by two identical springs of constant k. We have already seen that this harmonic approximation is relevant whatever the interaction potential, provided that the deviation from the equilibrium position is small. We use as generalized coordinates the displacements x_i of mass i with respect to its equilibrium position.

1. Write down the Euler-Lagrange equations. Set $\Omega^2 = k/m$ and $r = M/m$.

2. Find the proper angular frequencies and the corresponding proper modes.

3. A proper mode, known as "soft", has a null frequency. Explain the corresponding motion. In this case, the solution does not depend on time as does a harmonic oscillator. Find its time dependence. This mode corresponds to a conserved quantity. Which one and why?

4. Give a simple interpretation of the two other modes.

5. With a precise impulse, we give, at some initial time, a linear momentum p to the first mass. The molecule is initially at rest in its equilibrium position. Give the equations which describe the temporal evolution for the three masses. Interpret the result.

2.7. Elastic Transversal Waves in a Solid (F Waves) [Solution p. 88] ★ ★

Passing from a discrete to a continuous model for an elastic bar

Imagine an initially straight bar that is deformed transversely. This deformation will evolve in time, but how?

Consider a bar with section S and with mass density ρ. Virtually cut the bar into $N+1$ slices with thickness δ (Fig. 2.3) and place an identical mass at the center of mass of each slice, which we refer to as a node. Each of these nodes moves transversely by q_i with respect to a null deformation, and this last displacement is chosen as the coordinate relative to node i.

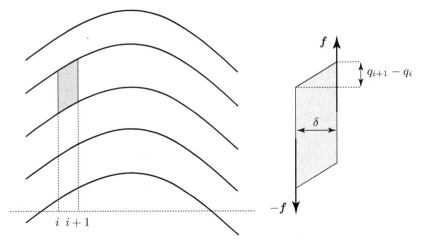

Fig. 2.3 Plane shearing waves in an elastic solid. The bar is cut into an infinite number of slices with vanishing thickness, which have a kinetic energy and an elastic potential energy corresponding to their shearing deformation

1. Give the expression for the kinetic energy of the bar.

    Each slice has an elastic potential energy: to get $q_i \neq q_{i+1}$, one must exert a shearing force f which is proportional to the deformation

    $$u = \frac{q_{i+1} - q_i}{\delta}$$

    and to the bar section S. Let $f = 2\mu u S$, where μ is the shearing modulus characteristic of the solid.

2. Determine the elastic energy of this sheared slice.

3. Deduce the total potential energy, then the Lagrangian.

4. Show, switching to a continuous treatment, that the speed for elastic transversal waves is $\sqrt{2\mu/\rho}$ (some km/s). In order to do this, it is useful to introduce a function $\varphi(x)$ such that $q_i = \varphi(x = i\delta)$. We then investigate the limit $\delta \to 0$ in Lagrange's equations which are finally transformed into a partial differential equation.

Note : Longitudinal waves, known as S waves (S for Second), can also be studied; they are also transmitted in fluids. They propagate more slowly and arrive after the F waves (F for First).

2.8. Lagrangian in a Rotating Frame
[Solution p. 89] ★ ★

Modification to the Lagrangian when one works in a rotating frame

A particle of mass m (or a set of particles) is studied in a frame which rotates around the Oz axis with an angular velocity $\dot{\phi}$.

1. Write down the kinetic energy choosing, in this frame, as generalized coordinates, first the Cartesian coordinates X, Y, then the polar coordinates ρ, ψ.

2. Deduce that, in the rotating frame, the inertial Coriolis and centrifugal forces arise from the potential $V = -\dot{\phi}L_z - \frac{1}{2}m\dot{\phi}^2\rho^2$, where L_z is the projection of the kinetic momentum on the rotation axis in the rotating frame, and ρ the distance of the particle to the rotation axis.

3. Consider the case of a rotating frame with an instantaneous rotation vector $\boldsymbol{\omega}$. Generalize the previous question to obtain the expression of the potential: $V = -\boldsymbol{\omega} \cdot \boldsymbol{L} - \frac{1}{2}m(\boldsymbol{\omega} \times \boldsymbol{r})^2$.

    Note: This expression can be used with an arbitrary choice of the coordinates.

4. With the help of this formula, find the expression of the Lagrangian for the bead on the hoop studied in Problem 1.4, Page 16.

5. Following on from the first question, point out the analogy with a particle in a frame that rotates with a small angular velocity, and the case of a charged particle in a Galilean frame embedded in a magnetic field $\boldsymbol{B}$ along Oz axis.

2.9. Particle Drift in a Constant Electromagnetic Field
[Solution and Figure p. 91]   ★ ★

Important example of motion in an electromagnetic field

A – Lagrangians and constants of the motion

In a frame with orthonormal axes, one wishes to study the motion of a particle of mass m and electric charge q_e, placed in an electric field $\boldsymbol{E} = (E, 0, 0)$ and a magnetic field $\boldsymbol{B} = (0, 0, B)$ uniform, constant and orthogonal to $\boldsymbol{E}$. As generalized coordinates, we choose the three Cartesian coordinates $\boldsymbol{r} = (x, y, z)$ of the particle. We will check that Expression (2.2) is indeed a generalized potential.

1. Give the expression of the Lorentz force, which is assumed to be the only force acting on the particle.

2. Calculate the scalar potential $U(\boldsymbol{r})$.

3. Show that one can choose a vector potential under the alternative forms $\boldsymbol{A} = (-yB, 0, 0)$ or $\boldsymbol{A} = (0, xB, 0)$ or the half sum $\boldsymbol{A} = -\frac{1}{2}(\boldsymbol{r} \times \boldsymbol{B})$. These formulae are very useful in practice for the physicist.

4. With the help of the first expression for $\boldsymbol{A}$, check that the generalized potential associated with the electromagnetic fields is written $V = q_e(\dot{x}yB - xE)$. Write the Lagrangian L of the particle in the electromagnetic field. Check that z is a cyclic coordinate and give the corresponding constant of the motion. Deduce the Lagrange equations relative to the variables x and y.

5. With the help of the second expression, check that the generalized potential associated with the electromagnetic field is $V = -q_e(x\dot{y}B + xE)$. Write the Lagrangian L' of the particle in the electromagnetic field. Check that y and z are cyclic coordinates and give the two corresponding constants of the motion. Could they be anticipated from the previous question?

6. Check that the difference between the Lagrangians L and L' is the total derivative with respect to time of a function depending on coordinates only.

B – Null electric field

For the moment, we assume a null electric field: $E = 0$.

7. From conserved quantities, deduce the first order in time differential equation for the three coordinates. Integrate these equations introducing the complex function $w(t) = x(t) + iy(t)$. It will be useful to introduce the

"cyclotron frequency" $\omega = q_e B/m$. Characterize the motion along this field and projected onto a plane perpendicular to the field.

This motion is known as cyclotron motion.

C – General case

Let us return to the general case $E \neq 0$.

8. Integrate the equations of motion. Show that, if one works in a frame moving with velocity $-E/B$ along the axis perpendicular to E and B, the particle follows the cyclotron motion studied in the previous question. This motion is known as a drift motion.

2.10. The Penning Trap
[Solution and Figure p. 94] ★ ★ ★

The drift in a radial electric field may be used to trap particles

Physics made considerable progress due to the fact that we are now able to work with a unique particle that is confined in space in a trap. For charged particles, such traps exist, for instance the Penning trap which is the subject of this problem and the Paul trap which will be presented in Chapter 4.

The principle is quite simple. We just saw in the previous Problem 2.9 that a charge in orthogonal electric and magnetic fields follows a cyclotron motion, associated with a drift perpendicular to the electric field with a drift speed equal to the ratio E/B.

Let us imagine a radial electric field, rather than a uniform electric field. The drift perpendicular to this field will occur around a circle, thus confining the charge in the plane perpendicular to the magnetic field. Such a radial field, from Poisson's law, is associated with a perpendicular electric field, which confines the charge in the vertical direction as well.

Let us study this trap in more detail.

With the help of two electrodes, one can create a scalar potential

$$U(x,y,z) = \frac{1}{4}k\left(2z^2 - x^2 - y^2\right)$$

with cylindrical symmetry around the Oz axis. In addition, we impose a uniform magnetic field, with intensity B, directed along this axis.

A particle of mass m and electric charge q_e is placed in this electromagnetic field.

1. Verify Poisson's equation with this type of potential. Plot schematically two equipotential curves with opposite signs. Using the gauge suggested in the previous problem $\boldsymbol{A} = -\frac{1}{2}\boldsymbol{r} \times \boldsymbol{B}$, write down the particle Lagrangian.

2. Give the three differential equations for motion on the three axes. Solve the equation of motion along Oz. It is convenient to introduce the cyclotron frequency $\omega_c = q_e B/m$ and the axial frequency $\omega_a = \sqrt{q_e k/m}$.

3. Use the trick of Problem 2.9 (introduction of a complex variable) to solve the two other equations. Under which condition concerning the electromagnetic field, is the charge confined in the three directions? Write down the stability condition for the Penning trap.

4. The motion in the plane xOy results from the composition of two uniform circular motions of frequencies ω_m (the smallest one) and ω_c. Interpret these two motions in the limit $\omega_c \gg \omega_a$.

   In order to do so, calculate the electric field in the plane $z = 0$, at a distance ρ from the revolution axis. What would be the drift velocity if the field is uniform. What would be the drift angular speed which is also called magnetron frequency.

5. The surroundings are time independent and there exists a constant of the motion: the particle energy. Calculate this energy as a function of the radii of the two circular motions.

2.11. Equinox Precession [Solution p. 97]   ★ ★ ★

Analysis of an important astronomical phenomenon with the Lagrangian method

In a first approximation, the trajectory of the Earth, of center O and mass $M_E = 5.974 \times 10^{24}$ kg, around the Sun, of center S and mass $M_S = 1.989 \times 10^{30}$ kg, is a circle of center S and radius $R = 1.496 \times 10^8$ km in a fixed plane (with respect to a Galilean frame attached to the stars), called the ecliptic. One considers in this ecliptic two fixed axes Sx and Sy, with unit vectors $\boldsymbol{u}_x, \boldsymbol{u}_y$ and the Sz axis, with unit vector $\boldsymbol{u}_z$, perpendicular to this plane. Thus the trihedron $Sxyz$ is orthonormal and Galilean. We denote by α the polar angle (Sx, SO) of the Earth on its orbit, which is accomplished in a year = 365.25 days.

On the other hand, the Earth is considered as a solid which retains its shape. We denote by $(OXYZ)$ a frame linked to the Earth (thus non Galilean) with unit vectors $\boldsymbol{u}_X, \boldsymbol{u}_Y, \boldsymbol{u}_Z$. The OZ axis is chosen as the pole axis (directed from south pole to north pole), OX and OY being two arbitrary perpendicular axes in the equatorial plane. The Earth rotates around its OZ axis with angular velocity Ω, assumed to be constant. It performs a complete revolution relative to the stars during a sidereal day which lasts 23 h 56 min 04 s. In a first approximation, the OZ direction is fixed relative to the stars; the angle $\theta = (\boldsymbol{u}_z, \boldsymbol{u}_Z) = 23°26'$ is called the declination angle with respect to the ecliptic. It is responsible for the seasons on the Earth.

Lastly, the Moon, of mass $M_M = 7.35 \times 10^{22}$ kg, rotates around the Earth along a circle, of center O and radius $r_M = 3.844 \times 10^5$ km. We assume in the following that the plane of this circle coincides with the ecliptic.

Because of the fact that the Earth is not exactly a sphere, the Sun's attraction manifests itself, in addition to a force exerted at the center of mass, by a torque which tends to make the Earth's rotation axis OZ perpendicular to the direction Sun-Earth SO. The consequence of this torque is a slow precession of the pole axis OZ around the fixed axis Sz which is known as the equinox precession; it has been measured experimentally with a high precision: a complete revolution needs 25,785 years.

Before looking at the consequences of this torque, we attempt to obtain a better understanding of its origin.

To this end we consider the Earth as a solid with a cylindrical symmetry axis OZ, and denote by I, I, I_3 the respective moments of inertia with respect to the inertial axes OX, OY, OZ. If the Sun is considered as spherical, its action on a mass dm in the Earth, located at $\boldsymbol{r}(x, y, z)$, produces a potential energy which is

$$dV = -\frac{G\,dm\,M_S}{\sqrt{(\boldsymbol{R}-\boldsymbol{r})^2}}$$

where $\boldsymbol{R}(X, Y, Z) = \boldsymbol{OS}$.

1. Expand this expression up to second order in r/R. Sum the contributions due to all the mass elements of the Earth. In the course of this derivation the inertia matrix

$$I_{ij} = \int dm\, x_i x_j$$

will be encountered. Simplify this expression considering that O is the center of mass of the Earth and that OX, OY and OZ are inertia axes. It is possible to take advantage of the freedom left in the determination of OX and OY to choose the OX axis in the plane SOZ (Fig. 2.4). We denote by ϕ the angle (OS, OZ). Express the potential energy not as a function of X, Y, Z but as a function of ϕ and R. The ϕ independent terms are responsible for the center of mass motion which is not interesting for the present purposes. The Earth is slightly flat at the poles. What would be the effect of the torque if there were no spin of the Earth?

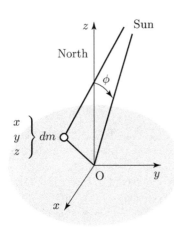 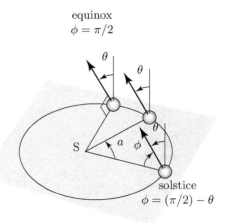

Fig. 2.4 Diagram for the calculation of the inertial tensor of the terrestrial geoid

Fig. 2.5 Revolution of the Earth around the Sun in the ecliptic plane

2. As is seen in Fig. 2.5, the angle ϕ varies during the revolution of the Earth around the Sun. Let us therefore admit that we should consider the potential energy averaged over one revolution. Express $\cos\phi$ as function of α and the declination angle θ. Deduce the average value of the potential as a function of θ.

3. At a given time t, one can choose the OX axis in the ecliptic plane. Let Ψ be the angle between $\boldsymbol{u}_x$ and $\boldsymbol{u}_X$. Give the expression of the instantaneous rotation vector $\boldsymbol{\omega}$ of the Earth in a reference system attached to the Earth. Note that $\boldsymbol{\omega}$ does not arise solely from the daily rotation.

4. Give the Earth's kinetic energy in the Galilean frame.

5. We are concerned now by the effect of the torque on the Earth's axis. Using the Lagrangian formalism, write down the Lagrange function and derive the equations of motion.

6. Show that an admissible solution of these equations is the absence of nutation $\theta = $ const. Deduce the precession angular velocity $\dot\Psi_S$ of the OZ axis along $\boldsymbol{u}_z$ due to the Sun.

7. The effect of the Moon due to the same phenomenon is far from negligible. Assuming that the Moon's orbit plane coincides with the ecliptic, calculate the ratio $\beta = \dot\Psi_M/\dot\Psi_S$ between the contributions of the Moon and the Sun. With the given data, calculate the numerical value of β.

Problem Statements

8. It can be shown that the Moon's contribution behaves in the same way as that of the Sun; hence, the precessional angular velocity measured experimentally is $\dot{\Psi} = \dot{\Psi}_M + \dot{\Psi}_S$. Deduce the expression of the Earth's "flatness" given by $(I_3 - I)/I_3$. Calculate its numerical value using the value $G=6,673 \times 10^{-11}$ MKSA for the gravitational constant.

9. One assumes that the Earth's shape is an ellipsoid of revolution with an equatorial radius R_e slightly greater than the polar radius R_p. Let us denote by $R_E = (R_e + R_p)/2 = 6,367$ km the Earth's mean radius. Calculate $(I_3 - I)/I_3$ as a function of R_e, R_p and R_E. From the value obtained previously, calculate the numerical values of R_e and R_p. Compare your results with the experimental values obtained by geodesic methods: R_e= 6,378.137 km and $R_p = 6,356.752$ km. Comments.

2.12. Flexion Vibration of a Blade
[Solution p. 102] ★ ★ ★

Why an embedded blade is able to vibrate?

Consider a homogeneous, uniform, thin blade, of mass M, length l and linear mass μ, which lies at rest in a horizontal position. Imagine for instance a knife or a flat ruler. If this blade is deformed, it attempts to return to equilibrium and starts moving, transforming its elastic potential energy into kinetic energy. It then overshoots its equilibrium position and the process repeats: there is vibration.

To study such vibrations, the Lagrangian formalism, based on a discontinuous model for the blade and ignoring the effect of gravity, will be used.

The discontinuous model

The blade is cut into N segments each of length δ and mass $\mu\delta$. The blade motion is assumed to take place always in the same vertical plane (absence of torsion). One denotes by A_i the middle of segment i (its center of mass) and one takes as generalized coordinates q_i the deviation of A_i from the equilibrium position along the vertical. Lastly, one assumes that the deformation is weak, and this imposes that the angle made by each segment with the horizontal remains small. Under these conditions, it is permissible to make the hypothesis that each segment has a fixed length δ, always equal to its horizontal projection during the deformation.

1. Give the expression of the kinetic energy of the blade.

   Let us consider now the elastic potential energy. From elasticity textbooks, one learns that the elastic energy, stored in a unit length, of a bent blade is $EI/(2R^2)$ where I is the so-called elastic moment of inertia, which is the product of the section area s by the square of the thickness e divided by 12 ($I = se^2/12$), R is the radius of curvature of the blade at the considered point and E, Young's modulus which is characteristic of the material.

2. One considers that the radius of curvature, R_i, for segment i is the radius of the circle passing through the three points A_{i-1}, A_i, A_{i+1} (see Fig. 2.6). Give an approximate expression, valid in the limit of weak flexions, of this radius in terms of the generalized coordinates q_{i-1}, q_i, q_{i+1}.

3. Ignoring the border effects, give the potential energy of the blade.

4. Write the Lagrangian for the blade and derive the Lagrange equations. You will notice that the equation for q_i requires the functions q_{i-2}, q_{i-1}, q_i, q_{i+1}, q_{i+2}.

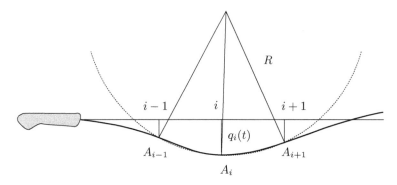

Fig. 2.6 Vibration of a blade in a vertical plane. Illustration of the curvature radius on an elementary segment

Passing to the continuous limits

Let $\varphi(x, t)$ be an interpolation function for the blade satisfying condition $\varphi(x = i\delta, t) = q_i(t)$ for all points A_i. Consider the limit $N \to \infty$, $\delta \to 0$; $N\delta = l = $ const.

1. Write down the partial differential equation that is satisfied by $\varphi(x, t)$.

2. Seek a solution of type $\varphi(x, t) = e^{-i\omega t}\psi(x)$. Find the differential equation satisfied by $\psi(x)$. As usual, one looks for a solution of the form $\psi(x) = Ae^{ikx}$. Solve this equation with the following limit conditions. The blade is embedded at one of its ends (knife handle): null displacement, null first derivative; the other end is free: infinite radius of curvature = null second derivative, null third derivative. Find the angular frequencies of the vibrations.

3. Estimate the vibrational frequency of your knife taking $E = 20 \times 10^{10}$ P.

Be careful: during the vibration, the blade does not have enough time to exchange heat with the surroundings. The flexion should therefore be considered as adiabatic, and one should use the corresponding Young's modulus.

Hints:

The 4th order derivative of a function can be approximated by

$$f^{(4)}(x) \cong \frac{f(x-2h) - 4f(x-h) + 6f(x) - 4f(x+h) + f(x+2h)}{h^4}.$$

For the solution of the 4th order differential equation, seek four independent solutions of real or complex type. Combine them to get the correct limit conditions.

2.13. Solitary Waves
[Solution and Figure p. 105] ★ ★

Waves that propagate without deformation, solutions of a non linear equation

The so-called sine-Gordon equation represents the passage to a continuous limit of a chain of simple pendulums coupled by a harmonic type force. The field $\varphi(x,t)$ describing the evolution of the system satisfies the partial differential equation (sine-Gordon equation)

$$\partial_{t^2}^2 \varphi - \frac{\lambda}{l^2} \partial_{x^2}^2 \varphi = -\omega^2 \sin\varphi$$

where λ is a constant parameter and l the common length of the pendulums.

The sine-Gordon equation possesses the property of exhibiting special solutions known as solitary waves or solitons, which move without deformation. Except for particular cases (electromagnetic, acoustic waves,...) the solutions of wave equations spread out sooner or later. One may think of the waves in the ocean. Existence of solitons is connected to the non linearity of the wave equation.[13]

The aim of this problem is to find a solution of this type for the sine-Gordon equation. These solitary waves are illustrated spectacularly by the tidal wave phenomenon which produces a bore, during flood tides, in the estuary of some large rivers, which moves up the river for several kilometers.

---

[13] The non linearity implies that the superposition of solutions is no longer a solution.

1. Show that, retaining generality, by a suitable change of space and time scales, the sine-Gordon equation can be written (we recall that $\varphi(x,t)$ represents the angular deviation from the downward directed vertical for the pendulum located at abscissa x at time t):

$$\frac{\partial^2 \varphi(x,t)}{\partial t^2} - \frac{\partial^2 \varphi(x,t)}{\partial x^2} = -\sin \varphi(x,t).$$

2. One seeks a solution of the form $\varphi(x,t) = \phi(u)$ with $u = \lambda(x - vt)$, the soliton speed v and the coefficient λ being undetermined for the moment. Show that this solution must fulfill the equation:

$$\frac{d^2 \phi(u)}{du^2} = \sin \phi(u)$$

if one chooses λ in a suitable way (be careful of the change of sign on the right hand side). Give the relationship between λ and v. What is the inequality that must be satisfied by the "speed" v?

We desire a solution which, for $|u| = \infty$, corresponds to a stable equilibrium situation for the chain of pendulums, for instance $\phi(u = \pm\infty) = 0$ (or 2π).

You will notice that the equation

$$\frac{d^2 \phi(u)}{du^2} = \sin \phi(u)$$

is the same as that corresponding to a pendulum with unit angular frequency, if u is taken as the time.

3. What is the constant of the motion associated with a translation of u? What value must be given to this function in order for the solution to satisfy $\phi(u = \pm\infty) = 0$? Show that, in this case, the equation to be solved is

$$\frac{d(\phi/2)}{du} = \pm \sin(\phi/2).$$

4. Show that the solutions of this equation are $\phi(u) = 4\arctan(e^{\pm u})$. Deduce two solitary solutions $\varphi_\pm(x,t)$. For several choices of the speed, plot these solutions schematically at a given time. You will find that the speed depends on the wave front stiffness.

2.14. Vibrational Modes of an Atomic Chain [Solution and Figure p. 107]   ★ ★ ★

Solvable model for a system composed of an infinite number of coupled oscillators

An infinite atomic chain is composed of two different types of atoms, disposed in an alternate way, the heavy ones of mass m and the light ones of mass $m' = rm$, with $r < 1$. The interaction between these atoms can be approximated by attractive forces represented by identical springs of constant k between each atom. In the following, we put $\Omega^2 = k/m$. It is possible to agree on an orientation of the chain and decide that the light atom numbered n comes before the heavy atom numbered n, following the orientation. As generalized coordinates, we use the deviations with respect to the equilibrium position which we denote for the nth light and heavy atoms respectively as u_n and v_n.

1. Write down the Euler-Lagrange equations.

2. Write the equations giving the proper modes.

3. The translational invariance for two atoms suggests that we search, for each type of atom, a solution of the form

$$u_n(t) = U \exp(in\phi + i\omega t)$$
$$v_n(t) = V \exp(in\phi + i\omega t)$$

   Give the two solutions $\omega(\phi)$. The largest one, $\omega_{\text{opt}}(\phi)$, is called the optical angular frequency, whereas the other one, $\omega_{\text{acou}}(\phi)$, is called the acoustical angular frequency.

4. Study the behaviour of $\omega(\phi)$ in the vicinity of $\phi = 0$. For each of these cases, study the sign of U/V. A positive sign corresponds to a motion in phase of all the atoms at low frequency, a negative sign corresponds to a motion with opposite phase for two neighboring atoms at high frequency. Give $\omega(\phi = \pi)$. Plot schematically $\omega_{\text{opt}}(\phi)$ and $\omega_{\text{acou}}(\phi)$.

5. Let us excite an atom with a wave of angular frequency ω. For what condition on ω does the perturbation propagate? Remark that this atomic chain is a filter with a stopping band or "gap".

6. Periodic conditions are imposed to the motion of the atoms. The motion of the group $N + 1$ is identical to that of group 1. Deduce the possible values for ϕ.

Problem Solutions

2.1. Disc on a Movable Inclined Plane
[Statement and Figure p. 57]

Let H be the contact point of the disc of center C on the inclined plane; one chooses $q = OH$ as generalized coordinate, O being an arbitrary origin chosen on the inclined plane. The rolling without slipping condition imposes $\dot{q} = -R\dot{\theta}$, where θ is the angle that specifies an arbitrary point B of the disc.

The velocity of the center of the disc is obtained easily adding the imposed velocity of the origin, which is $\dot{l}(t)$ along the horizontal and its relative velocity to the inclined plane which is $\dot{q}$ along the slope. As reference axes, let us choose the direction of the slope and the normal to the inclined plane; the velocity of the center is $\boldsymbol{v}_C = (\dot{q} + \dot{l}\cos\alpha, \dot{l}\sin\alpha)$. The kinetic energy of the disc is obtained after an application of Koenig's theorem as the sum of the center of mass kinetic energy $\frac{1}{2}mv_C^2$ and the rotational energy around the center $\frac{1}{2}I\dot{\theta}^2 = \frac{I}{2R^2}\dot{q}^2$.

1. The only force that performs work is the weight which arises from the potential $V = mgq\sin\alpha$ (up to an unimportant constant). After some algebra, the expression for the Lagrangian is obtained as

$$L(q, \dot{q}, t) = \frac{1}{2}m\left[\left(1 + \frac{I}{mR^2}\right)\dot{q}^2 + 2\dot{l}(t)\dot{q}\cos\alpha + \dot{l}(t)^2\right] - mgq\sin\alpha.$$
(2.13)

The Lagrange equation (2.4) provides the law giving the acceleration:

$$\left(1 + \frac{I}{mR^2}\right)\ddot{q} = -g\sin\alpha - \ddot{l}(t)\cos\alpha.$$

This is the equation of a movable object in a variable gravitational field. Let us integrate once in time to obtain the velocity:

$$\left(1 + \frac{I}{mR^2}\right)\dot{q} = -gt\sin\alpha - \dot{l}(t)\cos\alpha.$$

One can have $\dot{q} > 0$ – the disc can climb up the plane – if the condition $\dot{l}(t) < -gt\tan\alpha$ is fulfilled.

The time dependence of the motion is obtained by a further integration (up to a constant):

$$q(t) = -\left(1 + \frac{I}{mR^2}\right)^{-1}\left[\frac{g\sin\alpha}{2}t^2 + l(t)\cos\alpha\right].$$

Problem Solutions

2. The driven acceleration is horizontal: it is precisely that of the inclined plane, $a^{(e)} = \ddot{l}$. The horizontal component of $\boldsymbol{R}_{\text{cm}}$ in this frame is $q \cos \alpha$ and the inertial potential, as indicated in the statement, is written in this case $V_{\text{trans}} = mq\ddot{l} \cos \alpha$. The corresponding Lagrangian is:

$$L(q, \dot{q}, t) = \frac{1}{2} m \left(1 + \frac{I}{mR^2} \right) \dot{q}^2 - mgq \sin \alpha - mq\ddot{l} \cos \alpha.$$

It becomes identical to the previous one (2.13), if one adds the total time derivative of a function depending on coordinates only. The last term on the right hand side can be interpreted as an apparent weight (principle of equivalence: gravitational field – accelerated frame).

2.2. Painlevé's Integral [Statement p. 58]

1. Let O be a fixed origin, Ox a horizontal axis, Oz a vertical axis directed downwards and A the point of suspension of the pendulum, which moves along Ox according to the law $\overline{OA} = a(t)$. The only force that performs work is the weight which arises, following our conventions, from the potential $V = -mgz$. Hence, the system is a Lagrangian system.

2. Let q be the angle between the Oz axis and the pendulum direction; this is the generalized coordinate. The components of the pendulum (of mass m) position vector are $x = a + l \sin q$ and $z = l \cos q$. The kinetic energy is simply $T = \frac{1}{2}m(\dot{x}^2 + \dot{y}^2)$ and the potential energy $V = -mgl \cos q$. The Lagrangian is obtained by taking the difference between these quantities, that is in terms of the generalized coordinate:

$$L(q, \dot{q}, t) = \frac{1}{2} m \left(l^2 \dot{q}^2 + 2 l \dot{a}(t) \dot{q} \cos q + \dot{a}(t)^2 \right) + mgl \cos q.$$

There is an explicit time dependence through the $\dot{a}(t)$ function.

In this case, Lagrange's equation gives, introducing the usual angular frequency $\omega = \sqrt{g/l}$:

$$\ddot{q} + \omega^2 \sin q + \frac{\ddot{a}(t)}{l} \cos q = 0.$$

3. From the previous Lagrangian expression, it is easy to find that $\partial_t L = m\ddot{a}(l\dot{q} \cos q + \dot{a})$ which, in the case of a constant acceleration $\ddot{a}$, can be recast as $\partial_t L = m\ddot{a} \, d(l \sin q + a)/dt = dW/dt$, with the obvious definition of the function W:

$$W(q, t) = m\ddot{a}(l \sin q + a(t)).$$

4. The energy function is given, as usual, by $E = \dot{q}\,\partial_{\dot{q}}L - L$, or, performing the calculations,

$$E = \frac{1}{2}m(l^2\dot{q}^2 - \dot{a}^2) - mgl\cos q.$$

Generally, one always has $dE/dt = -\partial_t L$ which, in this particular case, takes the value $-dW/dt$. This implies that $d(E+W)/dt = 0$ so that the quantity $I = E + W$ is a constant of the motion, called Painlevé's integral.

Owing to the constant acceleration, which allows us to obtain, by integration, the velocity and the displacement laws, we are able to calculate Painlevé's integral which, after simplification and elimination of constant terms, is written:

$$I(q,\dot{q}) = \frac{1}{2}ml\left(l\dot{q}^2 - 2g\cos q + \ddot{a}\sin q\right).$$

5. In the frame where the point of suspension is at rest, the kinetic energy is simply $T = \frac{1}{2}ml^2\dot{q}^2$. To the gravitational potential energy, one must add the potential driven energy

$$V_{\text{trans}} = m\boldsymbol{a}^{(e)} \cdot \boldsymbol{R}_{\text{cm}} = m\ddot{a} \times (a + l\sin q).$$

Up to a time dependent function only, the Lagrangian is written:

$$L(q,\dot{q},t) = \frac{1}{2}m\left(l^2\dot{q}^2 + l\left(g\cos q - \ddot{a}(t)\sin q\right)\right),$$

which corresponds to a pendulum in an apparent gravitational field. When the acceleration is constant, this Lagrangian does not depend explicitly on time. There exists a constant of the motion, which is precisely Painlevé's integral I.

2.3. Application of Noether's Theorem
[Statement p. 58]

1. The Lagrangian $L = T - V$ is easily obtained:

$$L(x,y,\dot{x},\dot{y}) = \frac{1}{2}m(\dot{x}^2 + \dot{y}^2) - V(x - 2y).$$

Let $X = x + s$, $\dot{X} = \dot{x}$. In order to leave the kinetic energy unchanged, the new variable Y must fulfill $\dot{Y} = \dot{y}$, hence $Y = y + a$. The potential energy must be invariant as well, $x - 2y = X - 2Y$, which implies $a = s/2$. The transformation group that leaves the Lagrangian invariant is given by:

$$X = x + s; \quad Y = y + s/2.$$

2. Noether's theorem can be applied safely. There exists a constant of the motion

$$I = (\partial_{\dot{x}}L)\,(dX/ds)|_0 + (\partial_{\dot{y}}L)\,(dY/ds)|_0$$

or:
$$I = \dot{x} + \frac{y}{2} = \text{const.}$$

One can directly check this property starting from the two Lagrange's equations $m\ddot{x} = -V'$, $m\ddot{y} = 2V'$, and eliminating the potential derivative to get $\ddot{x} + \frac{1}{2}\ddot{y} = 0$ which, after integration, gives again the result of Noether's theorem.

2.4. Foucault's Pendulum
[Statement and Figure p. 59]

1. Let us adopt the axis conventions represented in Fig. 2.2. In the laboratory rotating frame, the point of suspension of the pendulum lies at altitude l. At equilibrium, the pendulum mass is placed at the origin. For a small deviation θ with respect to the vertical, the altitude is $z = l(1 - \cos\theta) \approx l\theta^2/2$. Let us perform an approximate calculation of the gravitational potential energy: $V = mgz \approx (mgl\theta^2)/2$. Furthermore,

$$\theta^2 \approx \sin^2\theta = \frac{OM^2}{l^2} = \frac{x^2 + y^2}{l^2}.$$

This leads to an approximate expression for the gravitational potential:

$$V(x,y) = mg\frac{x^2 + y^2}{2l}.$$

2. The pendulum coordinates in the frame attached to the Earth are by definition (x, y, z). The pendulum velocity in this frame is thus $(\dot{x}, \dot{y}, \dot{z})$. For a small deviation from equilibrium, the coordinates x and y are of order $l\theta$, whereas z is of order $l\theta^2$, hence negligible with respect to the horizontal components. It is then fully justified to consider that the motion takes place in the horizontal plane and that the relative velocity is given by:

$$\boldsymbol{v}_r = \dot{x}\,\boldsymbol{n} + \dot{y}\,\boldsymbol{w}.$$

3. The vector $\boldsymbol{n}$ is in the plane defined by the pole axis and the true vertical $\boldsymbol{z}$. As a consequence, the unit vector along the pole axis $\boldsymbol{Z}$ is in the plane formed by the vectors $(\boldsymbol{n}, \boldsymbol{z})$. A simple analysis based on the various projections shows that:

$$\boldsymbol{Z} = \cos\lambda\,\boldsymbol{n} + \sin\lambda\,\boldsymbol{z}.$$

4. The instantaneous rotation vector is directed along the pole axis: $\mathbf{\Omega} = \Omega \mathbf{Z}$. Let M_0 represent the coincident pendulum point at a given time. The driving velocity is simply expressed as $\mathbf{v}_e = \mathbf{\Omega} \times \mathbf{OM}_0$. Using the definition

$$\mathbf{OM}_0 = x\,\mathbf{n} + y\,\mathbf{w} + (R_E + z)\,\mathbf{z}$$

and the previous relation to express the instantaneous rotation vector, one obtains the equation which gives the driving velocity:

$$\mathbf{v}_e = -\Omega y \sin\lambda\,\mathbf{n} + \Omega\,(x\sin\lambda - (R_E + z)\cos\lambda)\,\mathbf{w} + \Omega y \cos\lambda\,\mathbf{z}. \quad (2.14)$$

5. The absolute velocity of the pendulum in the Galilean frame is obtained by summing the relative velocity and the driving velocity given in Questions 2 and 4: $\mathbf{v}_a = \mathbf{v}_r + \mathbf{v}_e$. The kinetic energy is obtained from $T = \frac{1}{2}m v_a^2$. Taking into account the small value $\Omega \approx 10^{-5} rad/s$, values around unity for x, y, the very small value of z and the very large value of $R_E \approx 10^6$ m, one must retain in the expression for T the terms Ωx, Ωy and $R_E \Omega^2$ (order 10^{-5}) but one can neglect the terms $\Omega^2 x^2$, $\Omega^2 y^2$, Ωz and $\Omega^2 xz$ (order 10^{-10}). Lastly, the Lagrangian L is the difference between the kinetic energy T and the potential energy as given in Question 1. It is of the form:

$$L = \frac{1}{2}m\left[\dot{x}^2 + \dot{y}^2 - 2\Omega \dot{x} y \sin\lambda + 2\Omega \dot{y}(x\sin\lambda - R_E \cos\lambda)\right.$$
$$\left. - R_E \Omega^2 x \sin(2\lambda)\right] - \frac{m\tilde{g}(x^2 + y^2)}{2l}$$

up to an uninteresting constant $m\Omega^2 R_E^2 \cos^2\lambda/2$ and in which we introduced the effective gravitational field $\tilde{g} = g - \Omega^2 R_E \cos^2\lambda$ modified by the centrifugal force.

6. We are concerned now by the Lagrange equations giving the motion in the horizontal plane. Starting from the previous Lagrangian and applying the traditional recipe (2.4), one obtains the equations of motion:

$$\ddot{x} - 2\Omega\dot{y}\sin\lambda + \frac{\tilde{g}}{l}x + \frac{1}{2}R_E\Omega^2\sin(2\lambda) = 0;$$

$$\ddot{y} + 2\Omega\dot{x}\sin\lambda + \frac{\tilde{g}}{l}y = 0.$$

7. Let define $\tilde{x} = x - x_e$ and substitute this value in the first Lagrange equation; the arbitrary value x_e is then chosen in order to cancel the constant term in the resulting equation. Owing to the fact that the value $R_E \Omega^2 \approx 10^{-3}\,m/s^2$ is very small as compared to $g \approx 10\,m/s^2$, it

Problem Solutions

is legitimate to approximate $\tilde{g} \cong g$, in which case the result takes the following form:

$$x_e = -\frac{R_E l \Omega^2 \sin(2\lambda)}{2g}.$$

The set $\tilde{x} = 0, y = 0$ is a solution of the equations of motion; indeed this is the equilibrium solution. Thus at equilibrium, the pendulum is not oriented along the true vertical, but along the apparent vertical, which makes an angle

$$\alpha \approx \sin\alpha = \frac{x_e}{l} = R_E \Omega^2 \sin\frac{2\lambda}{2g}$$

with respect to the true vertical. This deviation is due to the centrifugal force. It is maximum on the 45th parallel.

8. The coupled differential equations to be solved are rewritten:

$$\ddot{\tilde{x}} - 2\Omega \dot{y} \sin\lambda + \frac{\tilde{g}\tilde{x}}{l} = 0;$$

$$\ddot{y} + 2\Omega \dot{\tilde{x}} \sin\lambda + \frac{\tilde{g} y}{l} = 0.$$

Let us introduce the complex variable $u = \tilde{x} + iy$. Multiply the second Lagrange equation by i and add the first one; the auxiliary variable u occurs naturally in the unique differential equation:

$$\ddot{u} + 2i\Omega \dot{u} \sin\lambda + \frac{\tilde{g}}{l} u = 0.$$

Let us put $u(t) = U(t)e^{irt}$, substitute in the previous equation and choose $r = -\Omega \sin\lambda$ in order to get rid of the $\dot{U}$ term. Defining

$$\omega^2 = \frac{g}{l} + \Omega^2 \sin^2\lambda - \Omega^2 \frac{R_E}{l} \cos^2\lambda.$$

the resulting equation is written as $\ddot{U} + \omega^2 U = 0$.

One has $\Omega^2 \approx 10^{-9} \, s^{-1}$ while $\Omega^2(R_E/l) \approx 10^{-3} \, s^{-1}$. It is thus fully justified to neglect the second term as compared to the third one so that:

$$r = -\Omega \sin\lambda;$$

$$\omega^2 = \omega_0^2 - \Omega^2 \frac{R_E}{l} \cos^2\lambda,$$

where $\omega_0 = \sqrt{g/l}$ is the proper angular frequency of the pendulum in a Galilean frame. The solution of the equation $\ddot{U} + \omega^2 U = 0$ is trivial and gives $U = X + iY = Ae^{i\omega t} + Be^{-i\omega t}$.

It is always possible to choose the origin of time and the axis orientation in order to obtain $X(t) = A\cos\omega t$; $Y(t) = B\sin\omega t$. In this system of reference, the pendulum describes an ellipse with an angular frequency ω. In the complex plane, the multiplication by e^{irt} to switch from the set X, Y to the set x, y is just a rotation of angle rt. In other words, the axes of the ellipse turn slowly in time with the angular velocity $|r| = \Omega \sin\lambda$.

At the equator $\lambda = 0$ so that $r = 0$ and

$$\omega = \sqrt{\omega_0^2 - \Omega^2 \frac{R_E}{l}}.$$

The pendulum oscillates with an angular frequency slightly smaller than its proper value.

At the pole $\lambda = \pi/2$, then $r = -\Omega$ and $\omega = \omega_0$. The pendulum oscillates with its proper angular frequency and the ellipse axes make a complete revolution in one day (see Fig. 2.7).

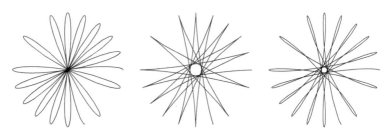

Fig. 2.7 Different types of trajectories for the ellipse drawn on the ground for three different initial release conditions. For these three cases $r/\omega = 1/10$. On the left hand side, the pendulum is released with a tangential velocity opposite to the driving velocity; in the middle, the pendulum is released with no initial velocity and on the right hand side one has a situation intermediate between the previous cases

2.5. Three-particle System [Statement p. 61]

A – Changing coordinates

1. In changing the origin $q_i' = q_i - a$, the velocities do not vary $\dot{q}_i' = \dot{q}_i$, neither do the relative distances $q_i' - q_j' = q_i - q_j$. The Lagrangian is invariant and one deduces the following constant of the motion (this is also a consequence of Noether's theorem):

$$P = \sum_i \partial_{\dot{q}_i} L,$$

Problem Solutions

or
$$P = m\,(\dot{q}_1 + \dot{q}_2 + \dot{q}_3) \quad \text{is a constant of the motion.}$$
This is the total momentum of the system.

2. Inverting the proposed relations, one gets: $q_1 = X + x/2 - y/(3\alpha)$; $q_2 = X - x/2 - y/(3\alpha)$; $q_3 = X + 2y/(3\alpha)$. It is easy to obtain the kinetic energy:
$$T = \frac{1}{2}m\,(\dot{q}_1^2 + \dot{q}_2^2 + \dot{q}_3^2) = m\left[\frac{3}{2}\dot{X}^2 + \frac{1}{4}\dot{x}^2 + \frac{1}{3\alpha^2}\dot{y}^2\right].$$
The desired α value must satify $3\alpha^2 = 4$, or
$$\alpha = \frac{2}{\sqrt{3}}$$
With the proposed change of variables, one is able to write the Lagrangian under the form:
$$L = \frac{3}{2}m\dot{X}^2 + \frac{1}{4}m\,(\dot{x}^2 + \dot{y}^2) - V_1(-\frac{\sqrt{3}}{2}y - \frac{1}{2}x) - V_2(\frac{\sqrt{3}}{2}y - \frac{1}{2}x) - V_3(x).$$
We notice that the X coordinate is cyclic. Consequently we find a constant of the motion $\partial_{\dot{X}}L = 3m\dot{X}$, which owing to the definition of the X variable, is precisely the quantity introduced in the first question. One can always choose a frame in which this quantity cancels. The Lagrangian depends on two degrees of freedom x and y.

3. In a cyclic permutation, the X variable remains unchanged while the new variables
$$x' = q_2 - q_3 \quad \text{and} \quad y' = \frac{2}{\sqrt{3}}(q_1 - \frac{1}{2}(q_2 + q_3))$$
are related to the old ones through
$$x' = -\frac{\sqrt{3}}{2}y - \frac{1}{2}x \quad \text{and} \quad y' = \frac{\sqrt{3}}{2}x - \frac{1}{2}y.$$
Replacing those values in $z' = x' + iy'$, one finds $z' = (x + iy)(-\frac{1}{2} + i\frac{\sqrt{3}}{2})$ or:
$$z' = z\exp(2i\pi/3).$$
Of course, one has $|z'|^2 = (x'^2 + y'^2) = |z|^2 = (x^2 + y^2)$.

It is just a matter of simple calculation to check that, with the new coordinates, the Lagrangian is written:
$$L = \frac{3}{2}m\dot{X}^2 + \frac{1}{4}m\,(\dot{x}'^2 + \dot{y}'^2) - V_1(x') - V_2(-\frac{\sqrt{3}}{2}y' - \frac{1}{2}x')$$
$$-V_3(\frac{\sqrt{3}}{2}y' - \frac{1}{2}x').$$

B – Harmonic approximation case

4. The harmonic potential is written, following the general expression,

$$\frac{1}{2}k\left[(\frac{\sqrt{3}}{2}y+\frac{1}{2}x)^2 + (\frac{\sqrt{3}}{2}y-\frac{1}{2}x)^2 + x^2\right],$$

which finally reduces to $\frac{3}{4}k(x^2+y^2)$. The corresponding Lagrangian can be recast, after some rearrangement, in the form:

$$\begin{aligned}L &= \frac{3}{2}m\dot{X}^2 + \frac{1}{4}(m\dot{x}^2 - 3kx^2) + \frac{1}{4}(m\dot{y}^2 - 3ky^2) \\ &= \frac{3}{2}m\dot{X}^2 + L_{\mathrm{ho}}(x,\dot{x}) + L_{\mathrm{ho}}(y,\dot{y}).\end{aligned}$$

The X coordinate is cyclic and we are brought back to the constant of the motion P already studied. This is not surprising, since this property is independent of the form of the potential. More astonishing is the fact that the Lagrangian can be decomposed into two decoupled parts. This property is specific to the harmonic potential. Since these parts are time independent, one can ascribe to each of them a constant of the motion analogous to the energy function $E_x(x,\dot{x}) = \dot{x}\,\partial_{\dot{x}}L_{\mathrm{ho}}(x,\dot{x}) - L_{\mathrm{ho}}(x,\dot{x})$ and a similar relation for the Lagrangian $L_{\mathrm{ho}}(y,\dot{y})$. After simple calculations, one arrives at the desired result:

$$P = 3m\dot{X};$$
$$E_{\mathrm{ho}}(x,\dot{x}) = \frac{1}{4}(m\dot{x}^2 + 3kx^2);$$
$$E_{\mathrm{ho}}(y,\dot{y}) = \frac{1}{4}(m\dot{y}^2 + 3ky^2)$$

are constants of the motion.

The proper mode in x corresponds to a vibration of the two first particles, the third one remaining at rest. The proper mode in y corresponds to a vibration of the third particle in opposition to the group formed by the others.

C – Hénon and Heiles potential

5. The Toda's potential

$$V(x,y) = V_0\left[\exp(-(\frac{1}{2}x + \frac{\sqrt{3}}{2}y)/a) + \exp((\frac{\sqrt{3}}{2}y - \frac{1}{2}x)/a) + \exp(x/a)\right]$$

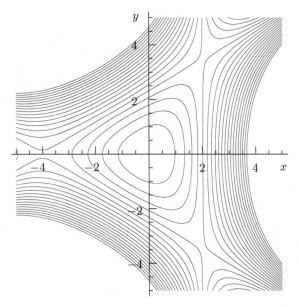

Fig. 2.8 Equipotential curves, for the Hénon and Heiles potential, obtained with the value $a = 1$. We remark the three-fold symmetry, the existence of a minimum and three saddle points

gives, for small distances, the Hénon's potential if the expansion is pursued to third order

$$V(x,y) = V_0 \left[3 + \frac{3}{4a^2}(x^2 + y^2) + \frac{1}{8a^3}(x^3 - 3xy^2) \right].$$

The equipotential curves are represented in Fig. 2.8. The corresponding Lagrangian is easily obtained:

$$L = \frac{1}{4}m(\dot{x}^2 + \dot{y}^2) - V_0 \left[3 + \frac{3}{4a^2}(x^2 + y^2) + \frac{1}{8a^3}(x^3 - 3xy^2) \right].$$

6. We already showed that the term $x^2 + y^2 = |z|^2$ is invariant under coordinate permutations. This is also the case of the term $x^3 - 3xy^2$ if one notices that it can be written as

$$x^3 - 3xy^2 = \frac{1}{2}(z^3 + \bar{z}^3) = \frac{1}{2}(z'^3 + \bar{z}'^3) = x'^3 - 3x'y'^2.$$

Hence:

$$V(x,y) = V(x',y').$$

Since passing from coordinates (x, y) to coordinates (x', y') corresponds in the plane to a rotation of $2\pi/3$, the previous invariance shows a three-fold symmetry for the potential. Moreover, $V(x = 2a, y) = 7V_0$; this proves that the straight line $x = 2a$ is an equipotential curve. Of course, the same property holds for the lines obtained by symmetry. These three lines intersect to form an equipotential equilateral triangle with the value $7V_0$. These properties are clearly seen in the Fig. 2.8.

2.6. Vibration of a Linear Triatomic Molecule: The "Soft" Mode
[Statement p. 63]

1. Let x_i be the deviation from equilibrium for atom i. The force acting on the first atom is given by $F_1 = k(x_2 - x_1)$, on the third one $F_3 = k(x_2 - x_3)$ and on the second one $F_2 = -F_1 - F_3 = k(x_1 + x_3 - 2x_2)$ since the molecule is at equilibrium. These forces arise from the potential $V = \frac{1}{2}k(x_2 - x_1)^2 + \frac{1}{2}k(x_3 - x_2)^2$ since one has $F_i = -\partial_{x_i} V$. The kinetic energy is easy to obtain and the Lagrangian is deduced as $L = T - V$:

$$L = \frac{1}{2}m\dot{x}_1^2 + \frac{1}{2}M\dot{x}_2^2 + \frac{1}{2}m\dot{x}_3^2 - \frac{1}{2}k(x_2 - x_1)^2 - \frac{1}{2}k(x_3 - x_2)^2.$$

The Lagrange equations are obtained by the classical recipe (2.4):

$$\ddot{x}_1 = \Omega^2(x_2 - x_1);$$
$$r\ddot{x}_2 = \Omega^2(x_1 + x_3 - 2x_2);$$
$$\ddot{x}_3 = \Omega^2(x_2 - x_3)$$

with the proposed definition of the parameters $\Omega = \sqrt{k/m}$ and $r = M/m$. These are coupled linear differential equations.

2. The proper modes are sought by imposing the same angular frequency ω on each coordinate, hence by seeking for a solution of the form $x_1 = Ae^{i\omega t}$, $x_2 = Be^{i\omega t}$, $x_3 = Ce^{i\omega t}$. Lagrange's equations become:

$$-\omega^2 A = -\Omega^2 A + \Omega^2 B;$$
$$-r\omega^2 B = \Omega^2 A + \Omega^2 C - 2\Omega^2 B;$$
$$-\omega^2 C = -\Omega^2 C + \Omega^2 B.$$

Let us simplify these equations dividing by Ω^2 and defining $\lambda = \omega/\Omega$. They can then be written in a matrix form:

$$\begin{pmatrix} 1 - \lambda^2 & -1 & 0 \\ -1 & 2 - r\lambda^2 & -1 \\ 0 & -1 & 1 - \lambda^2 \end{pmatrix} \begin{pmatrix} A \\ B \\ C \end{pmatrix} = \begin{pmatrix} 0 \\ 0 \\ 0 \end{pmatrix}.$$

Problem Solutions

This equation has a non trivial solution only if the matrix determinant vanishes. Performing the calculations, one finds the following three solutions: $\lambda^2 = 0$ with the eigenvector $A = B = C$, $\lambda^2 = 1$ with $A = -C, B = 0$ and $\lambda^2 = (r+2)/r = \lambda_3^2$ with $A = C, B = -2A/r$. Arbitrarily normalizing the eigenvectors, the final solution is obtained as:

$$\omega = 0 \quad \text{mode} \quad \begin{pmatrix} 1 \\ 1 \\ 1 \end{pmatrix}, \qquad \omega = \Omega \quad \text{mode} \quad \begin{pmatrix} 1 \\ 0 \\ -1 \end{pmatrix},$$

$$\omega = \sqrt{\frac{r+2}{r}}\Omega \quad \text{mode} \quad \begin{pmatrix} 1 \\ -2/r \\ 1 \end{pmatrix}.$$

3. In this particular case, we observe the existence of a soft mode $\omega = 0$ for which $A = B = C$. Substituting this solution in the Lagrange equations, results in $\ddot{x}_i = 0$, $\forall i$, which means $\dot{x}_1 = \text{const}, r\dot{x}_2 = \text{const}, \dot{x}_3 = \text{const}$. These conditions lead to the conservation of the total linear momentum

$$P = m\dot{x}_1 + M\dot{x}_2 + m\dot{x}_3 = \text{const}.$$

This property is the consequence of the Lagrangian invariance under a continuous space translation. The solution of the soft mode is given by:

$$x_1(t) = x_2(t) = x_3(t) = at + b.$$

In such a mode, the system moves as a bulk at constant speed. It is not interesting from the physical point of view, because one can always work in a Galilean frame where the total linear momentum vanishes. In such a frame, the soft mode corresponds to the equilibrium solution.

4. The mode $\omega = \Omega$ imposes $B = 0$ and $C = -A$. The mass M remains at rest whereas the two masses m vibrate with the same amplitude and opposite phases.

The mode $\omega = [(r+2)/r]^{1/2}\Omega$ imposes $C = A$ and $B = -(2/r)A$. The two masses m vibrate in phase, whereas the mass M vibrates with the opposite phase, with a reduced or enhanced amplitude depending on the r value.

5. The most general solution for the differential system is written

$$\begin{pmatrix} x_1(t) \\ x_2(t) \\ x_3(t) \end{pmatrix} = (at+b)\begin{pmatrix} 1 \\ 1 \\ 1 \end{pmatrix} + Be^{i\Omega t}\begin{pmatrix} 1 \\ 0 \\ -1 \end{pmatrix} + Ce^{i\lambda_3\Omega t}\begin{pmatrix} 1 \\ -2/r \\ 1 \end{pmatrix}.$$

Of course, one must consider the real part of this complex number.

Introducing the proposed initial conditions, a lengthy but straightforward calculation leads to the equations of motion:

$$x_1(t) = \frac{p}{2m+M}\left[t + \frac{2+r}{2\Omega}\sin(\Omega t) + \frac{r}{2\Omega\lambda_3}\sin(\lambda_3\Omega t)\right];$$

$$x_2(t) = \frac{p}{2m+M}\left[t - \frac{1}{\Omega\lambda_3}\sin(\lambda_3\Omega t)\right];$$

$$x_3(t) = \frac{p}{2m+M}\left[t - \frac{2+r}{2\Omega}\sin(\Omega t) + \frac{r}{2\Omega\lambda_3}\sin(\lambda_3\Omega t)\right].$$

2.7. Elastic Transversal Waves in a Solid
[Statement and Figure p. 64]

1. We place at each node a mass equal to that of a segment, or $m = \rho S \delta$. Since the node is free to oscillate only in the vertical direction, its velocity is $\dot{q}_i$ and its kinetic energy $T_i = \frac{1}{2}\rho S \delta \dot{q}_i^2$. The total kinetic energy is obtained by summing over nodes:

$$T = \frac{1}{2}\sum_i \rho S \delta \dot{q}_i^2.$$

2. The shear force on the slice i is $F_i = 2\mu S(q_{i+1} - q_i)/\delta$. The virtual work of this force on the node is $\delta W_i = F_i \delta q_i = 2\mu S(q_{i+1} - q_i)\delta q_i/\delta$. It appears that this work can be expressed as the negative of a potential function $\delta W_i = -dV_i$, this last being given by the expression:

$$V_i = \frac{\mu S}{\delta}(q_{i+1} - q_i)^2.$$

3. The total elastic potential energy is of course the sum of the energy for each slice, and the Lagrangian is the difference between the kinetic and potential energies. Thus:

$$L(q, \dot{q}) = \frac{1}{2}\sum_i \rho S \delta \dot{q}_i^2 - \sum_i \frac{\mu S}{\delta}(q_{i+1} - q_i)^2.$$

4. From this Lagrangian, one deduces $d(\partial_{\dot{q}_i} L)/dt = \rho S \delta \ddot{q}_i$ and $\partial_{q_i} L = 2\mu S(q_{i+1} + q_{i-1} - 2q_i)/\delta$ (to calculate the last term, don't forget to take into account a contribution coming from the slice labelled i, and another one coming from the slice labelled $i-1$). The Lagrange equations lead to coupled differential equations:

$$\rho \ddot{q}_i = 2\mu(q_{i+1} + q_{i-1} - 2q_i)/\delta^2.$$

Let us choose a field function $\varphi(x,t)$ which is identified with the node coordinates for each time: $\varphi(x = i\delta, t) = q_i(t)$. In this case, in the limit of small δ values, one has $\ddot{q}_i = \partial_{t^2}^2 \varphi|_{x=i\delta}$ and $(q_{i+1} + q_{i-1} - 2q_i)/\delta^2 = \partial_{x^2}^2 \varphi|_{x=i\delta}$. Taking the limit $\delta \to 0$, the Lagrange equations are represented by a unique partial differential equation:

$$\frac{1}{c^2}\frac{\partial^2 \varphi(x,t)}{\partial t^2} = \frac{\partial^2 \varphi(x,t)}{\partial x^2} \quad \text{with} \quad c = \sqrt{\frac{2\mu}{\rho}}.$$

This is precisely the equation for the propagation of a wave with speed c.

2.8. Lagrangian in a Rotating Frame
[Statement p. 65]

1. The Galilean frame is denoted $Oxyz$ and the Cartesian coordinates of the particle in this frame x, y, z. The rotating frame is denoted $OXYZ$ and the Cartesian coordinates of the particle in this frame X, Y, Z. It is easy to obtain the relation between the two sets of coordinates:

$$x = X \cos\phi - Y \sin\phi; \qquad y = X \sin\phi + Y \cos\phi; \qquad z = Z.$$

Differentiating these expressions and inserting them in the kinetic energy $T = \frac{1}{2}m(\dot{x}^2 + \dot{y}^2 + \dot{z}^2)$, one obtains the following expression:

$$T = \frac{1}{2}m(\dot{X}^2 + \dot{Y}^2 + \dot{Z}^2) + \frac{1}{2}m(X^2 + Y^2)\dot{\phi}^2 + m(X\dot{Y} - Y\dot{X})\dot{\phi}.$$

The first term $T_r = \frac{1}{2}m(\dot{X}^2 + \dot{Y}^2 + \dot{Z}^2)$ is just the kinetic energy in the rotating frame.

Instead of Cartesian coordinates, one can also use the polar coordinates (ρ, ψ, Z) in the same frame. With the traditional definition $X = \rho \cos\psi$, $Y = \rho \sin\psi$, the previous expression for the kinetic energy can be put into the form:

$$T = \frac{1}{2}m(\dot{\rho}^2 + \rho^2 \dot{\psi}^2 + \dot{Z}^2) + \frac{1}{2}m\rho^2 \dot{\phi}^2 + m\rho^2 \dot{\phi}\dot{\psi}.$$

2. Without loss of generality, the kinetic energy can be rewritten as $T = T_r - V$. For a free particle, Lagrange's equation is written $d(\partial_{\dot{q}} T)/dt - \partial_q T = 0$. In the rotating frame, one can identify T with a Lagrangian L for which T_r is the kinetic energy and V plays a role of a potential. In this case, $V = -\frac{1}{2}m\rho^2 \dot{\phi}^2 - m\rho^2 \dot{\phi}\dot{\psi}$. This is precisely the potential which gives rise to the inertial Coriolis and centrifugal forces. Moreover, in the rotating frame, the OZ component of the angular momentum $\mathbf{L} = \mathbf{OM} \times m\mathbf{V}_r$ is simply $L_z = -m(\dot{X}Y - \dot{Y}X) = m\rho^2 \dot{\psi}$.

Finally the potential is written:
$$V = -\dot\phi L_z - \frac{1}{2} m \rho^2 \dot\phi^2.$$

3. Generally, for the case of a pure rotation, the velocity addition law is $\boldsymbol{v} = \boldsymbol{v}_r + \boldsymbol{\omega} \times \boldsymbol{r}$. Substituting this expression into the kinetic energy $T = \frac{1}{2} m v^2$ and equating to $T_r - V$, one finds the expression for
$$V = -m\boldsymbol{v}_r \cdot (\boldsymbol{\omega} \times \boldsymbol{r}) - \frac{1}{2} m (\boldsymbol{\omega} \times \boldsymbol{r})^2.$$

The first term is a mixed product, which can be rewritten $-\boldsymbol{\omega} \cdot (\boldsymbol{r} \times m\boldsymbol{v}_r)$, that is $-\boldsymbol{\omega} \cdot \boldsymbol{L}$. Finally:
$$V = -\boldsymbol{\omega} \cdot \boldsymbol{L} - \frac{1}{2} m (\boldsymbol{\omega} \times \boldsymbol{r})^2.$$

4. In case of the bead moving on the hoop, $\boldsymbol{\omega} = \dot\phi \hat{\boldsymbol{z}}$, $\boldsymbol{v}_r = R\dot\theta \boldsymbol{u}_\theta$ whence $T_r = \frac{1}{2} m R^2 \dot\theta^2$ and the gravitational potential is $mgR\cos\theta$. To it, one must add the potential due to inertial forces $V = -\boldsymbol{\omega} \cdot \boldsymbol{L} - \frac{1}{2} m (\boldsymbol{\omega} \times \boldsymbol{r})^2$. The angular momentum in the hoop frame is perpendicular to its plane and thus $\boldsymbol{\omega} \cdot \boldsymbol{L} = 0$. We are left with the term $-\frac{1}{2} m (\boldsymbol{\omega} \times \boldsymbol{r})^2$, which is $-\frac{1}{2} m R^2 \sin^2\theta \, \dot\phi^2$. One deduces the Lagrangian of the system:
$$L = \frac{1}{2} m R^2 \dot\theta^2 + \frac{1}{2} m R^2 \sin^2\theta \, \dot\phi^2 - mgR \cos\theta.$$

5. If the system rotates with a small angular velocity, the centrifugal term $(\boldsymbol{\omega} \times \boldsymbol{r})^2$, which is of second order in ω, is negligible compared with the Coriolis term, which is of first order. Using the result of the first question, the kinetic energy is written simply:
$$T = \frac{1}{2} m(\dot X^2 + \dot Y^2 + \dot Z^2) + m(X\dot Y - Y\dot X)\dot\phi$$
and the Lagrange equations read: $m\ddot X = 2m\dot\phi \dot Y$; $m\ddot Y = -2m\dot\phi \dot X$.

Consider now a particle of mass m and charge q_e, placed in a constant magnetic field along Oz with amplitude B. The Lorentz force $q_e \boldsymbol{v} \times \boldsymbol{B}$ has the components $(\dot Y B, -\dot X B, 0)$ and the fundamental principle of dynamics leads to the equations of motion: $m\ddot X = q_e B \dot Y$, $m\ddot Y = -q_e B \dot X$. They are formally identical to those in a rotating frame, if one makes the substitution:
$$B = \frac{2m\dot\phi}{q_e}.$$

2.9. Particle Drift in a Constant Electromagnetic Field
[Statement p. 66]

1. Using the electric and magnetic fields, and the particle velocity $(\dot{x}, \dot{y}, \dot{z})$, the Lorentz force is easily calculated, $\boldsymbol{F} = q_e(\boldsymbol{E} + \boldsymbol{v} \times \boldsymbol{B})$:

$$F_x = q_e(E + B\dot{y});$$
$$F_y = -q_e B\dot{x};$$
$$F_z = 0.$$

2. The electric potential U follows from (up to a constant that can be taken with a null value) the equation: $\boldsymbol{E} = -\boldsymbol{\nabla}U$. With the previous expression, an integration provides at once:

$$U(x, y, z) = -Ex.$$

3. The equation $\boldsymbol{B} = \boldsymbol{\nabla} \times \boldsymbol{A}$ must be satisfied. With the form of the potential $\boldsymbol{A} = (-yB, 0, 0)$,

$$\boldsymbol{\nabla} \times \boldsymbol{A} = (0, 0, -\partial_y(-yB)) = (0, 0, B) = \boldsymbol{B}.$$

The equation is thus satisfied. With the alternative expression $\boldsymbol{A} = (0, xB, 0)$,

$$\boldsymbol{\nabla} \times \boldsymbol{A} = (0, 0, \partial_x(xB)) = (0, 0, B) = \boldsymbol{B},$$

the equation is again verified. The curl being a linear operator, the half sum of both expressions is also a solution.

4. Let us start with the first expression which leads to the generalized potential: $V = q_e(B\dot{x}y - Ex)$. With this expression, it is easy to check that $d(\partial_{\dot{x}_i}V)/dt - \partial_{x_i}V = F_{x_i}$, with the force components as given by question 1. The particle is subject to the electromagnetic field only and can thus be described in terms of the Lagrangian $L = T - V$, which is expressed in Cartesian coordinates:

$$L(\dot{x}, \dot{y}, \dot{z}, x, y, z) = \frac{1}{2}m(\dot{x}^2 + \dot{y}^2 + \dot{z}^2) + q_e(Ex - B\dot{x}y).$$

The z coordinate is cyclic; it follows that the constant of the motion $\partial_{\dot{z}}L = m\dot{z} = p_z$:

$$m\dot{z} = p_z = \text{const.}$$

The Lagrange equations for the two other variables give respectively:

$$m\ddot{x} - q_e B\dot{y} = q_e E;$$
$$m\ddot{y} + qB\dot{x} = 0.$$

5. The same study can be repeated, using the alternative expression for the vector potential that leads to the generalized potential $V = -q_e(B\dot{y}x + Ex)$. The new Lagrangian is:

$$L'(\dot{x}, \dot{y}, \dot{z}, x, y, z) = \frac{1}{2}m(\dot{x}^2 + \dot{y}^2 + \dot{z}^2) + q_e(Ex + Bx\dot{y}).$$

The z variable is still cyclic and this Lagrangian provides us with the same constant of the motion as before. This time, the y variable is cyclic as well and this leads to another constant of the motion: $\partial_{\dot{y}}L = m\dot{y} + q_eBx = p_y$,

$$m\dot{y} + q_eBx = p_y = \text{const}.$$

This last equation could have been derived by integration of the second equation of the previous question. The Lagrange equation for the x coordinate is identical to the corresponding equation coming from the Lagrangian of the previous question. It thus provides nothing new.

6. It is easy to check that $L' - L = q_eB(\dot{y}x + \dot{x}y)$, an expression that can be recast under the form:

$$L' = L + q_eB\frac{d(xy)}{dt}.$$

Both Lagrangians differ only by a total derivative with respect to time of the function $F(x, y) = q_eB\,xy$; as a consequence, they lead to the same set of Lagrange equations.

7. With the extra condition $E = 0$, the first Lagrange equation can be integrated and we are left with an additional constant of the motion $m\dot{x} - q_eBy = p_x$. To simplify the notation, let us put $p_x = a$, $p_y = b$, $p_z = c$. To summarize the results of the two previous questions, one must solve the following differential system:

$$m\dot{x} - q_eBy = a;$$
$$m\dot{y} + q_eBx = b;$$
$$m\dot{z} = c.$$

With a suitable choice of the time origin, the last equation can be integrated to give $z = (c/m)\,t$. Now by a suitable choice of the spatial origin, that is taken at $x_c = b/(q_eB)$ and $y_c = -a/(q_eB)$, it is possible to cancel the terms a, b. The two first equations form a coupled system that can be solved easily introducing the complex variable $w = x + iy$. This system is now equivalent to the unique differential equation $\dot{w} + i\omega w = 0$.

The condition for the initial time $t = 0$, imposes $w(0) = R$, and the equations of the motion become:

$$x(t) = R\cos(\omega t); \qquad y(t) = -R\sin(\omega t); \qquad z(t) = \frac{c}{m}t.$$

The unknown constant R is determined from the initial velocity v_0 via $R = mv_0/(q_e B)$.

Thus, the projection of the trajectory in a plane perpendicular to the magnetic field is just a circle of center $C(x_c, y_c)$ and radius R. The motion along Oz is performed at constant speed. Consequently, the particle motion in space is a helix.

This important result can be found as well by a reasoning "à la Newton", equating the centrifugal force with the Lorentz force.

8. Let us come now to the general case $E \neq 0$ and the equations of motion as stated in Questions 4 and 5. The integration concerning the z variable is performed at once $z(t) = (p_z/m)\,t$. If we make use of the cyclotron angular frequency $\omega = q_e B/m$, the equation for y with a null velocity at the origin provides the relation $\dot{y} = -\omega x$. Substituting this in the equation concerning x, one finds the differential equation $\ddot{x} + \omega^2 x = q_e E/m$, that can be integrated without any difficulty. Inserting this result in the equation $\dot{y} = -\omega x$ allows us to completely determine $y(t)$ with a single integration. To summarize, the general solution is given by:

$$x(t) = \frac{q_e E}{m\omega^2}(1 - \cos\omega t);$$

$$y(t) = \frac{q_e E}{m\omega^2}\sin\omega t - \frac{q_e E}{m\omega}t;$$

$$z(t) = \frac{p_z}{m}t.$$

If one works in the frame that drifts along Oy (thus perpendicular to $\boldsymbol{E}$ and $\boldsymbol{B}$) with the speed $-(q_e E)/m\omega = -E/B$, we see that the motion is still a helix, with axis Oz, characterized by the angular velocity ω. This is precisely the cyclotron motion studied in the previous question.

A few types of trajectories are represented in the Fig. 2.9.

9. The Lagrangian does not depend on time explicitly. The general theory predicts a constant of the motion – the energy – given by $\mathrm{E} = \dot{x}\partial_{\dot{x}}L + \dot{y}\partial_{\dot{y}}L + \dot{z}\partial_{\dot{z}}L - L$ (be careful to avoid confusion between the energy E and the electric field E). Using either form for the Lagrangian, an explicit calculation gives:

$$\mathrm{E} = \frac{1}{2}m(\dot{x}^2 + \dot{y}^2 + \dot{z}^2) - q_e E x = \frac{1}{2}mv_0^2.$$

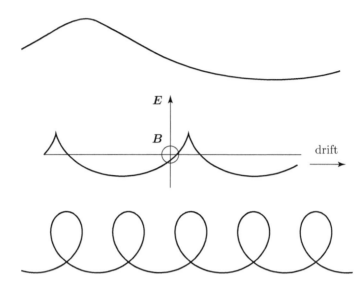

Fig. 2.9 Trajectories in the plane xy for a charged particle placed in a constant crossed electromagnetic field. The trajectory equations are calculated in Question 8. The drift direction is the axis $y'y$. This case corresponds to a null drift along $z'z$. The upper trajectory corresponds to a regime with a weak magnetic field, while the bottom one corresponds to a regime with a large magnetic field. The intermediate case corresponds to a special value of the magnetic field which highlights the transition between both types of regimes

2.10. The Penning Trap [Statement p. 67]

1. Since there is no charge in the region where the electric field is present, Poisson's equation reduces simply to the Laplace equation $\Delta U = 0$, an equation that is manifestly satisfied by the proposed form of the potential

$$\Delta U = \partial^2_{x^2} U + \partial^2_{y^2} U + \partial^2_{z^2} U = \frac{1}{2}k(-1 - 1 + 2) = 0.$$

The equipotential surfaces represented in Fig. 2.10 are hyperboloids of revolution. They give also the form of the electrodes which are held at fixed potentials.

The scalar potential is given by the expression $U = \frac{1}{4}k(2z^2 - x^2 - y^2)$ and the vector potential can be chosen as $\boldsymbol{A} = \frac{1}{2}(-yB, xB, 0)$. The generalized electromagnetic potential $V = q_e(U - \dot{\boldsymbol{r}} \cdot \boldsymbol{A})$ is written in terms of Cartesian coordinates as

$$V = \frac{1}{4}kq_e(2z^2 - x^2 - y^2) - \frac{1}{2}q_e B(\dot{y}x - \dot{x}y)$$

and the Lagrangian (difference between the kinetic and potential energies) is:

$$L = \frac{1}{2}m(\dot{x}^2 + \dot{y}^2 + \dot{z}^2) + q_e[\frac{1}{4}k(x^2 + y^2 - 2z^2) + \frac{1}{2}B(\dot{y}x - \dot{x}y)].$$

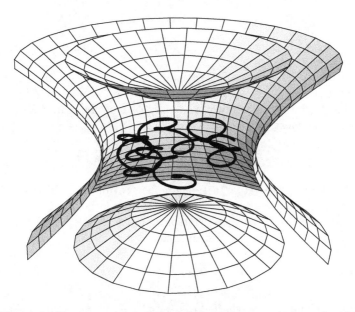

Fig. 2.10 Equipotential surfaces for the scalar potential employed in a Penning trap. A trajectory for the particle confined inside the trap is also shown in the figure

2. Applying (2.4), Lagrange's equations can be derived as:

$$\begin{aligned} m\ddot{x} &= q_e(B\dot{y} + \frac{1}{2}kx); \\ m\ddot{y} &= q_e(-B\dot{x} + \frac{1}{2}ky); \\ m\ddot{z} &= -q_e kz \end{aligned}$$

The equation for the z variable can be written $\ddot{z} + \omega_a^2 z = 0$, with the axial angular frequency ω_a given in the statement. This equation is integrated at once to give:

$$z(t) = a\,\cos(\omega_a t + \phi).$$

It reflects a sinusoidal behaviour.

3. The first two equations may be gathered into a single equation by introducing the complex variable $w = x + iy$; they lead to the differential equation $\ddot{w} + i\omega_c \dot{w} - \omega_a^2 w/2 = 0$ using the cyclotron and axial angular frequencies ω_c and ω_a given in the statement. One seeks a solution of the form $w = e^{irt}$. One then obtains the characteristic equation $r^2 + \omega_c r + \omega_a^2/2 = 0$ which has two solutions ω_m and $\tilde{\omega}_c$:

$$x(t) + iy(t) = R_c e^{-i\tilde{\omega}_c t} + R_m e^{-i\omega_m t};$$

$$\omega_m = \frac{1}{2}\left(\omega_c - \sqrt{\omega_c^2 - 2\omega_a^2}\right); \qquad \tilde{\omega}_c = \omega_c - \omega_m.$$

The particle oscillates around the plane $z = 0$. However, in order for it to remain confined in the trap, the coordinates x, y cannot be allowed to go to infinity which implies that the two roots must be real. This condition imposes the constraint $\omega_c^2 > 2\omega_a^2$ that is:

$$0 < k < \frac{q_e B^2}{2m}.$$

4. If the magnetic field is large enough, the previous condition is satisfied; moreover, one has $\omega_c \gg \omega_a$. In this case, the largest angular frequency may be identified with the cyclotron frequency $\tilde{\omega}_c \cong \omega_c$ and the smallest one, known as the magnetron frequency, is $\omega_m \cong \omega_a^2/(2\omega_c) = k/(2B)$. There is a superposition of two motions; one is a slow rotation with angular frequency ω_m, the other is a rapid rotation with angular frequency ω_c.

In the plane $z = 0$, the field is radial and its value at a distance R from the symmetry axis is $E = kR/2$. We saw in Problem 2.9 that in a constant electric field, the drift speed in a large magnetic field is $E/B = kR/(2B)$; moreover this velocity is perpendicular to both the electric and magnetic fields, so tangential in our case. The time needed to describe a complete circle of radius R is $\tau = 2\pi R/(kR/2B) = 4\pi B/k$. The drift frequency is thus $\omega_d = 2\pi/\tau$ or $\omega_d = \omega_m$. We thus recover the result obtained with a more complete calculation.

An example of a trajectory is shown on Fig. 2.10. The small circular motions are cyclotron motions, which roll up around the circular motion with larger frequency, corresponding to the magnetron motion.

5. Since the Lagrangian does not depend explicitly on time, the energy is a constant of the motion. Using the general relation (2.7), we obtain the value for energy (the energy E must not be confused with the electric field E): $E = \frac{1}{2}m(\dot{x}^2 + \dot{y}^2 + \dot{z}^2) + \frac{1}{4}q_e k(2z^2 - x^2 - y^2)$, which can be rewritten:

$$E = \frac{1}{2}m[(\dot{x}^2 + \dot{y}^2 + \dot{z}^2) + \frac{1}{2}\omega_a^2(2z^2 - x^2 - y^2)] = \text{const}.$$

Problem Solutions

Substituting the time law in this expression, we obtain, after an extensive but straightforward calculation, the value for the energy:

$$E = \frac{1}{2} m[(\tilde{\omega}_c^2 - \omega_a^2/2) R_c^2 + (\omega_m^2 - \omega_a^2/2) R_m^2 + \omega_a^2 a^2].$$

As $\tilde{\omega}_c > \omega_a$ and $\omega_m < \omega_a$, the contribution to the energy due to the cyclotron term is positive, whereas the contribution due to the magnetron term is negative. It can be noticed that if the particle loses its energy – for instance by collision processes with other particles – the magnetron radius increases and the particle can touch the trap wall.

In this type of trap, it is possible to store particles for very long times, typically of order of a year.

2.11. Equinox Precession
[Statement and Figures p. 68]

1. The Sun is considered as a spherical object; from the gravitational point of view, it behaves as a pointlike mass M_S located at its center S. Let O be an origin located at the center of mass of the Earth and $OXYZ$ a frame attached to it, with OZ along the pole axis. Let dm be a mass element of the Earth placed at position $\boldsymbol{r} = (x, y, z)$ with respect O and $\boldsymbol{R}(X, Y, Z) = \boldsymbol{OS}$ the position vector of the Sun in this frame. Obviously $SM = \sqrt{(\boldsymbol{R} - \boldsymbol{r})^2}$. The gravitational potential arising from the force due to the Sun on the mass element, dm, is

$$dV = -\frac{GM_S\, dm}{SM} = -\frac{GM_S dm}{\sqrt{R^2 + r^2 - 2\boldsymbol{R} \cdot \boldsymbol{r}}}.$$

On the other hand, one has the condition $r \ll R$ and it is legitimate to perform an expansion up to second order for the square root. We thus arrive at:

$$dV = -\frac{GM_S\, dm}{R}\left[1 + \frac{\boldsymbol{r} \cdot \boldsymbol{R}}{R^2} + \frac{3(\boldsymbol{r} \cdot \boldsymbol{R})^2 - r^2 R^2}{2R^4}\right].$$

The last term in the right hand side is called the quadrupole interaction.

To obtain the total gravitational potential, one must integrate the previous expression over the whole of the Earth's volume. The first term in brackets gives simply the mass M_E. Since O is chosen at the center of mass, one has

$$\int dm\, x_i = 0$$

(as usual, we will often write $x_1 = x$, $x_2 = y$, $x_3 = z$) and the second term cancels. The second order terms lead to moments and products of inertia. Let us choose a reference system with axes along the inertial axes. Since the Earth is a solid with cylindrical symmetry along OZ, one has two identical moments of inertia

$$I = \int dm(x^2 + z^2) = \int dm(y^2 + z^2)$$

and a third distinct moment $I_3 = \int dm(x^2 + y^2)$; of course the products of inertia are null by definition. With this choice (which is compatible with the choice of OZ along the pole axis), the potential can be written in a much simpler way:

$$V = -\frac{GM_S}{R}\left[M_E + (I_3 - I)\frac{X^2 + Y^2 - 2Z^2}{2R^4}\right].$$

Let us emphasize that for a spherical Earth $I_3 = I$ and the potential reduces to the usual expression $V = -(GM_S M_E/R)$.

Because of the cylindrical symmetry, the OX and OY axes can be chosen anywhere in the equatorial plane. Consequently, let us take OX in the plane SOZ and let us recall that ϕ is the angle between the pole axis OZ and the direction pointing to the Sun OS. With these conventions, we have the relations $X = R\sin\phi$, $Y = 0$, $Z = R\cos\phi$, which, after they have been inserted into the potential expression, lead to the desired formula:

$$V = -\frac{GM_S}{R}\left[M_E + \frac{(I_3 - I)}{2R^2}(1 - 3\cos^2\phi)\right].$$

Since the Earth is flattened at the pole, $I_3 > I$ and the potential is minimized when $\phi = \pi/2$; the torque coming from the second term in the potential tends to align the pole axis perpendicular to the direction of the Sun.

2. Let $Sxyz$ be a Galilean frame centered at the Sun, with the Sxy plane in the ecliptic plane. For instance, one can choose the Sx axis along the Earth-Sun direction when the Earth is at perihelion (in our particular case this does not make sense since the Earth's orbit is assumed to be circular). Since, during one revolution, the pole axis maintains a constant angle θ with respect to Sz, it is seen that the ϕ angle actually varies during a year.

Denoting by α the angle between Sx (the Earth's direction at the reference time) and SO (the Earth's direction at the considered time), a small graph convinces us that we have the relation:

$$\cos\phi = -\cos\alpha\sin\theta.$$

Inserting this expression in the potential expression and considering that θ does not vary during a revolution, it is legitimate to consider the potential as averaged over a full year:[14]

$$V(R,\theta) = (1/2\pi) \int_0^{2\pi} V(R,\theta,\alpha)\, d\alpha.$$

Finally, we arrived at:

$$V(R,\theta) = -\frac{GM_S}{R}\left[M_E + \frac{(I_3 - I)}{2R^2}\left(1 - \frac{3}{2}\sin^2\theta\right)\right].$$

3. Let $Oxyz$ be a frame attached to the Earth, but with axes which are fixed with respect to the stars so that the motion of $Oxyz$ with respect to the Galilean frame is just the revolution of the Earth around the Sun. Despite the fact that the Earth rotates around the Sun, this motion is purely translational (with speed $\dot{\alpha}$) since the axes of $Oxyz$ remain always parallel to the axes of the Galilean frame $Sxyz$. The only effect is a participation in the total kinetic energy of the Earth. The rotation concerns the link between $OXYZ$ rigidly attached to the Earth and $Oxyz$.

The instantaneous rotation vector $\boldsymbol{\omega}$ of the Earth originates from the superposition of several motions. There is first the daily rotation around the pole axis which contributes to $\Omega\boldsymbol{u}_Z$. There is also the precession of the plane XOZ around the Oz axis (responsible for the equinox precession) which contributes to $\dot{\psi}\boldsymbol{u}_z$ and there is lastly the nutation which separates the pole axis from the galactic vertical Sz and which contributes to $\dot{\theta}\boldsymbol{u}_X$. Thus, the instantaneous rotation vector is $\boldsymbol{\omega} = \Omega\boldsymbol{u}_Z + \dot{\psi}\boldsymbol{u}_z + \dot{\theta}\boldsymbol{u}_X$. It must be expressed in the frame attached to the inertial axes of the Earth; in order to do this we remark that $\boldsymbol{u}_z = -\sin\theta\,\boldsymbol{u}_Y + \cos\theta\,\boldsymbol{u}_Z$. Substituting this in the previous expression, we are led to the required result:

$$\boldsymbol{\omega} = \dot{\theta}\boldsymbol{u}_X - \dot{\psi}\sin\theta\,\boldsymbol{u}_Y + (\Omega + \dot{\psi}\cos\theta)\boldsymbol{u}_Z.$$

4. From Koenig's theorem, the kinetic energy of the Earth is the sum of the center of mass energy $\frac{1}{2}M_E R^2\dot{\alpha}^2$ (the angular velocity of revolution $\dot{\alpha}$ must be considered as constant) and the rotational energy $\frac{1}{2}(I\omega_X^2 + I\omega_Y^2 + I_3\omega_Z^2)$. Replacing the components of the rotation vector by those obtained previously, one writes the total kinetic energy of the Earth, T, under the form:

$$T = \frac{1}{2}M_E R^2 \dot{\alpha}^2 + \frac{1}{2}\left[I(\dot{\theta}^2 + \dot{\psi}^2\sin^2\theta) + I_3(\Omega + \dot{\psi}\cos\theta)^2\right].$$

---

[14] This crucial point can be justified by the perturbation theory developed in Chapter 7.

5. It is sufficient to take the difference between the kinetic energy T, just obtained, and the potential energy V from Question 2, to obtain the Earth Lagrangian (let us remove the constant terms $\frac{1}{2}M_E R^2 \dot\alpha^2$ and $GM_S M_E/R$ which are of no interest for the equations of the motion):

$$L = \left[I(\dot\theta^2 + \dot\psi^2 \sin^2\theta) + I_3(\Omega + \dot\psi \cos\theta)^2\right]$$
$$+ \frac{GM_S}{R}\left[\frac{(I_3 - I)}{2R^2}(1 - \frac{3}{2}\sin^2\theta)\right].$$

We remark that ψ is a cyclic variable and, thus, the conjugate momentum $p_\psi = \partial_{\dot\psi} L$ is a constant of the motion. Hence,

$$p_\psi = I \sin^2\theta\, \dot\psi + I_3 \cos\theta\, (\Omega + \dot\psi \cos\theta) = \text{const}.$$

The second Lagrange equation concerning θ is much more involved; it is written explicitly as:

$$I\ddot\theta = \frac{1}{2}(I - I_3)\dot\psi^2 \sin(2\theta) - I_3\Omega\dot\psi \sin\theta - \frac{3}{4}\frac{GM_S}{R^3}(I_3 - I)\sin(2\theta).$$

6. In fact, one can avoid a lot of unnecessary complications if one remarks that $\theta = \text{const}$ (no nutation) and $\dot\psi = \text{const}$ (precession with constant angular velocity) is a solution. The first Lagrange equation is automatically satisfied; as to the second one, it is also satisfied (neglecting the term in $\dot\psi^2$ as compared to $\Omega\dot\psi$, which is perfectly justified) providing that the following relation, which gives precisely the precession speed, is fullfiled:

$$\dot\psi = -\frac{3}{2}\frac{GM_S}{\Omega R^3}\frac{I_3 - I}{I_3}\cos\theta.$$

7. The Moon exerts on the Earth an analogous torque, which is far from negligible. Assuming that the revolution plane of the Moon is that of the ecliptic, one can perform again the same kind of calculations substituting simply the Sun's mass M_S by the Moon's mass M_M and the Earth-Sun distance R by the Moon-Earth distance r_M. After these changes, the expression for the precession due to the Moon is identical to that obtained before.

We thus obtain:

$$\beta = \frac{\dot\psi_M}{\dot\psi_S} = \frac{M_M}{M_S}\left(\frac{R}{r_M}\right)^3.$$

With the given data, one calculates $\beta = 2.178$.

8. Since the Moon's action reinforces that of the Sun, the total precession speed is the sum of both contributions, namely $\dot\psi = (1+\beta)\dot\psi_S$, or:

$$\dot\psi = -\frac{3}{2}\frac{GM_S}{\Omega R^3}(1+\beta)\frac{I_3 - I}{I_3}\cos\theta.$$

whence, after inversion, yields:

$$\frac{I_3 - I}{I_3} = \frac{2R^3\Omega|\dot\psi|}{3GM_S(1+\beta)\cos\theta}.$$

With the given data, one finds a flattening $(I_3 - I)/I_3 = 3.25 \times 10^{-3}$.

9. Let ρ be the constant mass density of the Earth. The shape of the Earth is assumed to be that of an ellipsoid of revolution whose equation is $(X^2 + Y^2)/a^2 + Z^2/c^2 = 1$ where $a = R_e$ is the equatorial radius and $c = R_p$ the polar radius of the Earth. The ellipsoidal volume is just $4\pi a^2 c/3$ and the mass density $\rho = 3M_E/(4\pi a^2 c)$.

We now calculate

$$I_{zz} = \int z^2\,dm = \rho\int z^2\,dx dy dz$$

by making the change of variables $u = x/a$, $v = y/a$, $w = z/c$ in order to transform the domain of integration into a sphere with unit radius. Then, one has

$$I_{zz} = \rho a^2 c^3 \int w^2\,du dv dw.$$

Introducing spherical coordinates, it can be shown that the value of the integral is $4\pi/15$. Consequently, $I_{zz} = M_E c^2/5$. By cyclic permutation, one also deduces $I_{xx} = I_{yy} = M_E a^2/5$. It is easy to obtain

$$I = I_{xx} + I_{zz} = M_E\frac{a^2 + c^2}{5} \quad \text{and} \quad I_3 = I_{xx} + I_{yy} = \frac{2}{5}M_E a^2.$$

The flattening is obtained as $(I_3 - I)/I_3 = (a^2 - c^2)/(2a^2)$. If we now introduce the mean radius of the Earth $R_E = (a+c)/2$ we find that, to first order in $(a-c)$, the flattening is approximately equal to $(a-c)/R_E$, or:

$$\frac{I_3 - I}{I_3} = \frac{R_e - R_p}{R_E}.$$

With the value of the flattening obtained in the previous question, and the mean radius of the Earth given in the statement, one calculates $R_e - R_p = 20.7$ km, which may be compared with the measured value 21.4 km. The agreement is quite remarkable.

2.12. Flexion Vibration of a Blade
[Statement and Figure p. 71]

The discontinuous model

1. The kinetic energy for segment i is simply $T_i = \frac{1}{2} dm\, v_i^2$. All the segments are identical and have the same mass $dm = \mu\delta$. Moreover, it is assumed that each segment is displaced in a vertical plane by a quantity $q_i(t)$ with respect to its equilibrium position. The velocity of the segment is therefore $v_i = \dot{q}_i$ and the corresponding kinetic energy is $T_i = \frac{1}{2}\mu\delta\, \dot{q}_i^2$. The total kinetic energy requires a sum over all segments:

$$T = \frac{1}{2}\mu\delta \sum_i \dot{q}_i^2.$$

2. In the vertical plane, the center of curvature O_i for segment i, with coordinates (X, Z), is located at the intersection of the medians of $A_{i-1}A_i$ and A_iA_{i+1} and the curvature radius R_i is the common value $R_i = O_iA_{i-1} = O_iA_i = O_iA_{i+1}$. Writing these relations, one has

$$(Z - q_{i-1})^2 + (X - x_i + \delta)^2 = R_i^2,$$
$$(Z - q_i)^2 + (X - x_i)^2 = R_i^2,$$
$$(Z - q_{i+1})^2 + (X - x_i - \delta)^2 = R_i^2;$$

multiplying by 2 the second equation and subtracting the other two, we are led (neglecting the terms in q^2 as compared to the terms in qZ) to: $Z(q_{i+1} + q_{i-1} - 2q_i) = \delta^2$. On the other hand, δ is small; one has $Z \approx R_i$, whence the desired relation:

$$\frac{1}{R_i} = \frac{|q_{i-1} + q_{i+1} - 2q_i|}{\delta^2}.$$

3. The elastic potential energy for the segment i is

$$V_i = \frac{EI}{2R_i^2}\delta$$

since E and I are characteristics of the material and not of the particular segment. Ignoring edge effects, the total elastic potential energy is obtained by summing the contribution of each segment. Using the value of the curvature radius calculated in the previous question, we are led to the result:

$$V = \frac{EI}{2\delta^3} \sum_i (q_{i-1} + q_{i+1} - 2q_i)^2.$$

4. From the kinetic energy calculated in question 1 and the potential energy calculated in Question 3, the Lagrangian $L = T - V$ is deduced to be:

$$L = \frac{1}{2}\mu\delta \sum_i \dot{q}_i^2 - \frac{EI}{2\delta^3} \sum_i (q_{i-1} + q_{i+1} - 2q_i)^2.$$

The Lagrange equations are obtained from (2.4). Be careful to take into account all the contributions to calculate $\partial_{q_i} L$. The reader is invited to rewrite the potential energy with a mute index j and to rely subsequently on the property $\partial_{q_i} q_j = \delta_{i,j}$. Finally, the Lagrange equations are written in the form:

$$\mu \ddot{q}_i = -\frac{EI}{\delta^4}(q_{i-2} - 4q_{i-1} + 6q_i - 4q_{i+1} + q_{i+2}).$$

The continuous model

5. With the definition of the continuous fields, it is seen that $\ddot{q}_i = \partial_{t^2}^2 \varphi(x_i, t)$ and, using the hint given in the statement,

$$\frac{q_{i-2} - 4q_{i-1} + 6q_i - 4q_{i+1} + q_{i+2}}{\delta^4} = \partial_{x^4}^4 \varphi(x_i, t).$$

Lagrange's equations are expressed in terms of the fields as

$$\mu \partial_{t^2}^2 \varphi(x_i, t) = -EI \partial_{x^4}^4 \varphi(x_i, t).$$

One assumes the continuous hypothesis which implies that this equation is true for any value of x, when the blade is cut in smaller and smaller pieces with the constraints $N \to \infty$, $\delta \to 0$, $N\delta = l$. Replacing the linear mass by its value $\mu = M/l$, one obtains the desired partial differential equation:

$$\frac{\partial^2 \varphi(x, t)}{\partial t^2} = -\frac{lEI}{M} \frac{\partial^4 \varphi(x, t)}{\partial x^4}.$$

6. Let seek a particular solution[15] in the separable form

$$\varphi(x, t) = e^{-i\omega t} \psi(x).$$

This leads to a differential equation for $\psi(x)$:

$$\frac{d^4 \psi(x)}{dx^4} = K^4 \psi(x) \text{ with } K^4 = \frac{\omega^2 M}{lEI}.$$

---

[15] The general solution is obtained with a Fourier transform. It is a combination of such particular solutions, each one presenting a spectral distribution.

Here again, following standard techniques, the solution is sought in the form $\psi(x) = e^{ikx}$, which leads to the simple equation $k^4 = K^4$. We have four roots $k = K, -K, iK, -iK$. The general solution therefore takes the form

$$\varphi(x,t) = e^{-i\omega t}\left[A_1 e^{iKx} + A_2 e^{-iKx} + A_3 e^{-Kx} + A_4 e^{Kx}\right].$$

We now consider the limit conditions. The blade is embedded at the origin which implies that $\varphi(0,t) = 0$ and $\varphi'(0,t) = 0$. The corresponding relationships between the integration constants are: $A_1 + A_2 + A_3 + A_4 = 0$ and $iA_1 - iA_2 - A_3 + A_4 = 0$. The other end of the blade is free. The curvature radius is infinite which means $\varphi''(l,t) = 0$ and $\varphi'''(l,t) = 0$. The corresponding relationships are: $-A_1 e^{iKl} - A_2 e^{-iKl} + A_3 e^{-Kl} + A_4 e^{Kl} = 0$ and $-iA_1 e^{iKl} + iA_2 e^{-iKl} - A_3 e^{-Kl} + A_4 e^{Kl} = 0$. This linear system of 4 equations with 4 unknowns has a non trivial solution provided that the determinant of the matrix vanishes, that is

$$\begin{vmatrix} 1 & 1 & 1 & 1 \\ i & -i & -1 & 1 \\ -e^{iKl} & -e^{-iKl} & e^{-Kl} & e^{Kl} \\ -ie^{iKl} & ie^{-iKl} & -e^{-Kl} & e^{Kl} \end{vmatrix} = 0.$$

The desired condition is very tedious to obtain, but, with some courage, we find the result $1 + \cos(Kl)\cosh(Kl) = 0$. Let us note $X = Kl$. We must solve the following transcendental equation:

$$\cos(X)\cosh(X) = -1; \quad X = \left(\frac{\omega^2 M l^3}{EI}\right)^{1/4},$$

which allows access to the vibrational angular frequency ω of the blade.

In fact, one has an infinite number of solutions X_i which are roots of the previous transcendental equation, which lead to the corresponding angular frequencies:

$$\omega_i = X_i^2 \sqrt{\frac{EI}{Ml^3}}.$$

The lower values are $X_1 = 1.875104$, $X_2 = 4.694091$, $X_3 = 7.854757$, $X_4 = 10.995541$, etc... and the fundamental angular frequency is:

$$\omega_{fund} = 3.516 \sqrt{\frac{EI}{Ml^3}}.$$

7. For a steel blade of thickness $e = 1$ mm and length $l = 10$ cm, one calculates a vibrational frequency of $f = \omega/(2\pi) \approx 75$ Hz.

2.13. Solitary Waves [Statement p. 73]

1. We put $x = \alpha\rho$, $t = \beta\tau$ and define the new field ψ by $\varphi(x,t) = \varphi(\alpha\rho, \beta\tau) = \psi(\rho, \tau)$. With the help of the usual rules of differentiation, it is easy to show that $\partial^2_{\rho^2}\psi = \alpha^2 \partial^2_{x^2}\varphi$ and $\partial^2_{\tau^2}\psi = \beta^2 \partial^2_{t^2}\varphi$. Inserting these relations in the sine-Gordon equation we find

$$\partial^2_{\tau^2}\psi = -\beta^2 \omega^2 \sin\psi + \frac{\beta^2 \lambda}{\alpha^2 l^2} \partial^2_{\rho^2}\psi.$$

Let us now take for the arbitrary variables α and β respectively the definitions $\alpha = \sqrt{\lambda}/(l\omega)$, $\beta = 1/\omega$. The wave equation simplifies to

$$\partial^2_{\tau^2}\psi - \partial^2_{\rho^2}\psi = -\sin\psi,$$

which is precisely the requested form. To conform to the notation proposed in the problem statement, let us rename the field ψ as φ, and ρ, τ as x, t. The simplified wave equation becomes:

$$\frac{\partial^2 \varphi(x,t)}{\partial t^2} - \frac{\partial^2 \varphi(x,t)}{\partial x^2} = -\sin\varphi(x,t).$$

2. Writing $\varphi(x,t)$ in the form $\phi(u)$ with $u = \lambda(x - vt)$, we have

$$\partial^2_{x^2}\phi(u) = \lambda^2 \phi''(u) \quad \text{and} \quad \partial^2_{t^2}\phi(u) = \lambda^2 v^2 \phi''(u).$$

The wave equation is expressed as $\lambda^2(v^2 - 1)\phi''(u) = -\sin\phi(u)$. For the moment, the parameters λ and v are arbitrary; let us impose the following relation:

$$\lambda^2 = \frac{1}{1 - v^2}.$$

The equation can then be recast in the simpler form:

$$\frac{d^2 \phi(u)}{du^2} = \sin\phi(u).$$

The parameters λ and v being real numbers, it is necessary that v, which has the dimension of a speed, satisfies the inequality $v < 1$.

3. There exists a constant of the motion analogous to the energy for Lagrange's equation. To see this, multiply the wave equation by $2\phi'$ to obtain

$$2\phi'(\phi'' - \sin\phi) = 0 = \frac{d(\phi'^2 + 2\cos\phi)}{du};$$

in other words $\phi'^2 + 2\cos\phi = $ const. For $u = \pm\infty$, we impose the limit conditions $\phi'(u) = 0$ and $\phi(u) = 0$ (pendulum equilibrium). These conditions allow us to determine the value of 2 for the constant of integration and, thus, we arrive at:

$$\phi'^2 + 2\cos\phi = 2.$$

Elementary trigonometry based on the half angle $\phi/2$ allows us to rewrite $\phi'^2/4 = \sin^2(\phi/2)$, or

$$\frac{d(\phi(u)/2)}{du} = \pm \sin(\phi(u)/2).$$

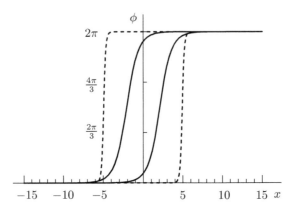

Fig. 2.11 Solitary waves $\varphi(x, t = -5)$ (left curve) and $\varphi(x, t = 5)$ (right curve), obtained for two different stiffness parameters λ. The steepest soliton (dotted line)) propagates more quickly than the softest solution (full line)

4. Let us check that the solution of the previous differential equation is $\phi(u) = 4\arctan e^{\pm u}$. After differentiation, one obtains

$$\frac{\phi'}{2} = \pm \frac{2e^{\pm u}}{1 + e^{\pm 2u}}.$$

Using $y = \tan(\phi/4) = \tan(\arctan e^{\pm u}) = e^{\pm u}$ as an intermediate variable, we obtain after some algebraic manipulation:

$$\frac{\phi'}{2} = \pm 2 \frac{\tan(\phi/4)}{1 + \tan^2(\phi/4)},$$

or $\phi'/2 = \pm \sin(\phi/2)$, which is precisely the equation we want to solve.

Problem Solutions

With an arbitrary choice for the signs, we obtain the two following solitary solutions (see Fig. 2.11):

$$\varphi_+(x,t) = 4\arctan e^{\lambda(x-vt)};$$
$$\varphi_-(x,t) = 4\arctan e^{-\lambda(x-vt)}.$$

2.14. Vibrational Modes of an Atomic Chain
[Statement p. 75]

1. The energy of an elementary chain consisting of light atoms with mass $m' = rm$ and heavy atoms with mass m, is obtained easily from the velocity of their respective displacements u_n and v_n:

$$T_n = \frac{1}{2}rm\dot{u}_n^2 + \frac{1}{2}m\dot{v}_n^2.$$

Between the nth light atom and the nth heavy atom, the net elongation of the spring is $v_n - u_n$ to which corresponds a potential energy $\frac{1}{2}k(v_n - u_n)^2$. Between the light atom $n+1$ and the heavy atom n, the net elongation of the spring is $u_{n+1} - v_n$ to which corresponds a potential energy $\frac{1}{2}k(u_{n+1} - v_n)^2$. The potential energy of the elementary mesh is the sum of these contributions. The total kinetic energy T and total potential energy V are obtained by summing over the elementary chains. The Lagrangian is the difference $L = T - V$, or

$$L = \sum_n \left[\frac{1}{2}m(r\dot{u}_n^2 + \dot{v}_n^2) - \frac{1}{2}k(v_n - u_n)^2 - \frac{1}{2}k(u_{n+1} - v_n)^2\right].$$

The Lagrange equations are derived using the usual techniques (2.4). Let us introduce the proper angular frequency of the springs $\Omega = \sqrt{k/m}$. Then, the Lagrange equations can be written in the form:

$$r\ddot{u}_n = \Omega^2(v_n + v_{n-1} - 2u_n);$$
$$\ddot{v}_n = \Omega^2(u_n + u_{n+1} - 2v_n).$$

2. To find the proper modes,[16] one seeks a solution for which all the atoms vibrate with the same frequency ω, i.e. in the form $u_n = A_n e^{i\omega t}$, $v_n = B_n e^{i\omega t}$. Introducing these expressions in Lagrange's equations and defining the quantity $\lambda = \omega^2/\Omega^2$, the Lagrange equations can be recast as:

$$(2 - r\lambda) A_n = B_n + B_{n-1};$$
$$(2 - \lambda) B_n = A_n + A_{n+1}.$$

---

[16] Here again, the general solution is obtained by a superposition of proper modes and leads to a Fourier series.

3. The previous equations form a linear system with infinite dimension; it is not easy to obtain a solution using this form. To go further, one takes advantage of translational invariance that suggests a solution of the form $A_n = U e^{in\phi}$, $B_n = V e^{in\phi}$. Inserting these definitions in the preceding equations, one obtains:

$$(2 - r\lambda)U = (1 + e^{-i\phi})V;$$
$$(2 - \lambda)V = (1 + e^{i\phi})U.$$

We are still faced with a linear system, but of dimension 2 only. It can be solved easily, for instance by multiplying both equations term by term and simplifying by UV. We are left with an equation of second order in λ:

$$r\lambda^2 - 2(r+1)\lambda + 4\sin^2(\phi/2) = 0.$$

Since λ is expressed as a function of the proper angular frequency ω, this equation gives the relation $\omega(\phi)$. The largest solution, ω_{opt}, is called the optical frequency, while the smallest one, ω_{acou}, is called the acoustical frequency. Explicitly, one has:

$$\omega^2_{\text{opt}}(\phi) = \Omega^2 \frac{1 + r + \sqrt{1 + r^2 + 2r\cos\phi}}{r};$$
$$\omega^2_{\text{acou}}(\phi) = \Omega^2 \frac{1 + r - \sqrt{1 + r^2 + 2r\cos\phi}}{r}.$$

4. The truncated Taylor expansion of the previous expression in the vicinity of $\phi = 0$ is not really a problem; nevertheless, we suggest writing the discriminant in the form $1 + r^2 + 2r\cos\phi = (1 + r)^2 - 4r\sin^2(\phi/2)$. Finally, we obtain (see Fig. 2.12):

$$\omega^2_{\text{opt}}(\phi) = \Omega^2 \frac{2(1+r)}{r}\left[1 - \frac{r}{(1+r)^2}\sin^2(\phi/2)\right];$$
$$\omega^2_{\text{acou}}(\phi) = \Omega^2 \frac{2}{1+r}\sin^2(\phi/2).$$

Relying on the expressions of the previous question, it is not difficult to obtain the angular frequencies for $\phi = \pi$:

$$\omega^2_{\text{opt}}(\phi = \pi) = \frac{2}{r}\Omega^2;$$
$$\omega^2_{\text{acou}}(\phi) = 2\Omega^2.$$

We have transformed a discrete matrix, which has an infinite number of eigenvalues into an infinity of continuous eigenvalues of two different types, labelled by the parameter ϕ.

In the vicinity of $\phi \approx 0$, one has $U/V \approx 1 - \lambda/2$. For the acoustical wave $\lambda \approx 0$ and $U/V \approx 1$. All the atoms vibrate in phase. For the optical wave $\lambda \approx 2(1+r)/r$ and $U/V \approx -1/r$. The light atoms vibrate with opposite phase with respect to the heavy ones, with an amplitude ratio which is inversely proportional to the mass ratio.

5. In order for a wave to propagate, it is necessary that its excitation frequency is precisely that of a proper mode. Thus, we must have $\omega = \omega_{\text{opt}}$ or $\omega = \omega_{\text{acou}}$. Comparing this value to the curves of each proper eigen frequency, the following conditions must be satisfied:

$$0 < \omega^2 < 2\Omega^2 \text{ for an acoustical wave;}$$
$$\frac{2\Omega^2}{r} < \omega^2 < \frac{2(1+r)\Omega^2}{r} \text{ for an optical wave.}$$

The waves such that $\omega^2 > 2(1+r)\Omega^2/r$ or those falling in the gap $2\Omega^2 < \omega^2 < 2\Omega^2/r$ are called evanescent waves.

6. If the atoms are subject to periodic boundary conditions such that $u_{N+1} = u_1$, $v_{N+1} = v_1$, there is an additional condition $e^{i(N+1)\phi} = e^{i\phi}$, which leads to $e^{iN\phi} = 1$. In this case, the ϕ parameter takes only the following discrete values:

$$\phi_l = \frac{2\pi l}{N}.$$

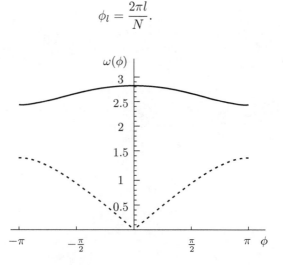

Fig. 2.12 Optical (full line) and acoustical (dotted line) angular frequencies as functions of the phase, obtained with a mass ratio $r = 0.3$

Chapter 3
Hamilton's Principle

Summary

3.1. Statement of the Principle

It is possible to state the laws of mechanics (Newton's or Lagrange's equations) under an equivalent and more concise form, but above all with a much broader impact, applicable to every domain of physics. This new formulation is named "Hamilton's principle" or "least action principle". It can be stated in the following form:

> For a given Lagrangian system, among all the imaginable time evolutions for the configuration $q(t)$ (we remind the reader that $q(t)$ is a simplified notation for the whole set of degrees of freedom $q(t) = (q_1(t), q_2(t), \ldots, q_n(t))$ which begin and end in a determined way (one speaks of a path), the actual evolution (we say a trajectory) is such that a quantity called **action**[1] is stationary.[2]

The task of the physicist is to propose a form of action, which is able to explain the observed physical phenomena.

---

[1] We note the close analogy with Fermat's principle or principle of least optical path. The role of time is played by the curvilinear abscissa, and the Lagrangian by the optical index. Just as there may exist several rays connecting the object to its image, there may exist several trajectories with fixed extremities, as you will see in the Problems 3.4 or 3.5.

[2] The term "stationary" implies that an infinitely small variation of the path, $q(t)+\varepsilon(t)$, produces no change in the action to first order in ε.

3.2. Action Functional

Given a path $q(t)$, the action S is calculated by a time integration of the Lagrange function L. It is a function of the function $q(t)$ and it is said to be a functional. More precisely, let us define:

$$S(q) = \int_{t_1}^{t_2} L(q(t), \dot{q}(t), t) \, dt$$

Its dimension is an energy × a time. In Nature, there exists a quantum of action (the smallest allowed action), which is the Planck constant, $\hbar$. Because this constant is so tiny, the quantification effect is only appreciable for microscopic systems.

Thanks to the calculus of variations, it can be shown that the prescription for stationarity of the action with fixed bounds:

$$\delta S = S(q+\varepsilon) - S(q) = 0 + O(\varepsilon^2)$$

is equivalent to Lagrange's equations

$$\frac{d}{dt}\left(\partial_{\dot{q}_i} L(q, \dot{q}, t)\right) = \partial_{q_i} L(q, \dot{q}, t).$$

Therefore, a unique condition on the functional is equivalent to n local conditions.

3.3. Action and Field Theory

When the number of degrees of freedom is infinite, a continuous function[3] $\varphi(x,t)$, known as a field, replaces the infinite number of functions, $q_i(t)$: the field φ which replaces the coordinates q represents the configuration of the system specified by the continuous variable x, which replaces the discrete index i. The Lagrange function, which depends on the system configuration, is an integral over x of a Lagrangian density $\ell(\varphi, \partial_t \varphi, \partial_x \varphi, x, t)$ which is a function of the field, its first time derivative, its partial space derivatives, and possibly of x and t. In this case, the action is a functional of the field, integral of the Lagrangian density in time and space

$$S(\varphi) = \int_{t_1}^{t_2} \int_{x_1}^{x_2} \ell(\varphi, \partial_t \varphi, \partial_x \varphi, \ldots, x, t) \, dx dt.$$

---

[3] At a given point, it could be the transverse deformation of a vibrating rope, or the value of the electric potential, etc.

Field theory treats the time and space variables on an equal footing. As a consequence, this theory is easily able to incorporate relativistic aspects.

For a continuous system, it can be proved that the stationarity of the action is equivalent to the Euler-Lagrange equations for the fields:

$$\frac{\partial \ell}{\partial \varphi} - \partial_x \left(\frac{\partial \ell}{\partial(\partial_x \varphi)} \right) - \partial_t \left(\frac{\partial \ell}{\partial(\partial_t \varphi)} \right) = 0.$$

3.4. Some Well Known Actions

Hamilton's principle is much more general than Newton's equations because it is a universal principle applicable to every domain in physics (provided we know the corresponding action), but even in mechanics because it allows a generalization to the relativistic case. Indeed, there exist invariance constraints on action which are due to the space-time properties (assumed up to now because they have been verified experimentally[4]), which severely restrict the forms for possible Lagrangians. It really seems that all Lagrangians present in Nature bestow on the action these invariance properties. By such considerations (and others[5]), particular forms of action are obtained in the following cases
– For one particle in a relativistic regime

$$S = -mc \int_{(1)}^{(2)} ds$$

where the infinitesimal distance element ds is given by

$$ds^2 = \sum_{\mu\nu} g_{\mu\nu} x^\mu x^\nu$$

($g_{\mu\nu}$ is the metric tensor), which, if gravitation is not effective, takes the simpler form (invariant for a change of Galilean frames):

$$ds^2 = c^2 dt^2 - dx^2 - dy^2 - dz^2.$$

---

[4] Invariance under time and space translations and reflections, isotropy for rotation, and relativistic covariance.
[5] Superposition principle, with which the results are obtained with the same elegance.

- For a charged relativistic particle embedded in an electromagnetic field $(U, \boldsymbol{A})$

$$S(\boldsymbol{r}) = \int_{(1)}^{(2)} [-mc\,ds + q_e\,(\boldsymbol{A}(\boldsymbol{r},t)\cdot d\boldsymbol{r} - U(\boldsymbol{r},t)\,dt)]$$

$$= \int_{(1)}^{(2)} \left[-mc^2\sqrt{1-\dot{\boldsymbol{r}}^2/c^2} + q_e\,(\boldsymbol{A}(\boldsymbol{r},t)\cdot \dot{\boldsymbol{r}} - U(\boldsymbol{r},t))\right] dt.$$

- For a scalar field [6] $\varphi(\boldsymbol{r}, t)$ representing a particle of mass μ in the presence of a source ρ

$$S(\varphi) = \int_{(1)}^{(2)} \left[\frac{1}{2}\left((\partial_t\varphi/c)^2 - (\boldsymbol{\nabla}\varphi)^2 - \mu^2\varphi^2\right) - \rho\varphi\right] d\boldsymbol{r}\,dt.$$

- For the electromagnetic field[7] $A = (U, \boldsymbol{A})$ in the presence of a charge density ρ and current density $\boldsymbol{j}$

$$S(A) = \int_{(1)}^{(2)} \left[\frac{1}{2}\varepsilon_0\left(\boldsymbol{E}^2 - c^2\boldsymbol{B}^2\right) - \rho U + \boldsymbol{j}\cdot\boldsymbol{A}\right] d\boldsymbol{r}\,dt$$

with $\boldsymbol{E} = -\boldsymbol{\nabla}U - \partial_t\boldsymbol{A}$ and $\boldsymbol{B} = \boldsymbol{\nabla}\times\boldsymbol{A}$.

3.5. The Calculus of Variations

Let us generalize somewhat Hamilton's principle, which is essentially of physical character, into a more mathematical framework. The trajectory $q(t)$ which, with fixed bounds, renders stationary a functional of the form $S(q) = \int_{(1)}^{(2)} L(q, \dot{q}, t)\,dt$ satisfies the differential equation

$$\frac{d}{dt}\left(\partial_{\dot{q}}L(q,\dot{q},t)\right) - \partial_q L(q,\dot{q},t) = 0.$$

---

[6] At each space-time point, an identical value for every observer. An example of a scalar field is the field at the origin of the nuclear interaction (the photon is replaced by the pion).

[7] Four quantities at a given point: the scalar potential $U = A^0/c$ and the vector potential $\boldsymbol{A}$, which transform, for a change of observer, as a quadrivector A^μ.

This result is the consequence of a general mathematical proof; $L(q, \dot{q}, t)$ need not necessarily be a Lagrangian, nor $q(t)$ a generalized coordinate, nor t a time.

For example, $S(q)$ may be the length $l(y)$ of a curve $y(x)$, in which case $L(q, \dot{q}, t)\, dt$ is the length differential element $\sqrt{dx^2 + dy^2} = \sqrt{1 + y'^2}\, dx$ and t the abscissa x. Thus, the shortest path $y(x)$ between two fixed points fulfills the equation[8]

$$\frac{d}{dx}\left(\partial_{y'}\sqrt{1+y'^2}\right) - \partial_y\sqrt{1+y'^2} = 0.$$

The mathematical theory, the purpose of which is the search for an extremum of some functional, is called the calculus of variations. It is mainly the work of the Swiss mathematician Euler, at the end of the XVIIIth century.

In some cases, situations occur for which the functions to be minimized are subject to constraints. There are essentially two types of constraint: constraints of integral type and constraints of holonomic type.

Constraints of integral form
Example: Which, in the plane, is the closed path of a given perimeter which encloses the largest area? The quantity to be maximized is the enclosed area; the constraint is a given value for the length of the path. The expression for the total length is obtained by an integral of the elementary length, whence the name of constraint of integral type.

Very generally, the following property can be proved:

The stationarity of $\int_{(1)}^{(2)} L(q, \dot{q}, t)\, dt$ with the constraint of integral type $\int_{(1)}^{(2)} c(q, \dot{q}, t)\, dt = \mathrm{const}$ is equivalent to making the functional

$$\int_{(1)}^{(2)} [L(q, \dot{q}, t) - \lambda c(q, \dot{q}, t)]\, dt$$

stationary.

The function $q_\lambda(t)$, obtained from Lagrange's equations, depends on λ (the Lagrange multiplier). This multiplier is in turn determined by the constraint

$$\int_{(1)}^{(2)} c(q_\lambda, \dot{q}_\lambda, t)\, dt = \mathrm{const}.$$

---

[8] Obviously this gives $y' = \mathrm{const}$, the straight line.

Holonomic constraints

One can imagine another type of constraint. For instance, let us search for the geodesic on a surface given as a parametric equation (τ being the parameter)

$$\phi(x(\tau), y(\tau), z(\tau)) = 0.$$

This is a holonomic constraint. The three functions x, y, z are not independent and, in principle, one could extract $z(x, y)$ from the constraint equation and restrict the search of the extremum to the two independent functions x, y.

In fact, it is often more elegant to use the following property of the variational method with holonomic constraints.

The stationarity of $\int_{(1)}^{(2)} L(q, \dot{q}, t)\, dt$ with a constraint of holonomic type $\phi(q, t) = 0$ is equivalent to making the functional

$$\int_{(1)}^{(2)} [L(q, \dot{q}, t) - \lambda(t)\phi(q, t)]\, dt$$

stationary.

The solution is a function of the Lagrange multiplier $\lambda(t)$ which is in turn determined, for each value of the integration variable t, by the constraint equation.

Problem Statements

3.1. The Lorentz Force [Solution p. 131] ★ ★

Direct application of Hamilton's principle

1. Check that the Lagrange equations resulting from the following Lagrangian

$$L(\boldsymbol{r}, \dot{\boldsymbol{r}}, t) = -mc^2 \sqrt{1 - \dot{\boldsymbol{r}}^2/c^2} + q_e\left(\boldsymbol{A}(\boldsymbol{r}, t) \cdot \dot{\boldsymbol{r}} - U(\boldsymbol{r}, t)\right)$$

lead to the equations of motion of a relativistic particle, of mass m and charge q_e placed in an electromagnetic field, described by the scalar potential U and the vector potential $\boldsymbol{A}$.

2. Calculate the energy of the system

$$E = \dot{\boldsymbol{r}} \cdot \partial_{\dot{\boldsymbol{r}}} L - L.$$

Show that, in general, it is not constant.

3.2. Relativistic Particle in a Central Force Field [Solution p. 132] ★ ★ ★

Example of a relativistic Lagrangian
A – Relativistic particle in an electrostatic field

Let us start from the action

$$S = \int \left(-mc^2\sqrt{1-\beta^2} - V(r)\right) dt,$$

where $\beta^2 = \dot{r}^2/c^2$ is the usual relativistic factor.

1. Use the rotational invariance to show that the motion takes place in a plane. One can for instance work with spherical coordinates and make use of the fact that the ϕ coordinate is cyclic.

2. Since the motion is planar, it is legitimate to employ the set of polar coordinates (ρ, ϕ). Give the constant of the motion σ associated with the cyclic coordinate ϕ (the angular momentum).

3. From now on, we focus on the determination of the trajectory (otherwise the problem is more involved). Thus, one seeks to replace the time derivative by the derivative with respect to the polar angle ϕ. In particular, we will denote $\rho' = d\rho/d\phi$.

    Express the angular velocity $\dot{\phi}$ as a function of σ, ρ, and β^2 (remember that β is also a function of ρ, ρ' and $\dot{\phi}$). Calculate v^2 and deduce the expression for the kinematical factor $\gamma = (1-\beta^2)^{-1/2}$.

4. Express the constant of the motion E (energy), associated with the time translational invariance.

5. The variable $u(\phi) = 1/\rho(\phi)$ is introduced. From the energy and the expression for γ given in Question 3, give the first order differential equation satisfied by $u(\phi)$.

6. Derive this equation with respect to ϕ, and deduce the second order differential equation (sometimes simpler to treat) that is satisfied by $u(\phi)$. Compare to Binet's equation (2.10).

B – Relativistic particle in a scalar field

We start now from the action $S = \int \left(-mc^2 - V(r)\right) \sqrt{1-\beta^2}\, dt.$

Examine again all the questions of the previous case.

3.3. Principle of Least Action?
[Solution p. 135] ★ ★ ★

Why we speak of least action whereas Hamilton's principle speaks only of stationary action?

In many textbooks, Hamilton's principle is often referred to as a least action principle. In this problem, we will understand that, if the trajectory extremities are close enough (the problem will make this notion more precise), the action which is stationary (Hamilton's principle) does indeed exhibit a minimum (least action principle).

1. A particle of mass m, referred to by its coordinate q, is subject to a force that arises from a potential $V(q)$. Let us start from the Lagrangian

$$L(q, \dot{q}) = \frac{1}{2} m \dot{q}^2 - V(q)$$

and imagine a path $q(t) = \tilde{q}(t) + \varepsilon(t)$ close to the trajectory $\tilde{q}(t)$. Show that the second order variation of the action between the time $t = 0$ and the time T can be written :

$$\delta^2 S \cong \frac{1}{2} \int_0^T \left[m \dot{\varepsilon}(t)^2 - \varepsilon(t)^2 \, V''(\tilde{q}(t)) \right] \, dt.$$

It is always possible to choose a time T sufficiently small so that $V''(\tilde{q})$ maintains a constant sign.

If, over the whole range, $V''(\tilde{q})$ is always negative or null (case of a gravitational field), then it is clear that, owing to the fact that $\delta^2 S > 0$, we are in the presence of a minimum for the action.

In the other more complex case, it is possible to minimize the term containing the potential by replacing it by $V''_{\max} \int_0^T \varepsilon(t)^2 \, dt$, where $V''_{\max}$ is the maximum of the second derivative along the trajectory. The function $\varepsilon(t)$ can be defined up to a multiplicative constant, without changing the order of inequalities. One can take advantage of this freedom to fix a value for the integral $\int_0^T \varepsilon(t)^2 \, dt$.

2. Using the variational method show, with the constraint of a fixed value for $\int_0^T \varepsilon(t)^2 \, dt$, that the variation compatible with the boundary conditions and which minimizes the kinetic energy integral is $\varepsilon(t) = \nu \sin(\pi t/T)$.

3. Deduce that, if the time range is sufficiently small: $T^2 < m\pi^2/V''_{\max}$, we are dealing with a minimum.

3.4. Minimum or Maximum Action?
[Solution p. 137] ★ ★ ★

A stationary action may sometimes be a maximum

Usually, one speaks of the principle of least action. The more appropriate word would be stationary action. In this problem, we will see that the Lagrange equations lead, depending on the type of variation, either to minima or to maxima for the action.

Let us take for example the case of a one dimensional harmonic oscillator and, to simplify the notation (the price to pay is that dimensional analysis is not easy to employ), write the Lagrange function as

$$L(q, \dot{q}) = \frac{1}{2} \left(\dot{q}^2 - q^2 \right).$$

1. Write and solve the Lagrange equation giving the trajectory $\tilde{q}(t)$.
2. Imagine a path $q(t)$ close to the trajectory: $q(t) = \tilde{q}(t) + \varepsilon(t)$, such that

$$\varepsilon(0) = \varepsilon(T) = 0.$$

Show that the second order variation of the action between a time 0 and an arbitrary time T is:

$$\delta^2 S(\varepsilon) = \frac{1}{2} \int_0^T \left(\dot{\varepsilon}(t)^2 - \varepsilon(t)^2 \right) dt.$$

3. Consider the case of a very simple function for the variation $\varepsilon(t) = \sin(\pi t/T)$, satisfying the limit conditions. Show that, if T is greater than a value that should be determined, the variation of the action is negative so that the action presents a maximum.
4. Let us choose a trajectory passing through the point q_0 at time $t = 0$ and the point q_1 at time t_1. Show that if $t_1 \neq n\pi$ (n integer), there exists one and only one trajectory satisfying the conditions. Show that if $t_1 = n\pi$ and $q_1 \neq (-1)^n q_0$ there exists no possible trajectory, while if $q_1 = (-1)^n q_0$ there exists an infinite number of possible trajectories. Check that, in this last case,

$$\delta^2 S(\varepsilon) = 0,$$

for the previous function $\varepsilon(t)$ and for $n = 1$. The point q_1 is called the conjugate point to q_0.

3.5. Is There Only One Solution Which Makes the Action Stationary?
[Solution and Figure p. 138] ★ ★ ★

Several trajectories may pass through two given points

How many functions (trajectories) with fixed bounds make the action stationary ?

Let us take a free particle in a homogeneous infinite space. One knows that there is only one solution: the straight line covered at uniform speed. However there exist more amusing situations. For instance, let us consider a particle of mass m in a one dimensional space, located by its coordinate q, free in the domain $[0, L]$ and subject to harmonic forces outside this interval:

$$V(q) = \begin{cases} \frac{1}{2}m\omega^2 q^2, & \text{if } q < 0, \\ 0, & \text{if } 0 < q < L, \\ \frac{1}{2}m\omega^2(q-L)^2, & \text{if } q > L. \end{cases}$$

$T = 2\pi/\omega$ is the period of the motion in the harmonic potential.

1. Show that, if $t_2 - t_1 > T/2$, there exist at least three trajectories $q(t)$ starting in the domain $[0, L]$ from the position q_1 at time t_1, ending in the same domain at the position q_2 at time t_2, and which all make the action stationary.

   Indication: You should find the uniform motion, plus two trajectories that bounce on the potentials on the right and on the left.

   For greater time separations, it is possible to have more solutions with several bounces.

2. Give the expression of the action for each of these trajectories. It is interesting to use the following property: for a harmonic oscillator, the average in time for the kinetic energy and for the potential are equal. This is just a consequence of the virial theorem.

   One of these trajectories minimizes the action. What can be said for the others? To study this point, let us imagine a variation with one bounce. To make things easier, consider a very close path, with the same shape, but which corresponds to a smaller period $T_<$.

3. Show that the variation of second order is negative. Nevertheless it is neither a minimum nor a maximum, because if we consider the solution with a greater period, the action increases. We say that the trajectory corresponds to a saddle point action.

3.6. The Principle of Maupertuis
[Solution p. 141]   ⋆ ⋆

A principle less general than Hamilton's principle which gives the trajectory directly

We will see in Chapter 5 that the trajectory of a particle of mass m in a conservative force field makes extremal the "reduced action" which is defined by $S_0(q) = \int_{(1)}^{(2)} ds \sqrt{2m\,(E - V(q))}$, where ds is the length element along the trajectory. We remark the analogy with Fermat's principle. It allows us to obtain the form of the trajectory, without being worried by the temporal evolution; in many cases, this is sufficient.

Let us employ the calculus of variations on this reduced action to determine the trajectory equation in a plane for two particular cases:

First case: gravitational field

The potential under consideration, $V(z) = mgz$, corresponds to a uniform gravitational field g along the vertical Oz. We seek a solution in the vertical plane xOz in the form $x(z)$. The reduced action is, in this case, the functional $S(x)$.

1. Show that the x variable is cyclic and that there exists a constant of the motion.

2. By integration of the previous equation, show that the trajectory is a parabola.

Second case: central field

The potential considered is central. We know that the motion takes place in a plane. In this plane, we use the polar coordinates (ρ, ϕ).

1. We study first the trajectory in the form $\phi(\rho)$. The reduced action is a functional $S_0(\phi)$. Show that the ϕ variable is cyclic and that there exists a constant of the motion which is identified with the angular momentum σ (more exactly its opposite). Deduce the integral expression for the trajectory $\phi(\rho)$.

2. At present, we study the trajectory under the form $\rho(\phi)$. The reduced action is a functional $S_0(\rho)$. Show that there exists also a constant of the motion which is again identified as σ. To simplify this expression, we make the change of function $u(\phi) = 1/\rho(\phi)$. After derivation, one obtains a second order differential equation in u, called Binet's equation. Give the corresponding equation and check the result in the Expression (2.10).

3.7. Fermat's Principle
[Solution and Figure p. 144] ★★

Application of Hamilton's principle in a domain different from mechanics

Fermat's principle stipulates that "light rays" make stationary the optical path $\int n\,ds$, where n is the index of refraction and ds the differential length element. It was stated a long time ago by Fermat in the XVIIth century, but it is of course much later that it was recognized that it could be connected to "least action" type problems. We will use here what we learned on the variational method to find the equation of rays in a two dimensional space (x, z), for which the refraction index $n(z)$ depends only on the z coordinate.

To seek the minimum of the optical path (Fermat's principle), we consider two methods:

1. Write the expression of the optical path as an integral over the z variable. Use the Lagrange equations to determine the curve $x(z)$. Notice that there exists a constant of the motion. Interpret the corresponding relation in the framework of geometrical optics.

2. Write the expression of the optical path as an integral over the x variable. Use the Lagrange equations to determine the curve $z(x)$. Notice that we find the same constant of the motion.

3. Find the light trajectory for a linear variation of the index $n(z) = n_0 + \lambda z$. One imposes the initial conditions $z(0) = 0$, $z'(0) = 0$ on the solution.

   Give a plausible explanation of the "mirages" observed in deserts.

   It is possible to make a complete analogy between the previous study and the propagation of a sound wave: normally, the temperature decreases with the altitude and so does the sound speed. What conclusion can be drawn concerning the propagation of noise?

3.8. The Skier Strategy
[Solution and Figure p. 146] ★★★

A very classical problem for application of the calculus of variations

On a snow covered inclined plane with uniform slope, making an angle α with respect to the horizontal, in a constant vertical gravitational field g, a skier of mass m starts from a gate O with a null speed, and wishes to arrive with the fastest time at the gate A below. The optimal trajectory (the quickest) is not necessarily the straight line OA.

Explain rapidly why, assuming no dissipation (friction with the snow and the air), assumption which is of course physically far from reality.

We take in the plane a frame with the origin at O, a horizontal axis Ox and the axis Oy in the direction of the steepest downward slope. The zero for the gravitational potential energy is chosen at O.

1. Express the mechanical energy of the skier at a given time. What is its value?
2. Express the time interval dt between two positions (x, y) and $(x+dx, y+dy)$ for the skier.
3. Deduce the time needed for the skier to go from O to A. This time will be given as a functional $T(x)$ where $x(y)$ represents the trajectory of the skier. The optimal trajectory minimizes this time.
4. Show that the x coordinate is cyclic and deduce a constant of the motion.
5. Obtain the equation of the optimal trajectory under a parametric form $x(\theta), y(\theta)$, θ being an angle. It is strongly suggested to carry out a change of variable such that y is proportional to $\sin^2 \theta$.
6. Now, a more tricky question: must we consider trajectories that pass below the end point? Compare the course times for the optimal trajectory, for a straight trajectory and for a descent along the steepest slope followed by a horizontal segment.

3.9. Free Motion on an Ellipsoid
[Solution p. 150] ★ ★

Calculus of variations with a holonomic constraint

A particle, of mass m, is free to move on an ellipsoid with equation:

$$(x/a)^2 + (y/b)^2 + (z/c)^2 = 1$$

One wishes to determine the reaction force that maintains the particle on the surface.

Let us first remark that the number of degrees of freedom is 2 and, if we use as generalized coordinates the Cartesian coordinates (x, y, z), one must take into account the constraint.

1. Write the equation which expresses Hamilton's principle corresponding to this holonomic constraint. Deduce the Lagrange equations.
2. Check that the reaction force is normal to the surface.

3. Check the conservation of energy (kinetic in this particular case).

4. Give the expression for the intensity of the reaction force, R, as a function of coordinates and velocities, but eliminating accelerations. Check your result for a sphere.

5. The motion on an ellipsoid is "integrable" that is, as we will see in Chapter 6, there exist as many constants of the motion as the number of degrees of freedom. Check that

$$\dot{x}^2 + \frac{(x\dot{y} - \dot{x}y)^2}{(a^2 - b^2)} + \frac{(x\dot{z} - \dot{x}z)^2}{(a^2 - c^2)},$$

together with the cyclic permutations, are constants of the motion.

3.10. Minimum Area for a Fixed Volume
[Solution p. 152] ⋆ ⋆

Calculus of variations with an integral constraint

With a given volume, among the pear, the apple or the orange, which is the fruit which has less peel?

One considers a solid with a cylindrical symmetry around the Oz axis. This solid is specified by the function $r(z)$ – the distance from the symmetry axis as function of the altitude – between the altitudes 0 and h (example on Fig. 3.1). One assumes a solid "without circumvolution"; this means that the function $r(z)$ is single valued (for each altitude z there corresponds a unique value of r). Moreover, the Oz axis passes inside the solid, and the lowest and highest points belong to the solid, so that $r(0) = 0 = r(h)$. Given these conditions, one seeks the solid which has a minimum surface for a given volume.

1. Cutting the solid into slices of thickness dz, give the volume functional: $V(r)$ and the area functional: $A(r)$.

2. Write down the functional to be minimized, introducing the Lagrange multiplier λ.

3. Rather than solving the differential equation resulting from the Euler-Lagrange equation, it is judicious to remark the existence of a constant of the motion corresponding to the translation along the Oz axis (equivalent to time in mechanics). Determine the value of this constant when located at $z = 0$. Give the corresponding first order differential equation.

4. Express $r' = dr/dz$ as a function of r. Integrate this equation (for instance r^2 could be used as an auxiliary variable). Deduce that the solution we seek is a sphere. What is the interpretation of the Lagrange multiplier?

5. From the previous study, solve in a very simple way the inverse problem: what is the shape of a solid with a maximum volume for a given surface?

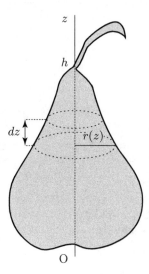

Fig. 3.1 Area and volume of a solid with cylindrical symmetry

3.11. The Form of Soap Films
[Solution and Figures p. 154] ★ ★ ★

The Hamilton principle in a bubble

What is the underlying basis for the form of the surface of a soap film which is characterized by one or several closed curves? It is, of course, the surface tension forces that tend to minimize the area of the surface. Thus, for determining the form of the minimum surface based on a given curve, it is not necessary to appeal to high-performance numerical codes; it is enough to dip an iron stem which has precisely the form of this curve in a detergent solution and look at the resulting soapy film. Its form gives to you the answer!

In this problem, we look for the equation of the surface of a film formed by two identical hoops with the same axis, assuming the natural hypothesis of cylindrical symmetry. Searching for a solution possessing the symmetry of the problem is known in physics as the Curie principle.

1. The OZ axis can be chosen as the symmetry axis, and one can define the surface by $r(z)$, giving the radius at altitude z. The two hoops are located at altitudes z_1 and z_2. Write down the integral expression of the area, which is a functional of the curve $r(z)$.

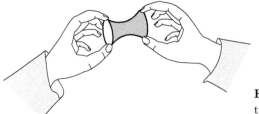

Fig. 3.2 Form of a soap film between two hoops

2. To obtain an extremum for this area, we should, in principle, write and solve the second order differential equation resulting from Lagrange's equation. Nevertheless, one can do better. Find a constant of the motion; in other words find a quantity, function of $r(z)$ and $r'(z) = dr/dz$, which does not depend on z.

3. Check that the solution of the previous first order differential equation is of the form $r(z) = \rho \cosh((z-h)/\rho)$ (Fig. 3.2). Plot schematically this function and interpret the constant parameters h and ρ.

4. For brevity, let us take two identical hoops of radius R located at $z_1 = -H$ and $z_2 = H$. Give the relation which allows the determination of the constant ρ as function of R and H.

5. Find an inequality obeyed by R/H in order that the physical problem admits at least one solution. If this inequality is fulfilled, there exist two mathematical solutions, but only one physical solution. Explain why? What happens if we separate the two hoops up to the limit of the inequality? Discuss briefly what happens after.

   Indication: Plot, as a function of $x = H/\rho$, $\cosh(x)$ and the straight line Rx/H.

6. What would you do to determine the force that maintains the cohesion of the film as a function of the distance $2H$ between the hoops? Remember that the capillarity potential energy is proportional to the area and that the force is the gradient of the potential energy.

   The result is simple, but only a posteriori!

   Hint: Find the expression of the area

   $$A = 2\pi H^2 \left((\cosh(x)\sinh(x))/x^2 + 1/x\right)$$

   and differentiate it with respect to H.

3.12. Laplace's Law for Surface Tension
[Solution p. 158]   ★ ★ ★

The Hamilton principle applied to hydrostatics

Let us consider an incompressible liquid (mass density ρ) lying, under the influence of a vertical constant gravitational field g directed downwards, in a parallelepipedic channel along an infinite horizontal axis $y'y$ (to avoid the boundary effect in that direction). Let O be an arbitrary origin on $y'y$, and Ox a horizontal axis perpendicular to Oy; the upward vertical is Oz. The xOz plane is thus a section plane and the form of the surface is a curve $z(x)$. This form is determined, in a static way, by a minimization of the gravitational potential energy and the surface potential energy TS. The surface tension of the liquid in contact with the air is T, supposed to be constant, and S the air-liquid interface area. Because of the translational invariance along the $y'y$ direction, we can reason using a slice with unit thickness in that direction. The edges of the channel are taken at the abscissas $x = 0$ and $x = l$ (Fig. 3.3).

The liquid-air-wall interface points are assumed to be fixed at $z(0) = h = z(l)$. We wish to minimize the total potential energy with fixed bounds. Moreover, we have in addition the constraint of a constant volume.

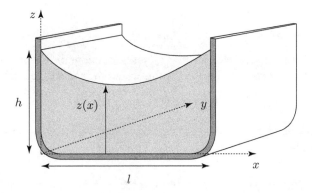

Fig. 3.3 Channel containing an incompressible liquid. In a slice of liquid, one defines a system of axes xOz. The form of the meniscus is given by the curve $z(x)$

1. Give the expression of the functional of the total potential energy $V_P(z)$.

2. Express in the same way the functional of the volume $V(z)$.

3. Give the Euler-Lagrange equation constrained by a given volume, which allows the determination of the curve $z(x)$. It is useful to introduce the curvature radius $R(x)$ at each point.

3.13. Chain of Pendulums [Solution p. 160] ★ ★

Hamilton's principle for a continuous system

One considers a chain of N identical pendulums of length l and mass m, all suspended from a horizontal axis. Each pendulum is separated from the next by a distance δ. In addition to gravity, the pendulums are subject to a harmonic force between nearest neighbours $\frac{1}{2}k(q_i - q_{i-1})^2$, where q_i is the angle between the vertical and the direction of the pendulum numbered i and k is a constant characteristic of the system which takes into account the effect of restoring forces.
1. Write down the Lagrangian of the system.

2. Using this Lagrangian and passing to a continuous description, calculate the Lagrangian density of the system assuming that $k\delta^2/m = \lambda$ is a constant.

3. Applying the Euler-Lagrange equation to this system, recover the wave equation studied in the Problem 2.13. We remind the reader that this equation is linear only for small motions.

3.14. Wave Equation for a Flexible Blade
[Solution p. 161] ★ ★

Hamilton's principle for a continuous system

We revisit Problem 2.12 concerning a metallic blade, of linear mass density μ, allowed to vibrate in a plane. Again, we assume weak flexions.
1. Passing to a continuous limit, give the expression of the Lagrangian density for the system.

2. With the help of the Euler-Lagrange equations, recover the wave equation proposed in the Problem 2.12.

3.15. Precession of Mercury's Orbit
[Solution p. 162] ★ ★ ★

Hamilton's principle can be directly applied in the framework of general relativity

In classical Newtonian theory, the trajectory of a planet of mass m, subject only to the attraction of the Sun of mass M, is an ellipse with semi-major axis a and eccentricity e, with the Sun located at one focus.

Problem Statements

In reality, due to miscellaneous perturbations (influence of the other planets among others), the true trajectory is not closed; the symmetry axes of this ellipse rotate very slowly between two revolutions. This phenomenon is called precession of perihelia. The precession concerning Mercury is most important.

In spite of all refinements imagined in the framework of celestial mechanics during the nineteenth century, there remained a disagreement between the predicted and measured values. The explanation of this disagreement using Einstein's theory of general relativity was one of its most fantastic achievements.

The aim of this problem is to present a way for addressing this disagreement. Taking into account relativistic effects for celestial mechanics makes sense in the theory of general relativity. Don't give up faced with these impressive words; read the remainder, you will see that the problem is well within reach.

The action has the traditional relativistic form $S = -mc \int ds$ with, in general, the differential length element ds given by the metric tensor $g_{\mu\nu}(x)$, $(\mu, \nu = 0, 1, 2, 3)$, but with a value depending on the position quadrivector $x = (x_0 = ct, x_1 = x, x_2 = y, x_3 = z)$:

$$ds^2 = \sum_{\mu,\nu} g_{\mu\nu}(x)\, dx_\mu\, dx_\nu.$$

The gravitational attraction by a body of mass M with spherical symmetry[9] exhibits a metric, known as Schwarzschild's metric, which, using spherical coordinates, is written:

$$ds^2 = e(r)c^2 dt^2 - r^2(d\theta^2 + \sin^2\theta\, d\phi^2) - dr^2/e(r)$$

with

$$e(r) = 1 - \frac{2GM}{c^2 r} = 1 - \frac{r_0}{r} \tag{3.1}$$

where G is the universal gravitational constant, c the speed of light and $r_0 = 2GM/c^2$ the Schwarzschild radius of the Sun.

As for any central interaction, the trajectory is situated in a plane, that we choose as equatorial ($\theta = \pi/2$).

1. Write the action as a time integral and deduce the Lagrange function.

2. The ϕ variable is cyclic; deduce the associated constant of the motion σ (the angular momentum).

---

[9] S. Weinberg, *Gravitation and Cosmology: Principles and Applications of the General Theory of Relativity*, John Wiley and sons, New York, 1972.

3. Deduce the expression of the angular velocity as a function of the radius r and its derivative $r' = dr/d\phi$.

   We focus now on the study of the trajectory. One introduces the constant of the motion associated with the translational time invariance, namely the energy E.

4. Give its expression.

5. Use the result of the previous question to obtain the differential equation giving the trajectory. Make the usual change of variable $u = 1/r$ and give the corresponding differential equation.

6. Differentiating this equation with respect to ϕ, obtain a second order differential equation for the u variable.

7. Find the inverse of the radius $u_c = 1/R_c$ for circular orbits.

8. Let $v = u - u_c$. Solve the differential equation assuming a small value for v (linearization of the equation of motion). Show that one solution is stable. The corresponding trajectory is approximately an ellipse. Calculate the angular shift between successive perihelia for the revolution of the planet around the Sun. Give an estimation of this shift for the planet Mercury (we remind you of the relationship between the angular momentum and the characteristics of the orbit $\sigma^2/(GMm^2) = a(1-e^2)$, where a is the length of the semi-major axis of the orbit and e the corresponding eccentricity). The measured value of the shift is 572″ per century; 42.6″ remained unexplained by classical mechanics.

   What about the precession of perihelia for the Earth (in this case the unexplained difference was 4.6″ per century)? Conclusion.
   Take the values:

   G (gravitational constant) $= 6.673 \times 10^{-11}$ MKSA
   M (mass of the Sun) $= 1.989 \times 10^{30}$ kg
   c (speed of light) $= 2.998 \times 10^8$ m/s
   a (Mercury) $= 57.9 \times 10^6$ km
   e (Mercury) $= 0.2056$
   T (Mercury) $= 87.969$ days
   a (Earth) $= 149.6 \times 10^6$ km
   e (Earth) $= 0.0167$
   T (Earth) $= 365.25$ days.

Problem Solutions

3.1. The Lorentz Force [Statement p. 116]

1. The Lagrange equations are written generally

$$\frac{d}{dt}(\partial_{\dot{r}}L) = \partial_r L.$$

For the proposed Lagragian:

$$\partial_{\dot{r}}L = \gamma m \dot{\boldsymbol{r}} + q_e \boldsymbol{A},$$

where $\gamma = \left(1 - \dot{r}^2/c^2\right)^{-1/2}$ is the usual relativistic factor. Let us introduce the relativistic linear momentum of the particle $\boldsymbol{\pi} = \gamma m \dot{\boldsymbol{r}}$; it then follows that $\partial_{\dot{r}}L = \boldsymbol{\pi} + q_e \boldsymbol{A}$. Taking the total time derivative of this expression leads to

$$\frac{d(\partial_{\dot{r}}L)}{dt} = d\boldsymbol{\pi}/dt + q_e \left[\partial_t \boldsymbol{A} + \dot{x}\partial_x \boldsymbol{A} + \dot{y}\partial_y \boldsymbol{A} + \dot{z}\partial_z \boldsymbol{A}\right]$$

and moreover

$$\partial_r L = q_e \left[\dot{x}\partial_r A_x + \dot{y}\partial_r A_y + \dot{z}\partial_r A_z - \boldsymbol{\nabla} U\right].$$

The Lagrange equations imply:

$$\frac{d\boldsymbol{\pi}}{dt} = q_e[-\boldsymbol{\nabla} U - \partial_t \boldsymbol{A} + \dot{x}\left(\partial_r A_x - \partial_x \boldsymbol{A}\right)$$
$$+ \dot{y}\left(\partial_r A_y - \partial_y \boldsymbol{A}\right) + \dot{z}\left(\partial_r A_z - \partial_z \boldsymbol{A}\right)].$$

On the right hand side, one recognizes the electric field

$$\boldsymbol{E} = -\boldsymbol{\nabla} U - \partial_t \boldsymbol{A}.$$

Let us focus on the term $\partial_r A_x - \partial_x \boldsymbol{A}$ which has three components. The x component is $\partial_x A_x - \partial_x A_x = 0$; the y component is $\partial_y A_x - \partial_x A_y = -(\boldsymbol{\nabla} \times \boldsymbol{A})_z = -B_z$, and the z component is $\partial_z A_x - \partial_x A_z = (\boldsymbol{\nabla} \times \boldsymbol{A})_y = B_y$. The components of $\partial_r A_y - \partial_y \boldsymbol{A}$ and $\partial_r A_z - \partial_z \boldsymbol{A}$ are obtained similarly by cyclic permutations.

Thus $d\pi_x/dt = q_e \left[E_x + \dot{y}B_z - \dot{z}B_y\right] = q_e \left[\boldsymbol{E} + \boldsymbol{v} \times \boldsymbol{B}\right]_x$ with two similar equations for the y and z components. Therefore, one can write with vector notation

$$\frac{d\boldsymbol{\pi}}{dt} = q_e \left[\boldsymbol{E} + \boldsymbol{v} \times \boldsymbol{B}\right]$$

which is nothing more than the equation of motion for a charged particle submitted to the Lorentz force.

2. From the general formula, the energy E (not to be confused with the electric field E) is expressed as $E = \dot{r} \partial_{\dot{r}} L - L$. With the already proven relation $\partial_{\dot{r}} L = \pi + q_e A$, one obtains

$$E = \gamma m \dot{r}^2 + q_e A \cdot \dot{r} + mc^2/\gamma - q_e A \cdot \dot{r} + q_e U,$$

which simplifies to:

$$E(r, \dot{r}, t) = \gamma mc^2 + q_e U(r, t).$$

One clearly distinguishes the contributions of the relativistic energy and the electric energy. The magnetic field does not contribute to the energy since the Lorentz force, always perpendicular to the velocity, performs no work. For a time dependent scalar potential, the energy itself depends on time and thus is not a constant of the motion.

3.2. Relativistic Particle in a Central Force Field [Statement p. 117]

Case of a particle in an electromagnetic field

1. Let us use spherical coordinates to write the Lagrangian:

$$L(r, \theta, \phi, \dot{r}, \dot{\theta}, \dot{\phi}) = -mc^2 \sqrt{1 - \beta^2} - V(r)$$

where $\beta^2 = \dot{r}^2/c^2 = (\dot{r}^2 + r^2 \dot{\theta}^2 + r^2 \sin^2\theta\, \dot{\phi}^2)/c^2$.

The ϕ coordinate is cyclic; consequently there exists a constant of the motion

$$p_\phi = \partial_{\dot{\phi}} L = mr^2 \sin^2\theta\, \dot{\phi}/\sqrt{1-\beta^2} = \gamma mr^2 \sin^2\theta\, \dot{\phi} = \text{const}.$$

It is easy to check that p_ϕ is the z component of the angular momentum of the particle: $p_\phi = \sigma_z$, where $\sigma = r \times \pi$ and $\pi = \gamma mv$. The Oz axis was chosen arbitrarily; a similar reasoning can be made by a choice of two other mutual orthogonal axes which implies $\sigma_x = \text{const} = \sigma_y$, or equivalently $\sigma = \text{const}$. The angular momentum vector being constant, the same property holds for a plane perpendicular to it; this plane contains both vectors r and v, that is the trajectory, i.e. The trajectory is plane

Problem Solutions   133

2. Let us work in this plane and employ the polar coordinates (ρ, ϕ). We have $\beta^2 = (\dot{\rho}^2 + \rho^2 \dot{\phi}^2)/c^2$ and the constant of the motion resulting from the rotational invariance is the modulus of the angular momentum $p_\phi = \partial_{\dot{\phi}} L = \gamma m \rho^2 \dot{\phi} = \sigma$.

$$\gamma m \rho^2 \dot{\phi} = \sigma.$$

3. In the expression of $v^2 = \dot{\rho}^2 + \rho^2 \dot{\phi}^2 = \dot{\phi}^2(\rho'^2 + \rho^2)$ substitute the expression for $\dot{\phi}$ obtained in the previous question; let us calculate $1 - \beta^2 = 1 - v^2/c^2$ from this equation, and then $\gamma = (1-\beta^2)^{-1/2}$. After simple calculations, we obtain:

$$\gamma = \sqrt{1 + \frac{\sigma^2(\rho'^2 + \rho^2)}{m^2 c^2 \rho^4}}.$$

4. The Lagrangian does not depend explicitly on time. The energy is conserved:

$$E = \dot{\rho}\, \partial_{\dot{\rho}} L + \dot{\phi}\, \partial_{\dot{\phi}} L - L = \text{const.}$$

Performing the calculation, we obtain the traditional expression (the relativistic kinetic energy plus the potential energy):

$$E = \gamma m c^2 + V(\rho).$$

5. We extract γ from this last expression and replace γ^2 by the expression of Question 3 thereby obtaining a differential equation concerning ρ. Now, let us perform the proposed change of variable $u(\phi) = 1/\rho(\phi)$. We are led to the following first order differential equation in u ($u' = du/d\phi$):

$$u'^2 + u^2 = \frac{m^2 c^2}{\sigma^2} \left[\left(\frac{E - V(1/u)}{mc^2} \right)^2 - 1 \right].$$

6. Let us differentiate this expression with respect to ϕ, then simplify by $2u'$. We are left with a second order differential equation in u.

$$u'' + u = -\frac{E - V(1/u)}{\sigma^2 c^2} \frac{d}{du} V(1/u).$$

One could qualify this equation as the "relativistic Binet equation" for a particle in an electromagnetic field. For specific cases, it is more convenient than the first order differential equation given in the previous question (it avoids the cumbersome square root).

Case of a particle in a scalar field

1. The proof follows exactly the same reasoning as in the previous case and results from the rotational invariance. Nevertheless, remember that the angular momentum $\boldsymbol{\sigma} = \boldsymbol{r} \times \boldsymbol{p}$ is not identical with the kinetic momentum since the generalized momentum $\boldsymbol{p} = \gamma(m + V(r)/c^2)\boldsymbol{v}$ is different from the linear momentum $\boldsymbol{\pi} = \gamma m \boldsymbol{v}$, where, as usual, $\gamma = 1/\sqrt{1-\beta^2}$.

2. With polar coordinates, the Lagrangian is now written

$$L(\rho, \phi, \dot{\rho}, \dot{\phi}) = -\left(mc^2 + V(\rho)\right)\sqrt{1-\beta^2},$$

with a similar expression for β^2 as in the preceding case. The ϕ variable is still cyclic and leads to a constant of the motion which is precisely the angular momentum: $p_\phi = \partial_{\dot{\phi}} L = \sigma$. A short calculation gives:

$$\gamma\left(m + \frac{V(\rho)}{c^2}\right) \rho^2 \dot{\phi} = \sigma.$$

3. A calculation completely analogous to the one developed previously provides the relation:

$$\gamma = \sqrt{1 + \frac{\sigma^2 c^2 \left(\rho'^2 + \rho^2\right)}{\left(mc^2 + V(\rho)\right)^2 \rho^4}}.$$

4. In this case also, the Lagrangian is time-independent and the energy is a constant of the motion: $E = \dot{\rho}\, \partial_{\dot{\rho}} L + \dot{\phi}\, \partial_{\dot{\phi}} L - L = \text{const}$.

Performing the calculations, we obtain (this expression is not as natural as the previous one):

$$E = \gamma\left(mc^2 + V(\rho)\right).$$

In contrast to the preceding case, we remark that it is enough to replace mc^2 by $mc^2 + V(\rho)$, $V(\rho) \to 0$.

5. The calculations require the same steps as before. Finally, one obtains the first order differential equation in u:

$$u'^2 + u^2 = \frac{E^2 - \left(mc^2 + V(1/u)\right)^2}{\sigma^2 c^2}.$$

6. Let us differentiate this expression with respect to ϕ, then simplify by $2u'$. We are left with the second order differential equation in u:

$$u'' + u = -\frac{mc^2 + V(1/u)}{\sigma^2 c^2} \frac{d}{du} V(1/u).$$

One could qualify this equation as the "relativistic Binet equation" for a particle in an scalar field.

3.3. Principle of Least Action?
[Statement p. 118]

1. From the Lagrangian $L(q,\dot{q}) = \frac{1}{2}m\dot{q}^2 - V(q)$ and from Lagrange's equations, we derive the equation of motion: the trajectory $\tilde{q}(t)$ satisfies the differential equation $m\ddot{\tilde{q}} = -V'(\tilde{q})$. The action functional is defined, as usual, by

$$S(q) = \int_0^T L(q(t), \dot{q}(t))\, dt.$$

Let us choose a path $q(t)$ slightly different from the trajectory $q(t) = \tilde{q}(t) + \varepsilon(t)$, still imposing fixed bounds $\varepsilon(0) = 0 = \varepsilon(T)$, and calculate the variation of the action $\Delta S = S(q) - S(\tilde{q})$, to second order in ε. The first order vanishes since the trajectory fulfills Hamilton's principle. We end up with the desired expression:

$$\delta^2 S = \frac{1}{2} \int_0^T \left[m\dot{\varepsilon}(t)^2 - \varepsilon(t)^2 V''(\tilde{q}(t)) \right]\, dt.$$

2. It is always possible to assume a bound T which is small enough so as to maintain a constant sign for V'' over the whole integration interval. If, in the small interval $[0,T]$, $V'' \le 0$, then the second term in the action integral is always positive or null; since the first one has the same property, one always has $\delta^2 S > 0$ and the action corresponds to a minimum.

Let us assume now the opposite relation $V'' > 0$ and let $V''_{\max}$ be the maximum value of V'' in the interval: $0 < V''(\tilde{q}) < V''_{\max}$. Of course,

$$\delta^2 S > F(\varepsilon) = \frac{1}{2} \int_0^T \left[m\dot{\varepsilon}(t)^2 - \varepsilon(t)^2 V''_{\max} \right]\, dt.$$

The quantity $$F(\varepsilon) = \frac{1}{2} \int_0^T f(\varepsilon, \dot{\varepsilon})\, dt$$
is still a functional. If one requires a function $\varepsilon(t)$ which makes it extremum, one sees that there is no reason that the condition $\varepsilon(0) = 0 = \varepsilon(T)$ should still hold, since these conditions have been imposed for the functional $\delta^2 S(\varepsilon)$ which is different from the functional $F(\varepsilon)$.

We must modify our conditions, noting that the function $\varepsilon(t)$ can be defined up to a multiplicative constant, which does not change the inequalities.

Let us profit from this property to impose the further condition that $\int_0^T \varepsilon^2\, dt$ retains a fixed value; obviously the same property holds for

$$-\frac{1}{2} V''_{\max} \int_0^T \varepsilon^2\, dt.$$

Consequently, it is sufficient to search for a function $\varepsilon(t)$ which minimizes $\int_0^T \dot{\varepsilon}^2\, dt$ with a given value for $\int_0^T \varepsilon^2\, dt$. This goal is achieved by introducing a Lagrange multiplier λ and requiring that

$$\int_0^T (\dot{\varepsilon}^2 - \lambda \varepsilon^2)\, dt$$

be a minimum. Hamilton's principle applied to this functional leads to the equation $\ddot{\varepsilon} + \lambda \varepsilon = 0$, the solution of which is $\varepsilon(t) = \nu \sin(\sqrt{\lambda} t + \phi)$. The Lagrange multiplier is determined by imposing boundary conditions: $\varepsilon(0) = 0$ leads to $\phi = 0$ and $\varepsilon(T) = 0$ to $\sqrt{\lambda} = \pi/T$. It is always possible to assume $\nu = 1$ since ν is a multiplicative constant.

Thus the function which minimizes our quantity is simply:

$$\varepsilon(t) = \sin(\pi t/T).$$

3. Using this expression, it is easy to calculate the value

$$\frac{1}{2} \int_0^T \left[m\dot{\varepsilon}(t)^2 - \varepsilon(t)^2 V''_{\max} \right] dt = \left[m\pi^2 - T^2 V''_{\max} \right]/(4T).$$

Thereby we prove very important inequalities

$$\delta^2 S > \frac{1}{2} \int_0^T \left[m\dot{\varepsilon}(t)^2 - \varepsilon(t)^2 V''_{\max} \right] dt > \left[m\pi^2 - T^2 V''_{\max} \right]/(4T).$$

If T is chosen sufficiently small, explicitly if

$$T^2 < \frac{m\pi^2}{V''_{\max}},$$

the inequality $\delta^2 S > 0$ holds and the action is a minimum.

In conclusion, we always have a principle of least action provided the action is computed between two sufficiently close instants. This property historically gave the name "principle of least action" to Hamilton's principle.

3.4. Minimum or Maximum Action?
[Statement p. 119]

1. With the Lagrangian
$$L(q,\dot{q}) = \frac{1}{2}(\dot{q}^2 - q^2),$$
the Lagrange equation leads to the differential equation $\ddot{q} + q = 0$ whose general solution is:
$$\tilde{q}(t) = a \sin t + b \cos t.$$

2. Let us use the result of the previous problem with $m = 1$, $V(q) = q^2/2$. In the particular case of a harmonic potential this result is exact (not only valid up to second order in ε)
$$\delta^2 S = \frac{1}{2} \int_0^T \left[\dot{\varepsilon}(t)^2 - \varepsilon(t)^2\right] dt.$$

3. Let consider the function $\varepsilon(t) = \sin(\pi t/T)$, which obviously fulfills the boundary conditions. One easily deduces that
$$\int_0^T \varepsilon(t)^2 \, dt = \frac{T}{2} \quad \text{and} \quad \int_0^T \dot{\varepsilon}(t)^2 \, dt = \frac{\pi^2}{2T},$$
whence:
$$\delta^2 S = \frac{T}{4}\left(\frac{\pi^2}{T^2} - 1\right).$$
It is seen at once that if $T > \pi$, $\delta^2 S < 0$ so that the action is a maximum, and that we can make the opposite conclusion if $T < \pi$.

$T > \pi$ the action is maximum
$T < \pi$ the action is minimum.

4. The real solution was determined in the first question: $\tilde{q}(t) = a \sin t + b \cos t$. The condition $\tilde{q}(0) = q_0$ imposes $b = q_0$ and the additional condition $\tilde{q}(t_1) = q_1$ imposes $q_1 = a \sin t_1 + q_0 \cos t_1$.

Several cases must be investigated.
- If $t_1 \neq n\pi \Leftrightarrow \sin t_1 \neq 0$, then it is possible to find one and only one value for a, namely $a = (q_1 - q_0 \cos t_1)/\sin t_1$. Thus, through the two given points, there passes one and only one trajectory.
- If $t_1 = n\pi \Leftrightarrow \sin t_1 = 0$, then the required condition becomes $q_1 = (-1)^n q_0$. Two cases must be then considered:

- $q_1 \neq (-1)^n q_0$, the condition is not satisfied and there exists no trajectory passing through the two points. Indeed, after an even number of half periods, the system recovers its state, and for an odd number a symmetric state.

- $q_1 = (-1)^n q_0$, the condition is always satisfied whatever the value of a. There exists an infinite number of trajectories (which differ by the values of a) passing through the two points.

This situation is illustrated in the Fig. 3.4.

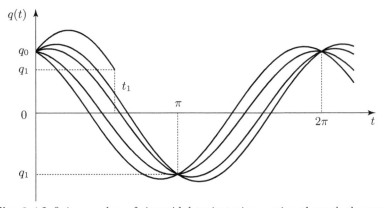

Fig. 3.4 Infinite number of sinusoidal trajectories passing through the conjugate points q_0 and q_1

3.5. Is There Only One Solution Which Makes the Action Stationary?
[Statement p. 120]

1. We are interested in the domain such that $0 < |q_1| < |q_2| < L$.
   - A first trajectory always lies in the interval $0 < q < L$. In this case, the Lagrangian is simply $\ell = \frac{1}{2}m\dot{q}^2$ and the equation of motion $\ddot{q} = 0$ which implies a linear solution $q(t) = vt + u$.

     Since, in the domain influenced by the harmonic potential, the solutions of the equation of motion are sinusoidal curves with angular frequency ω and period T, the time spent by the particle inside this zone is $T/2$ (the particle enters and leaves with the same value of q) for each exploration. Thus, if $t_2 - t_1 > T/2$ the particle has physically enough time to remain for at least a semi-period in the zone corresponding to the harmonic potential. We are led to two other trajectories.

- From t_1 to t_3, the particle stays in the free zone with a motion described by the equation $q(t) = vt + u$. At time t_3 it enters the border zone $q(t_3) = L$ with velocity v. Between t_3 and $t_3 + T/2$, it stays in the harmonic zone governed by the solution $q(t) - L = A\sin(\omega t + \phi)$; it leaves this zone at time $t_3 + T/2$: $q(t_3 + T/2) = L$ with the velocity $-v$. Between $t_3 + T/2$ and t_2, it remains in the free zone with the time law $q(t) = -vt + w$.

- Similarly, we have the possibility of a trajectory that explores the left harmonic zone. From t_1 to t_4 the particle stays in the free zone with a motion described by $q(t) = -vt + u$. At time t_4, it enters the border zone $q(t_4) = 0$ with velocity $-v$. Between t_4 and $t_4 + T/2$, it remains in the harmonic zone with the equation of motion $q(t) = A\sin(\omega t + \phi)$; it leaves this zone at time $t_4 + T/2$: $q(t_4 + T/2) = 0$ with the velocity v. Between $t_4 + T/2$ and t_2, it is to be found in the free zone with the temporal law $q(t) = vt + u$ (see Fig. 3.5).

If $t_2 - t_1 > T$, in addition to the three previous trajectories, we have two supplementary trajectories for which, in each zone of the harmonic potential, the particle executes half a period before leaving. Every time the difference in time increases by T, two more trajectories are possible for which the particle spends half a period more in the harmonic zones.

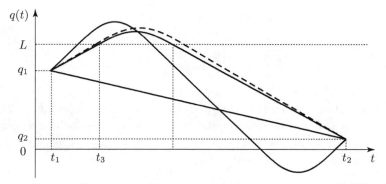

Fig. 3.5 Various possible trajectories between two points (full lines). Infinitely close paths are indicated by dashed lines

2. The action is given by $S = \int_{t_1}^{t_2} \ell(q, \dot{q})\, dt$.

- For the first trajectory, the equation is $q(t) = vt + u$ and the speed is constant $v = (q_2 - q_1)/(t_2 - t_1)$.

Since in this case $\ell = \frac{1}{2}m\dot{q}^2 = \frac{1}{2}mv^2$ remains constant, the action is simply $S = \frac{1}{2}mv^2(t_2 - t_1)$, or, replacing the speed by its value:

$$S_1 = \frac{m}{2}\frac{(q_2 - q_1)^2}{(t_2 - t_1)}.$$

- The time spent in the free zone is simply $t_2 - t_1 - T/2$ and the speed remains constant. The distance for the first part of the path is $L - q_1$ and for the second part $L - q_2$, so that the total path length in the free zone is $2L - q_2 - q_1$ and the speed is

$$|v| = \frac{2L - q_2 - q_1}{t_2 - t_1 - T/2}.$$

With the same argument as before the contribution to the action from this part of the trajectory is

$$\frac{m}{2}\frac{(2L - q_2 - q_1)^2}{t_2 - t_1 - T/2}.$$

The action in the harmonic zone vanishes because the integral on the kinetic energy just cancels the integral on the potential energy; this is a consequence of the virial theorem (of course, one can perform a rigorous calculation!). Thus the total action is restricted to its contribution from the free zone:

$$S_2 = \frac{m}{2}\frac{(2L - q_2 - q_1)^2}{t_2 - t_1 - T/2}.$$

- For the third trajectory, the treatment is exactly the same but, in this case, the distance travelled by the particle in the free zone is $q_2 + q_1$. The corresponding action is therefore:

$$S_3 = \frac{m}{2}\frac{(q_2 + q_1)^2}{t_2 - t_1 - T/2}.$$

3. Let us consider a small variation in the path close to the first trajectory. It occurs in a zone free of potential. The variation of the action to second order is thus

$$\delta^2 S = \frac{1}{2}m\int_{t_1}^{t_2}\dot{\varepsilon}^2\,dt > 0.$$

The action is always a minimum.

Problem Solutions

Let us consider now the second trajectory. In this case, we choose a path which is still half a sinusoid in the harmonic zone but with a period slightly smaller $T_< < T$ and with a linear motion (at constant speed) in the free zone. The distance travelled by the particle in the free zone is still $2L - q_2 - q_1$, but within a time $t_2 - t_1 - T_<$ and the action is still null in the harmonic zone. The action along this new path is thus

$$S'_2 = \frac{m}{2} \frac{(2L - q_2 - q_1)^2}{t_2 - t_1 - T_</2}.$$

One sees that $S'_2 < S_2$. On the contrary, if one chooses a path with a period in the harmonic zone which is a little greater $T_> > T$, we arrive at the opposite conclusion $S''_2 > S_2$. Thus the action S_2 is stationary, and corresponds neither to a minimum nor to a maximum. We are dealing with a saddle point.

For the third trajectory, one obtains of course a similar conclusion: the action corresponds to a saddle point.

Remark: The existence of several classical trajectories is at the origin of important implications for quantum mechanics. This, however, is another story!

3.6. The Principle of Maupertuis
[Statement p. 121]

First case

1. One investigates a trajectory beginning at the origin O and ending at A in the plane xOz; let $x(z)$ be the equation for the trajectory. The reduced action is written

$$S_0(x) = \int_O^A ds \sqrt{2m(E - V(z))},$$

or, with $V(z) = mgz$ and $ds^2 = dx^2 + dz^2$

$$S_0(x) = \int_O^A \sqrt{2m(E - mgz)}\sqrt{1 + x'(z)^2}\, dz.$$

We must apply Hamilton's principle with the equivalent of a "Lagrangian"

$$L(x, x', z) = \sqrt{2m(E - mgz)}\sqrt{1 + x'(z)^2}.$$

One sees that the x variable is cyclic and, consequently, there exists a constant of the motion $\partial_{x'} L = $ const. Calculating the partial derivative, we find:

$$\frac{\sqrt{2m(E-mgz)}\, x'(z)}{\sqrt{1+x'(z)^2}} = \text{const.}$$

2. Let us define by H the maximum altitude, obtained with a value $x = x_m$; at this point the potential energy is mgH and the kinetic energy is denoted mgA; thus $E = mg(H+A)$. The differential equation can be recast as:

$$H + A - z = \frac{1}{C^2}\left[1 + \left(\frac{dz}{dx}\right)^2\right]$$

where a new constant C was introduced. When $(dz/dx) = 0$, one has $z = H$. From this, one obtains $A = (1/C)^2$. The differential equation is separable and gives $dz/\sqrt{H-z} = C dx$ which can be integrated to provide, with the conditions, the result $\pm C(x - x_m) = 2\sqrt{H-z}$ or, equivalently:

$$z - H = -(C/2)^2 (x - x_m)^2.$$

One recognizes the classical parabolic equation, with the imposed conditions. It was not necessary to consider the temporal characteristics. This is precisely the essence of the principle of Maupertuis.

Second case

1. For a central potential, the motion lies in a plane and we may employ the polar coordinates (ρ, ϕ); the potential being central, it can be written as $V(\rho)$. For a trajectory beginning at A and ending at B, the reduced action is:

$$S_0(\phi) = \int_A^B ds\, \sqrt{2m(E-V(\rho))},$$

with $ds^2 = d\rho^2 + \rho^2 d\phi^2$. The reduced action functional is put into the traditional form with a "Lagrangian"

$$L(\phi, \phi', \rho) = \sqrt{2m(E-V(\rho))}\sqrt{1+\rho^2 \phi'(\rho)^2}.$$

The ϕ variable is obviously cyclic and, consequently, there exists a constant of the motion $\partial_{\phi'} L = \sigma$.

Problem Solutions

The angular momentum σ is written explicitly as

$$\sigma = \rho^2 \phi' \frac{\sqrt{2m(E-V(\rho))}}{\sqrt{1+\rho^2\phi'^2}}.$$

Simple algebraic manipulation allows us to recast the differential equation as

$$\phi' = \frac{d\phi}{d\rho} = \frac{\sigma}{\left[m\rho^2 \sqrt{(2/m)\left(E-V(\rho)-\sigma^2/(2m\rho^2)\right)}\right]}.$$

This equation can be integrated to give the trajectory in the traditional form:

$$\phi(\rho) = \int \frac{\sigma\, d\rho}{m\rho^2 \sqrt{\frac{2}{rm}\left(E-V(\rho)-\frac{\sigma^2}{2m\rho^2}\right)}}.$$

2. Let us take now ϕ as the independent variable. The new "Lagrangian" is written

$$L(\rho,\rho',\phi) = \sqrt{2m(E-V(\rho))}\sqrt{\rho^2 + \rho'(\phi)^2}.$$

The independent variable does not appear explicitly in the Lagrangian. There exists a constant of the motion, analogous to the energy, namely

$$\sigma = \rho^2 \frac{\sqrt{2m(E-V(\rho))}}{\sqrt{\rho^2+\rho'^2}}.$$

The constant, σ, is identified with the angular momentum introduced previously. Isolating ρ'^2, we obtain the equivalent equation:

$$\rho^4\left[2m(E-V(\rho))\right] = \sigma^2\rho^2 + \sigma^2\rho'^2.$$

Instead of working with the function $\rho(\phi)$, let us use $u(\phi) = 1/\rho(\phi)$. The previous equation becomes:

$$2m\left(E-V(1/u)\right) = \sigma^2(u^2+u'^2).$$

Differentiating with respect to ϕ and simplifying by $2u'$, we are left with:

$$u'' + u = -\frac{m}{\sigma^2}\frac{d}{du}V\left(\frac{1}{u}\right)$$

which is nothing more than the famous Binet equation, mentioned in (2.10).

3.7. Fermat's Principle [Statement p. 122]

Consider the plane xOz; the optical index depends on z only: $n(z)$. The iso-index surfaces are straight lines parallel to the Ox axis. To go from the origin O to the point A, light takes the path that minimizes the optical path

$$\int_O^A n(z)\,ds = \int_O^A n(z)\,\sqrt{dx^2 + dz^2}.$$

1. We seek the curve $x(z)$ and we thus keep z as the integration variable. The optical path is described in terms of a functional

$$S(x) = \int_O^A n(z)\,\sqrt{1 + x'(z)^2}\,dz$$

in which we purposely employ the notation relative to the action; with these conventions, the "Lagrangian" is

$$L(x, x', z) = n(z)\,\sqrt{1 + x'(z)^2}.$$

We notice that it does not depend on x; there is "translational invariance" since $L(x - a, x', z) = L(x, x', z)$. One deduces that the "momentum" is constant: $p = \partial_{x'} L$. (One can consider also that x is cyclic in which case we are led to the same conclusion.)

In other words
$$\frac{n(z) x'(z)}{\sqrt{1 + x'(z)^2}} = \text{const.}$$

The medium can be considered as being formed by an infinite number of plane diopters parallel to Ox. Let $i(z)$ be the incident angle of light rays on these diopters. One can easily check that $\sin i(z) = dx/\sqrt{dx^2 + dz^2}$, so that $\sin i(z) = x'(z)/\sqrt{1 + x'(z)^2}$. The previous invariance equation is thus written more simply:

$$n(z)\sin i(z) = \text{const.}$$

This is the expression of the well known Snell–Descartes law (for an illustration of the angle of incidence, see Fig. 3.6).

2. Let us now consider the situation in terms of the independent variable x. In this case, the optical path is written:

$$S(z) = \int_O^A n(z)\,\sqrt{1 + z'(x)^2}\,dx.$$

Problem Solutions

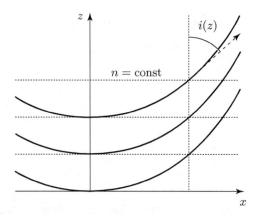

Fig. 3.6 Curves followed by the light rays in a transparent medium the index of which varies with the altitude. $i(z)$ is the angle between the light ray and the normal to the iso-index surface (angle of incidence)

The "Lagrangian" $L(z, z', x) = n(z)\sqrt{1 + z'(x)^2}$ does not depend on the independent variable x. There exists a "constant of the motion" analogous to the energy : $pz' - L =$ const where, this time,

$$p = \partial_{z'} L = \frac{nz'}{\sqrt{1 + z'(x)^2}}.$$

Consequently

$$\text{const} = \frac{nz'^2}{\sqrt{1 + z'^2}} - n\sqrt{1 + z'^2} = -\frac{n}{\sqrt{1 + z'^2}}.$$

Moreover we have $\sin i = 1/\sqrt{1 + z'^2}$. The previous relation is thus equivalent to

$$n(z) \sin i(z) = \text{const}.$$

which is again the Snell–Descartes law. This treatment thus adds nothing new.

3. Let us start from the previous equation

$$\frac{n(z)}{\sqrt{1 + z'^2}} = a,$$

where a is an arbitrary constant. This differential equation can be recast under the form $z'^2 = (n/a)^2 - 1$, or, using the proposed value for the index,

$$z' = \sqrt{[(n_0 + \lambda z)/a]^2 - 1}.$$

Let $(n_0 + \lambda z(x))/a = \cosh u(x)$. The differential equation is now

$$\frac{a}{\lambda} u' \sinh u = \sinh u,$$

or, even more simply, $u' = du/dx = \lambda/a$, which can easily be integrated to give $u(x) = (\lambda/a)x + b$. The constants a and b are determined from the initial conditions $z(0) = 0$ and $z'(0) = 0$. One obtains $a = n_0$ and $b = 0$. Thus the trajectory of the light ray is:

$$z(x) = \frac{n_0}{\lambda}\left[\cosh\left(\frac{\lambda}{n_0}x\right) - 1\right].$$

When $\lambda > 0$, that is when the index increases with the altitude, the rays are curved upwards. The same type of argument can be employed with sound waves. The temperature decreases with altitude. Therefore the speed of sound, which behaves as $\sqrt{T}$ also decreases. The refraction index $n = v_0/v$ must thus increase with altitude. In our model, this corresponds to a positive value for λ. Therefore, "noises rise".

The wave aspect of light and sound propagation allows us to easily recover these conclusions. Let us assume that the speed decreases with the altitude. This means that the wave length decreases and thus that the wave surfaces tighten more and more with the altitude. The rays, which are perpendicular to the wave surfaces, are curved upwards.

3.8. The Skier Strategy
[Statement p. 122]

The straight line from O to A indeed corresponds to the shortest path. However the time needed depends also on the speed along the trajectory. Along this line, the slope is followed slantwise, thus with a speed, which is smaller than in the direction of steepest slope. One can imagine a path which, although longer, profits from the slope to obtain a larger speed.

1. The kinetic energy of the skier is given by: $T = \frac{1}{2}m(\dot{x}^2 + \dot{y}^2)$ and the gravitational potential energy, with our conventions for the axes, by $V = -mgy \sin \alpha$. The mechanical energy $E = T + V$ is a constant of the motion. At the beginning, the speed of the skier is null and he is positioned at the origin; at that time his kinetic energy as well as the potential energy vanish and the same is true for the mechanical energy. Thus

$$E = \frac{1}{2}m(\dot{x}^2 + \dot{y}^2) - mgy \sin \alpha = 0.$$

Problem Solutions

2. Writing $\dot{y} = dy/dt$ and $\dot{x} = dx/dt = x'\dot{y}$ in the equation for the energy, we obtain

$$dt^2 = (1 + x'^2)\, dy^2/(2gy \sin \alpha).$$

We deduce the infinitesimal time between two close points:

$$dt = \frac{\sqrt{1 + x'(y)^2}}{\sqrt{2gy \sin \alpha}}\, dy.$$

3. The total time necessary for the skier to make the run is obtained by the integration of the previous expression between the start and finish points, namely:

$$T(x) = \frac{1}{\sqrt{2g \sin \alpha}} \int_0^A \sqrt{\frac{1 + x'(y)^2}{y}}\, dy.$$

This expression is a functional of the curve $x(y)$ performed by the skier. To minimize this time is a classical problem in the calculus of variations.

4. Identifying the functional with an "action", the corresponding "Lagrangian" (up to an unimportant multiplicative constant) is

$$L(x, x', y) = \sqrt{(1 + x'^2)/y}.$$

One notices that the x coordinate is cyclic, so that there exists an associated constant of the motion, the momentum:

$$\partial_{x'} \sqrt{(1 + x'^2)/y} = x'/\sqrt{(1 + x'^2)y} = \text{const}.$$

This equality shows that the slope of the trajectory cannot be reversed (there is no possibility to go back up again to the end point!). The preceding equation can be rewritten (introducing a new ad-hoc constant C) as:

$$\sqrt{y(x)(1 + y'(x)^2)} = \sqrt{C}.$$

5. A simple algebraic manipulation allows us to recast this equation as $dx = \sqrt{y/(C-y)}\, dy$. Let introduce the new variable θ through $y = C \sin^2 \theta$ or $y = C(1 - \cos 2\theta)/2$. This allows us to calculate dx as:

$$dx = 2C \sin^2 \theta\, d\theta = C(1 - \cos 2\theta)\, d\theta,$$

which can be easily integrated with the help of the initial conditions: $x = C(2\theta - \sin 2\theta)/2$.

Finally, we obtain the required curve, called a brachistochrone, under a parametric form:

$$x(\theta) = \frac{C}{2}(2\theta - \sin 2\theta);$$
$$y(\theta) = \frac{C}{2}(1 - \cos 2\theta).$$

One recognizes the equation of a cycloid, for which only one branch must be retained, namely $0 \leq \theta \leq \pi/2$.

6. *Let us be a little more curious; is there really an important gain of time?*
   - We first calculate the time along the brachistochrone. Using the θ variable, the integral gives $T(\theta_f) = \sqrt{2H/(g \sin \alpha)}\, \theta_f / \sin \theta_f$ where H is the ordinate of the end point (the point A specified parametrically by the angle θ_f), the horizontal distance L (the abscissa) being expressed as a function of θ_f by

   $$L = H \frac{2\theta_f - \sin(2\theta_f)}{1 - \cos(2\theta_f)}.$$

   - Let us imagine another choice: the skier takes the steepest slope then turns at a right angle to cover the horizontal part with the largest possible speed. The speed at A is the largest, but the total distance is longer. The time required to make the run in this case is:

   $$T'(L) = \sqrt{\frac{2H}{g \sin \alpha}}\left(1 + \frac{L}{2H}\right).$$

   - It is also possible to calculate the time if one follows the shortest path, namely the straight line from O to A.

   $$T''(L) = \sqrt{\frac{2H}{g \sin \alpha}}\sqrt{\left(1 + \frac{L^2}{H^2}\right)}.$$

   The three corresponding times are compared on Fig. 3.7. One finds $T(\theta_f(L)) < T''(L) < T'(L)$, but the gain seems not to be very important.

   The limitation to a single branch implies that there is no possible cycloid if $L > \pi H/2$. What is the best trajectory under those conditions? Let us notice that there exists another solution to the equation $\sqrt{y(x)(1 + y'(x)^2)} = \sqrt{C}$: this is the horizontal path $y = C$ at constant speed. Thus, the best choice is the cycloid with its top at the bottom of the slope, followed by a horizontal path (see Fig. 3.8).

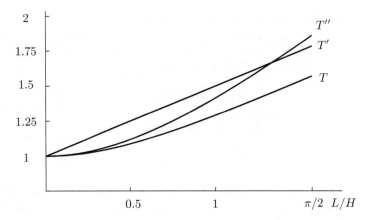

Fig. 3.7 Comparison of the times as a function of L/H (on the abscissa). In order of increasing performance, one has first the brachistochrone, then the straight line and lastly the descent along the steepest slope followed by a turn at a right angle. The time unit is the time needed to perform the descent with the same difference of altitude, but in the direction of the steepest slope

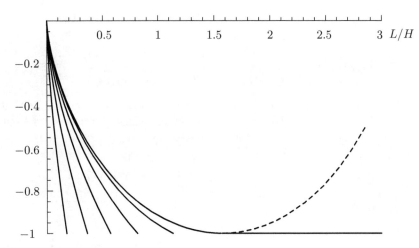

Fig. 3.8 Different ideal trajectories for the skier with the starting point at the same point and the same difference of altitude, but with different horizontal distances L/H. For a too big horizontal distance, the trajectory is a cycloid followed by a straight segment

3.9. Free Motion on an Ellipsoid
[Statement p. 123]

1. Since the particle is free, its Lagrangian is

$$L(\dot{x}, \dot{y}, \dot{z}) = \frac{1}{2}m(\dot{x}^2 + \dot{y}^2 + \dot{z}^2);$$

it is constrained to move on an ellipsoid of equation

$$\left(\frac{x}{a}\right)^2 + \left(\frac{y}{b}\right)^2 + \left(\frac{z}{c}\right)^2 = 1.$$

The constraint is holonomic, since one of the coordinates can be extracted from the others with the help of this equation; the system therefore has two degrees of freedom.

To find the equations of motion, one begins with with the action constrained by a Lagrange multiplier $\lambda(t)$: $S = \int F(x, y, z, \dot{x}, \dot{y}, \dot{z}, t)\,dt$ with

$$F(x, y, z, \dot{x}, \dot{y}, \dot{z}, t) = L(\dot{x}, \dot{y}, \dot{z}) - \lambda(t)\left[\left(\frac{x}{a}\right)^2 + \left(\frac{y}{b}\right)^2 + \left(\frac{z}{c}\right)^2 - 1\right].$$

The equations of motion are given by $d(\partial_{\dot{x}}F)/dt = \partial_x F$, ... together with circular permutations. With the previous expression for F, we obtain:

$$m\ddot{x} = -2\frac{\lambda(t)x}{a^2}; \quad m\ddot{y} = -2\frac{\lambda(t)y}{b^2}; \quad m\ddot{z} = -2\frac{\lambda(t)z}{c^2}.$$

2. The contact force is given by Newton's equation $\boldsymbol{R} = m\ddot{\boldsymbol{r}}$ and the normal $\boldsymbol{n}$ by the gradient of the surface equation. Explicitly for the component along Ox: $R_x = m\ddot{x}$. As for the component n_x of the normal, it is proportional to $\partial_x[(x/a)^2 + (y/b)^2 + (z/c)^2] = 2x/a^2$.

Using the equations of motion, one can check that $R_x \propto \lambda n_x$. Of course, we have similar equations for the other two components; thus, one can write the result in the vectorial form:

$$\boldsymbol{R} \propto \lambda \boldsymbol{n}.$$

Therefore, the force is normal to the surface, as it should be for contact without friction. Moreover, the Lagrange multiplier is proportional to the modulus of the force.

3. The term $F(x, y, z, \dot{x}, \dot{y}, \dot{z}, t)$ appearing in the action allows us to define the "energy" $E = \dot{x}p_x + \dot{y}p_y + \dot{z}p_z - F$. Owing to the equations of motion, we must have $dE/dt = -\partial_t F$. But

$$-\partial_t F = \lambda'(t)[(x/a)^2 + (y/b)^2 + (z/c)^2 - 1] = 0$$

because of the surface condition. Thus $dE/dt = 0$ and the energy remains constant on the surface. With $p_x = m\dot{x}$, and using again the equation of the surface, it is easy to calculate the expression of the energy $E = \frac{1}{2}m(\dot{x}^2 + \dot{y}^2 + \dot{z}^2)$. This is precisely the kinetic energy, since the particle is free (the contact force does not produce work):

$$E = \frac{1}{2}mv^2 = \text{const.}$$

4. Differentiating the constraint equation twice with respect to time; one obtains:

$$(\dot{x}/a)^2 + (\dot{y}/b)^2 + (\dot{z}/c)^2 + x\ddot{x}/a^2 + y\ddot{y}/b^2 + z\ddot{z}/c^2 = 0.$$

The terms containing accelerations are eliminated using the equations of motion proposed in the first question; one obtains an equation giving λ as a function of the coordinates and velocities. Substituting this value in the first equation of motion, one obtains $\ddot{x}$, then $R_x = m\ddot{x}$. Lastly the modulus of the force follows from $R = \sqrt{R_x^2 + R_y^2 + R_z^2}$. The calculations are rather involved, but without any major difficulty. Finally, one finds:

$$R = m\frac{[(\dot{x}/a)^2 + (\dot{y}/b)^2 + (\dot{z}/c)^2]}{\sqrt{(x/a^2)^2 + (y/b^2)^2 + (z/c^2)^2}}.$$

For a sphere $a = b = c = r$, one finds $R = mv^2/r$, or $R = mA$, where A is the acceleration of the particle. This equation is simply the expression of the fundamental principle of dynamics.

5. Let $\quad h(x, y, z, \dot{x}, \dot{y}, \dot{z}) = \dot{x}^2 + \dfrac{(x\dot{y} - \dot{x}y)^2}{(a^2 - b^2)} + \dfrac{(x\dot{z} - \dot{x}z)^2}{(a^2 - c^2)}.$

Calculate the total derivative with respect to time:

$$\frac{dh}{dt} = \dot{x}\partial_x h + \dot{y}\partial_y h + \dot{z}\partial_z h + \ddot{x}\partial_{\dot{x}} h + \ddot{y}\partial_{\dot{y}} h + \ddot{z}\partial_{\dot{z}} h.$$

Performing somewhat cumbersome calculations, one arrives at

$$\frac{dh}{dt} = 2\dot{x}\ddot{x} + \frac{2(x\dot{y} - \dot{x}y)(x\ddot{y} - \ddot{x}y)}{(a^2 - b^2)} + \frac{2(x\dot{z} - \dot{x}z)(x\ddot{z} - \ddot{x}z)}{(a^2 - c^2)}.$$

Expressing the second derivatives in terms of the Lagrange multiplier, we can write alternatively

$$dh/dt = -4\lambda/(ma^2)\left[x\dot{x} + xy(x\dot{y} - \dot{x}y)/b^2 + xz(x\dot{z} - \dot{x}z)/c^2\right].$$

After some algebra involving the use of the surface equation, one arrives at the equivalent equation

$$\frac{dh}{dt} = -4\frac{\lambda x^2}{ma^2}\left[\frac{x\dot{x}}{a^2} + \frac{y\dot{y}}{b^2} + \frac{z\dot{z}}{c^2}\right].$$

The term in brackets is nothing more than the derivative of the surface equation which cancels because its value is the constant 1. Then one has $dh/dt = 0$ and thus $h = \text{const}$

$$\dot{x}^2 + \frac{(x\dot{y} - \dot{x}y)^2}{(a^2 - b^2)} + \frac{(x\dot{z} - \dot{x}z)^2}{(a^2 - c^2)} = \text{const}.$$

Of course, the same property holds for the two other equations obtained by permutation.

Notice that the sum of these three constants of the motion gives the energy, another constant of the motion, which is thus not independent.

3.10. Minimum Area for a Fixed Volume
[Statement p. 124]

1. Let us find quickly and intuitively the expressions for the volume and the area of the "object". To this end, we cut it into infinitesimal slices with thickness δz. Consider the slice whose faces are specified by $z + \delta z/2$ and $z - \delta z/2$. Its volume is comprised between $\pi[r(z + \delta z/2)]^2 \delta z$ and $\pi[r(z - \delta z/2)]^2 \delta z$, both quantities taking on the same value $\pi r(z)^2 \delta z$ to first order. The sum of all these contributions, which is the required volume, tends towards the integral:

$$V(r) = \pi \int_0^h r(z)^2 \, dz.$$

The width of the lateral strip of the slice is, to first order, $\sqrt{dr^2 + dz^2}$ (Pythagoras' theorem) and, completely spread out, its length is comprised between $2\pi r(z + \delta z/2)$ and $2\pi r(z - \delta z/2)$. The area to first order is thus $2\pi r(z)\sqrt{dr^2 + dz^2}$. The sum of all these contributions, which is the required area, tends towards the integral:

$$A(r) = 2\pi \int_0^h r(z)\sqrt{1 + r'(z)^2} \, dz.$$

2. We must minimize the functional $A(r)$ with the integral constraint $V(r) =$ const. We introduce the Lagrange multiplier λ and minimize the quantity:

$$S(r) = \pi \int_0^h \left(2r(z)\sqrt{1+r'(z)^2} - \lambda r(z)^2\right) dz$$

3. This functional looks similar to an "action", with a "Lagrangian"

$$L(r, r', z) = 2r(z)\sqrt{1+r'(z)^2} - \lambda r(z)^2$$

(we forget about the factor π which is unimportant for our purpose). This "Lagrangian" does not depend explicitly on z; one deduces the existence of a constant of the motion, analogous to the "energy": $r'\partial_{r'}L - L = \text{const}$. Performing the corresponding simple calculations, one obtains the result:

$$\lambda r^2 - \frac{2r}{\sqrt{1+r'^2}} = \text{const}.$$

Since the surface cuts the symmetry axis, the value $r=0$ is possible (one can imagine surfaces with a shape like a torus surrounding a cylinder). Calculating the constant for these particular points, one finds $\text{const} = 0$ and the resulting differential equation is then written:

$$\lambda r = \frac{2}{\sqrt{1+r'^2}}.$$

4. To simplify, let us put $\mu = \lambda/2$. Simple algebra transforms this equation into the form: $\mu r r' = \sqrt{1-\mu^2 r^2}$. The integration is performed separating the variables

$$\frac{\mu d(r^2)}{\sqrt{1-\mu^2 r^2}} = 2dz$$

to give, with the boundary conditions $r(0) = 0 = r(h)$, $(z-h/2)^2 + r^2 = h^2/4$ and $\mu = 2/h$. In the plane (r, z), this is the equation for a circle with radius $h/2$ and a center situated halfway between the extremities of the object. By symmetry, it generates a sphere. The Lagrange multiplier is twice the inverse of the radius.

5. The inverse problem is expressed at present by the minimization of the quantity

$$S(r) = \pi \int_0^h \left(r(z)^2 - \sigma\, 2r(z)\sqrt{1+r'(z)^2}\right) dz,$$

where we introduced the Lagrange multiplier σ.

Rewriting the "action" as

$$S(r) = -\pi\sigma \left[\int_0^h \left(2r(z)\sqrt{1+r'(z)^2} - \frac{1}{\sigma} r(z)^2 \right) dz \right],$$

it appears immediately that the problem amounts to the previous one if one makes the identification $\sigma = 1/\lambda$. The solution is therefore known; it is a sphere. In this case, the Lagrange multiplier represents half the radius.

3.11. The Form of Soap Films
[Statement p. 125]

1. The surface of the film has cylindrical symmetry around Oz. Let us first calculate the elementary area comprised between altitudes z and $z + dz$. Developing this strip one obtains, to first order in z, a rectangle of length $2\pi r(z)$ and width $dl(z)$. Using Pythagoras' theorem it is seen that $dl^2 = dz^2 + dr^2 = (1 + r'^2)\,dz^2$. The elementary area is thus $dA = 2\pi r \sqrt{1 + r'^2}\,dz$ and the total area is obtained by an integration over the altitudes:

$$A(r) = 2\pi \int_{z_1}^{z_2} r(z) \sqrt{1 + r'(z)^2}\, dz.$$

This is a functional of the shape of the curve (see also Question 1 in Problem 3.10).

2. Formally, the expression for the area looks like an "action" for which the Lagrangian is $L(r, r', z) = 2\pi r \sqrt{1 + r'^2}$. As it does not depend explicitly on z, the Lagrange equations, resulting from Hamilton's principle, imply the existence of a constant of the motion, analogous to the "energy": $r'\,\partial_{r'}L - L = \text{const}$. With this form for the Lagrangian, this quantity is equivalent to the first order differential equation:

$$r(z) = \rho \sqrt{1 + r'(z)^2}$$

where ρ is a real constant that can be identified with the minimal radius.

3. One can check at once that $r(z) = \rho \cosh((z - h)/\rho)$ is the only possible solution. The integration constant h is the altitude for which the surface is closest to the axis, and ρ is the value of the radius at that point.

The shape of the soap film, $r(z)$, in the symmetry plane, for two identical hoops located at altitudes H and $-H$, is depicted in Fig. 3.9.

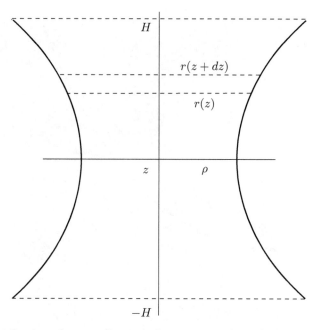

Fig. 3.9 Section of a soap film in a plane passing through the symmetry axis. The hoops are located at altitudes $-H$ and $+H$. The radius of the circle corresponding to the neck is ρ

4. One must fulfill the conditions $r(H) = r(-H) = R$, which imply $h = 0$ and therefore

$$R = \rho \cosh \frac{H}{\rho}.$$

This is the relationship which gives the unknown constant ρ as a function of the physical quantities R and H.

5. We define $x = H/\rho$, so that the previous relation becomes $\cosh x = (R/H)\,x$. This is a transcendental equation that gives x and thus ρ as a function of the ratio R/H. The solutions are the intersections of the curve $\cosh x$ with the straight line $(R/H)\,x$. Plotting those curves (see Fig. 3.10), one remarks that if R/H is small there is no solution whereas two solutions exist if R/H is large. The critical value x_c corresponds to the situation for which the straight line is tangential to the hyperbolic curve, which implies $\sinh x_c = R/H$. Coupled with the relation $\cosh x_c = (R/H)\,x_c$, this last equation provides $x_c = \sqrt{1 + H^2/R^2}$ which leads to the transcendental equation for the intersection:

$$\sinh\left(\sqrt{1 + H^2/R^2}\right) = \frac{R}{H}.$$

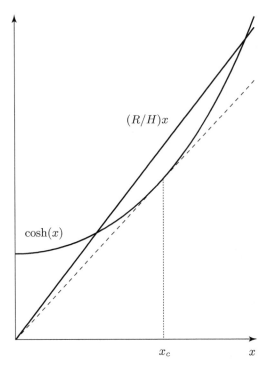

Fig. 3.10 Graphical construction to find the radius at the neck; it is necessary to calculate the intersection of a straight line with a hyperbolic cosine, as explained in the text

The solution is $(H/R)_c = 0.6627429....$ Therefore, there exists a solution only in the case for which:

$$\frac{H}{R} < \left(\frac{H}{R}\right)_c = 0.6627429...$$

As we saw, if this inequality is satisfied, there exist two solutions. To obtain a better understanding, it is useful to study numerically the area of the surface formed by the revolution of the hyperbolic cosine which passes through the two hoops, as a function of the radius at the neck ρ. This area possesses two extrema, a minimum for the largest value of the neck radius – and this is the physical solution – and a non-physical maximum for the lowest value of the neck radius. For the critical value H/R these two extrema are identical and the solution corresponds to an unstable saddle point. The area can still decrease up to a conical form for the surface; for a further decrease, the cone will degenerate into two separated films based on the two hoops. This discussion is illustrated graphically in Fig. 3.11.

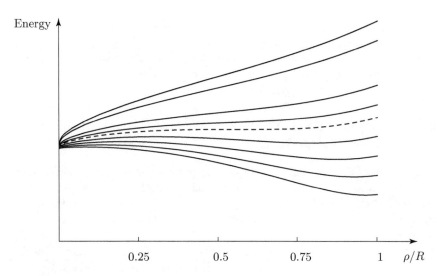

Fig. 3.11 Surface tension energy, proportional to the surface, for a film with a catenoidal shape as a function of the neck radius. The point at the origin corresponds to the two layers of a cone based on the hoops. The various curves correspond to different values of H/R. For a value H/R less than a critical value (symbolized by the dashed line) there exists a minimal energy solution; for a larger value there is no minimum

6. The surface potential energy is $V = TA$, where T is the coefficient of surface tension and A the area of the film. The calculus of the area is obtained by computing the integral given in Question 1 with the expression of the film shape as given in Question 3. A long calculation, without particular difficulty, gives: $A = 2\pi\rho^2 [x + \cosh x \sinh x]$, where $x(R/H)$ is the physical solution of our transcendental equation $\cosh x = (R/H) x$. Remember that $\rho = H/x(R/H)$ is a function of R **and** H.

Note in passing that this area is a increasing function of $x(R/H)$; this justifies the choice of the largest solution.

A rather involved calculation, performed with care, allows us to obtain the variation of the area as function of the degree of stretching (of course the radius of the hoop R remains unchanged) $\partial A/\partial H = 4\pi\rho$. The force due to the surface tension is defined as the derivative of the potential energy with respect to the stretching $F = \partial V/\partial(2H) = (T/2)\partial A/\partial H$ that is

$$F(H) = 2\pi T \rho(H)$$

This result is in fact easy to obtain.

Let us imagine a cut at the neck. One must remember that T, the coefficient of surface tension, is the force per unit length (perpendicular to the cut) and, obviously, at the neck the cohesion forces, all parallel, are exerted over a distance $2\pi\rho(H)$, the perimeter of the neck, whence the expression of the force derived above.

3.12. Laplace's Law for Surface Tension
[Statement p. 127]

The section of the free liquid surface is given by a curve $z(x)$ between point A on the Oz axis and point B, located at a distance l on the horizontal axis. We impose the conditions $z(0) = h = z(l)$. The required curve minimizes the total energy, which is the sum of the gravitational potential energy and the surface energy.

1. Let us take a small basic rectangle with sides dx and dy centered at the point (x, y). The small parallelepiped built on this rectangle has a volume $dV = z(x)\,dx\,dy$ and a mass $dm = \rho dV = \rho z(x)\,dx\,dy$. Its center of mass lies at the altitude $z/2$; therefore, its gravitational potential energy is $dV_G = dm\,g\,(z/2)$, or $dV_G = \rho g\,(z^2/2)\,dx\,dy$. The total potential energy is obtained by integrating this expression over the whole surface. Because of the translational invariance along Oy, the y variable does not play any practical role and one can reason using a slice with thickness $dy=1$. Thus the gravitational potential energy is $dV_G = \rho g\,(z^2/2)\,dx$.

With identical conditions, the element of the free surface is $ds(x)\,dy = ds(x)$, where $ds(x)$ is the elementary length element at point x. Obviously, $ds^2 = dx^2 + dz^2$ or $ds = dx\sqrt{1+z'^2}$. The free surface has an area $dA = dx\sqrt{1+z'^2}$ and the corresponding potential energy is $dV_T = T dA$, where T is the coefficient of surface tension. Therefore $dV_T = T\sqrt{1+z'^2}\,dx$. The total potential energy is the sum of the contributions due to gravity and surface tension, that is

$$dV_P = dV_G + dV_T = \rho g \frac{z^2}{2}\,dx + T\sqrt{1+z'^2}\,dx.$$

The total potential energy for the slice of unit thickness is obtained by integrating this quantity along Ox:

$$V_P(z) = \int_0^l \left[\rho g\,(z(x)^2/2) + T\sqrt{1+z'(x)^2}\right]dx.$$

It is a functional of the shape of the meniscus.

Problem Solutions

2. We already saw that the elementary volume is $dV = z(x)\,dx\,dy$, or (for a slice of unit thickness) $dV = z(x)\,dx$. The total volume is obtained by integrating this quantity along Ox:

$$V(z) = \int_0^l z(x)\,dx.$$

It is also a functional of the shape of the meniscus.

3. Since the liquid is incompressible, its volume remains constant. The equations of motion minimize the potential energy. We are faced with a variational principle subject to an integral constraint. The usual method consists of introducing a Lagrange multiplier λ and then searching for the minimum of the functional

$$S(z) = \int_0^l \left[\rho g\, (z(x)^2/2) + T\sqrt{1+z'(x)^2} - \lambda z(x) \right] dx.$$

We have an expression analogous to an "action" with a "Lagrangian"

$$L(z, z') = \rho g\, (z(x)^2/2) + T\sqrt{1+z'(x)^2} - \lambda z(x)$$

The Euler-Lagrange equation is written as: $\partial_z L = d(\partial_{z'} L)/dx$. In this case

$$\partial_z L = \rho g z - \lambda, \quad \partial_{z'} L = \frac{T z'}{\sqrt{1+z'^2}} \quad \text{and} \quad \frac{d}{dx}(\partial_{z'} L) = \frac{T z''}{(1+z'^2)^{3/2}}.$$

In this last expression, we can identify the radius of curvature of the meniscus at point x: $R(x) = (1+z'^2)^{3/2}/|z''|$ (be very careful, in our case $z'' < 0$); therefore $d(\partial_{z'} L)/dx = -T/R(x)$. The Euler-Lagrange equation leads to:

$$\rho g z(x) + \frac{T}{R(x)} = \lambda$$

If one sets $Z(x) = z(x) - \lambda/\rho g$, the equation for the meniscus is $\rho g Z(x) + T/R(x) = 0$, which does not depend on the Lagrange multiplier (i.e. on the volume).

Fortunately, one can recover this equation much more elegantly

To simplify, let us take a meniscus with a concavity oriented upwards. The horizontal is an equipotential curve, i.e. an isobar. The pressure is the sum of the atmospheric pressure, of the hydrostatic pressure $\rho g Z(x)$ and the difference of pressure on opposite sides of the meniscus (Laplace's equation) $T/R(x)$.

3.13. Chain of Pendulums [Statement p. 128]

1. For the atom numbered i, the kinetic energy is $T_i = \frac{1}{2}ml^2\dot{q}_i^2$, the gravitational potential energy (with our conventions for the orientation of the axes) $V_i = -mgl\cos q_i$ and the harmonic potential energy $V'_i = \frac{1}{2}k(q_i - q_{i-1})^2$. The total Lagrangian is the difference between the kinetic energy and the potential energies, after summation over all the point-like masses. Therefore:

$$L(q,\dot{q}) = \sum_{i=1}^{N} \left[\frac{1}{2}ml^2\dot{q}_i^2 + mgl\cos q_i - \frac{1}{2}k(q_i - q_{i-1})^2\right].$$

2. Passing to a continuous treatment, the distance between the pendulums becomes infinitesimal $\delta = dx$, the masses become infinitesimal $dm = \mu\, dx$ (μ is the linear mass density), the coordinates become fields $q_i(t) = \varphi(x_i, t)$ and the summations become integrals. In consequence, passing to this limiting case, the kinetic energy becomes $T = \frac{1}{2}\mu l^2 \int dx\, (\partial_t\varphi)^2$, the gravitational potential energy $V = -\mu gl \int dx \cos\varphi$ and the harmonic potential energy $V' = \frac{1}{2}kdx \int dx\, (\partial_x\varphi)^2$. With the proposed value at this limit $kdx = \mu\lambda$, the Lagrangian is written:

$$L = \int \left[\frac{1}{2}\mu l^2 (\partial_t\varphi)^2 - \frac{1}{2}\mu\lambda(\partial_x\varphi)^2 + \mu gl\cos\varphi\right] dx$$

which can be written in the form $L = \int \ell\, dx$, by introducing the Lagrangian density, ℓ. It thus appears that the Lagrangian density is expressed as:

$$\ell\left(\varphi, \frac{\partial\varphi}{\partial x}, \frac{\partial\varphi}{\partial t}\right) = \frac{1}{2}\mu l^2 \left(\frac{\partial\varphi}{\partial t}\right)^2 - \frac{1}{2}\mu\lambda\left(\frac{\partial\varphi}{\partial x}\right)^2 + \mu gl\cos\varphi.$$

3. The Euler-Lagrange equations for the continuous fields are written:

$$\partial_\varphi\ell - \partial_x\left(\frac{\partial\ell}{\partial(\partial_x\varphi)}\right) - \partial_t\left(\frac{\partial\ell}{\partial(\partial_t\varphi)}\right) = 0.$$

With the Lagrangian density obtained previously

$$\partial_\varphi\ell = -\mu gl\sin\varphi;\quad \partial_x\left(\frac{\partial\ell}{\partial(\partial_x\varphi)}\right) = -\mu\lambda\partial_{x^2}^2\varphi;\quad \partial_t\left(\frac{\partial\ell}{\partial(\partial_t\varphi)}\right) = \mu l^2\partial_{t^2}^2\varphi.$$

Introducing the proper angular frequency $\omega^2 = g/l$, the Euler-Lagrange equations lead to the non-linear wave equation, of soliton type (see Problem 2.13 for the solution of this equation):

$$\frac{\partial^2\varphi(x,t)}{\partial t^2} = \frac{\lambda}{l^2}\frac{\partial^2\varphi(x,t)}{\partial x^2} - \omega^2\sin\varphi(x,t).$$

3.14. Wave Equation for a Flexible Blade
[Statement p. 128]

1. We saw in the Problem 2.12 that the kinetic energy of the blade is

$$T = \frac{1}{2}\mu\delta \sum_{i=1}^{N} \dot{q}_i^2$$

and the elastic potential energy

$$V = \frac{EI}{2\delta^3} \sum_{i=1}^{N} (q_{i-1} + q_{i+1} - 2q_i)^2.$$

Passing to a continuous treatment, the coordinates become fields $q_i(t) = \varphi(x_i, t)$, $\delta \to 0$ and $\delta \sum_{i=1}^{N} \to \int_0^L dx$. Thus, the kinetic energy transforms into $T = \frac{1}{2}\mu \int_0^L (\partial_t \varphi)^2\, dx$. For the potential energy, one uses

$$\frac{q_{i-1} + q_{i+1} - 2q_i}{\delta^2} \to \partial_{x^2}^2 \varphi \quad \text{and} \quad V = \frac{1}{2} EI \int_0^L (\partial_{x^2}^2 \varphi)^2 \, dx.$$

Consequently the Lagrangian is expressed as

$$L = \int_0^L \left[\frac{1}{2}\mu (\partial_t \varphi)^2 - \frac{1}{2} EI (\partial_{x^2}^2 \varphi)^2\right] dx,$$

which can be written as $L = \int \ell \, dx$, where ℓ is the Lagrangian density. Explicitly, this density is expressed as:

$$\ell\left(\varphi, \frac{\partial\varphi}{\partial x}, \frac{\partial^2\varphi}{\partial x^2}, \frac{\partial\varphi}{\partial t}\right) = \frac{1}{2}\mu \left(\frac{\partial\varphi}{\partial t}\right)^2 - \frac{1}{2} EI \left(\frac{\partial^2\varphi}{\partial x^2}\right)^2.$$

2. The Euler-Lagrange equations for the continuous fields are written:

$$\partial_\varphi \ell - \partial_x \left(\frac{\partial \ell}{\partial(\partial_x \varphi)}\right) - \partial_t \left(\frac{\partial \ell}{\partial(\partial_t \varphi)}\right) + \partial_{x^2}^2 \left(\frac{\partial \ell}{\partial(\partial_{x^2}^2 \varphi)}\right) = 0.$$

With the previous Lagrangian density

$$\partial_\varphi \ell = 0, \qquad \partial_x \left(\frac{\partial \ell}{\partial(\partial_x \varphi)}\right) = 0,$$

$$\partial_t \left(\frac{\partial \ell}{\partial(\partial_t \varphi)}\right) = \mu \partial_{t^2}^2 \varphi, \qquad \partial_{x^2}^2 \left(\frac{\partial \ell}{\partial(\partial_{x^2}^2 \varphi)}\right) = -EI \partial_{x^4}^4 \varphi.$$

Introducing the mass M and the length L of the blade ($\mu = M/L$), in the wave equation, one obtains the final expression:

$$\frac{\partial^2 \varphi}{\partial t^2} = -\frac{EIL}{M} \frac{\partial^4 \varphi}{\partial x^4}.$$

3.15. Precession of Mercury's Orbit
[Statement p. 128]

1. The action can be rewritten as $S = \int L\, dt$ where, with the proposed expression, the Lagrangian is expressed as $L = -mc\, ds/dt$. For the Schwarzschild metric, one has explicitly:

$$L = -mc^2 \sqrt{e(r) - (r^2\dot\theta^2 + r^2 \sin^2\theta\, \dot\phi^2)/c^2 + \dot r^2/(e(r)c^2)}.$$

This Lagrangian is rotationally invariant; this property implies the constancy of the angular momentum which implies a planar motion. It is therefore natural to work in this plane and to adopt the polar coordinates (r, ϕ); this is carried out simply by substituting $\theta = \pi/2$ in the previous expression. The form of the Lagrangian is then:

$$L(r, \phi, \dot r, \dot\phi) = -mc^2 \sqrt{e(r) - \frac{1}{c^2}(r^2\dot\phi^2 + \dot r^2/(e(r)))}.$$

2. One notices that the ϕ variable is cyclic; therefore there exists a constant of the motion, the conjugate momentum; this last quantity is nothing more than the angular momentum. Performing all the necessary calculations in $\sigma = p_\phi = \partial_{\dot\phi} L$, we obtain the expression:

$$\sigma = \frac{mr^2\dot\phi}{\sqrt{e(r) - \frac{1}{c^2}(r^2\dot\phi^2 + \dot r^2/(e(r)))}}.$$

3. Since we are concerned by the trajectory rather than the temporal evolution, it is judicious to work with $r' = dr/d\phi$ rather than $\dot r = dr/dt$. Substituting $\dot r = r'\dot\phi$ in the previous expression and isolating $\dot\phi$, we arrive at the alternative expression:

$$\dot\phi^2 = \frac{e(r)}{\frac{1}{c^2}\left(r^2 + \frac{r'^2}{e(r)}\right) + \frac{m^2 r^4}{\sigma^2}}.$$

4. Since the Lagrange function does not depend explicitly on time, the energy $E = \dot r p_r + \dot\phi p_\phi - L$ is also a constant of the motion. In order to simplify the notation, let us set

$$D = \sqrt{e(r) - \frac{1}{c^2}(r^2\dot\phi^2 + \dot r^2/(e(r)))}.$$

Then $L = -mc^2 D$ and one can calculate $p_r = \partial_{\dot r} L = m\dot r/(e(r)D)$ and $p_\phi = \partial_{\dot\phi} L = mr^2\dot\phi/D$. One deduces the value of the energy $E = mc^2 e(r)/D$.

Problem Solutions

Employing now the D value from the second question, expressed as a function of σ, the energy can be written in the form:

$$E = \frac{\sigma c^2 e(r)}{r^2 \dot{\phi}}.$$

5. Let us square this expression and replace $\dot{\phi}^2$ by the value obtained in Question 3. A little algebra finally leads to the differential equation giving the trajectory:

$$e(r)\left(\frac{1}{r^2} + \frac{m^2 c^2}{\sigma^2}\right) + \frac{r'^2}{r^4} = E^2/(c^2\sigma^2).$$

As is often the case, it is more astute to work with the inverse of the radius $u(\phi) = 1/r(\phi)$. Substituting this change of variable in the differential equation and using the explicit expression of $e(u) = 1 - r_0 u$, we arrive at the differential equation of first order in u:

$$u'^2 = \frac{E^2}{c^2\sigma^2} - (1 - r_0 u)\left(u^2 + \frac{m^2 c^2}{\sigma^2}\right).$$

6. Let us differentiate this equation with respect to ϕ and simplify by u'; we now obtain a differential equation of second order in u:

$$u'' + u = \frac{3}{2} r_0 u^2 + \frac{m^2 c^2 r_0}{2\sigma^2}.$$

This equation could be called the relativistic Binet equation obtained with the Schwarzschild metric. It can be written alternatively as $u'' = \frac{3}{2} r_0 (u - u_c)(u - u'_c)$, after the introduction of the two roots u_c and u'_c for the second degree equation. These latter quantities obey the condition $u'' = 0$. Explicitly, one has $u_c = \left(1 - \sqrt{1-d}\right)/3r_0$ and $u'_c = \left(1 + \sqrt{1-d}\right)/3r_0$ with $d = 3\left(mcr_0/\sigma\right)^2$.

7. For a circular orbit, one has $u''(\phi) = 0$, a condition that is possible for $R'_c = 1/u'_c$ and another circle with larger radius $R_c = 1/u_c$. There exist orbits around those two particular solutions.

8. Let us set $v = u - u'_c$; in terms of the new variable v the differential equation becomes:

$$v'' + \frac{3}{2} r_0 v \left(v - 2\sqrt{1-d}/3r_0\right) = 0.$$

Let us now assume that $v \ll 2\sqrt{1-d}/3r_0$; the linearized equation is even simpler $v'' - \omega^2 v = 0$.

Its solution increases exponentially so that the approximation is no longer valid. There are no stable orbits close to this radius. The situation is completely different close to u_c for which the equation becomes $v'' + \omega^2 v = 0$. This equation can be integrated easily to give $v = A\cos(\omega\phi + \phi_0)$. Returning to the r variable and defining $\omega = (1-d)^{1/4}$, one obtains:

$$r(\phi) = \frac{1}{u_c + A\cos(\omega\phi + \phi_0)}.$$

If $\omega = 1$, one recovers the usual equation of an ellipse. Owing to the fact that ω deviates slightly from 1, the ellipse is slightly distorted: its axes rotate slowly from one revolution to the other. This phenomenon is known as the precession of perihelia.

For example, one can say that the major axis rotates by $\Delta\phi$ after one revolution. To fulfill this, the length of the major axis must be unchanged and this leads to the equation $\cos(\omega\phi) = \cos(\omega(2\pi + \phi + \Delta\phi))$, or $2\pi + \Delta\phi = 2\pi/\omega$. Substituting the value of ω previously obtained, we deduce the value of the precession $\Delta\phi = 3\pi m^2 c^2 r_0^2/2\sigma^2$. Finally, expressing the Schwarzschild radius in terms of fundamental constants, the new expression for the precession of perihelia is obtained:

$$\Delta\phi = 6\pi \frac{G^2 M^2 m^2}{c^2 \sigma^2}.$$

One can eliminate the angular momentum, σ, by expressing it in terms of the characteristics of the classical Kepler orbit (semi-major axis a and eccentricity e) $\sigma^2/(GMm^2) = a(1-e^2)$; the final expression is written:

$$\Delta\phi = 6\pi \frac{GM}{c^2 a(1-e^2)}.$$

- With the characteristic data relative to Mercury's orbit, one finds $\Delta\phi = 0.502 \times 10^{-6}$ rd for one revolution, that is $\Delta\phi = 0.1035''$ per revolution. During a century Mercury performs $100 \times 87.969/365.25 = 415.203$ revolutions. Its perihelion thus advances by $415.203 \times 0.1035'' \cong 43''$ per century. This results agrees remarkably well with the experimental observations.

- An analogous calculation for the Earth leads to an advance of $0.0384''$ per revolution, that is $3.84''$ per century; here again the agreement is very good.

Chapter 4
Hamiltonian Formalism

Summary

4.1. Generalized Momentum

In Chapter 2, we introduced the Lagrange function (or Lagrangian) $L(q, \dot{q}, t)$, which depends on generalized coordinates q and generalized velocities $\dot{q}$, considered as independent variables, and, possibly, on time. The Hamiltonian formalism is an alternative to the Lagrangian formalism for the description of a mechanical system. Instead of generalized velocities employed in Lagrangian formalism, it relies on **generalized momenta**[1] defined as:

$$p_i = \frac{\partial}{\partial \dot{q}_i} L(q, \dot{q}, t). \tag{4.1}$$

Theoretically, one could return to generalized velocities $\dot{q}(q, p, t)$ from these generalized momenta,[2] simply by inverting the relations (4.1). In this new framework, the state of a system is defined using its n generalized coordinates q and its n generalized momenta p, that is by a point (q, p) in a $2n$ dimensional space, called **phase space**.

---

[1] The word "conjugate variable of the generalized coordinate" is also of common use. Note that we must distinguish between generalized momentum and linear momentum.

[2] Let us remark that the product of a coordinate by a momentum has always the dimension of an action.

4.2. Hamilton's Function

The Lagrange functions which derive from a Lagrangian are fully equivalent to Newton's equations. For a system with n degrees of freedom (without additional constraints), they form a set of n coupled second order differential equations. In the new formalism, one defines a function, known as **Hamilton's function** (or **Hamiltonian**), which depends on the generalized coordinates q and momenta p (which are considered as independent variables), and possibly on time, as the Legendre transform of the Lagrangian:

$$H(q, p, t) = p \cdot \dot{q}(q, p, t) - L(q, \dot{q}(q, p, t), t). \tag{4.2}$$

In this expression, all the generalized velocities are expressed in terms of coordinates and momenta $\dot{q}(q, p, t)$, after inversion of relation (4.1). Remember that $p \cdot \dot{q}$ means explicitly

$$p \cdot \dot{q} = \sum_{i=1}^{n} p_i \dot{q}_i.$$

Consequently, the Hamiltonian depends on a point in phase space, and possibly on time.

Mechanics can be formulated equivalently using either the Lagrangian or the Hamiltonian. If we know the Hamiltonian, the Lagrangian is easily recovered from the inverse Legendre transform. One begins by determining the velocities thanks to $\dot{q} = \partial_p H(q, p, t)$, which can be inverted to give $p(q, \dot{q}, t)$. The Lagrangian is easily deduced from the Legendre transform:

$$L(q, \dot{q}, t) = \dot{q} \cdot p(q, \dot{q}, t) - H(q, p(q, \dot{q}, t), t).$$

Some examples[3] **in common use.**

- *A one-dimensional particle subject to a force arising from a potential*

$$L(q, \dot{q}, t) = \frac{m\dot{q}^2}{2} - V(q, t) \leftrightarrow H(q, p, t) = \frac{p^2}{2m} + V(q, t).$$

- *The composite pendulum*

$$L(\theta, \dot{\theta}) = \frac{I\dot{\theta}^2}{2} + mg \cos \theta \leftrightarrow H(\theta, p) = \frac{p^2}{2I} - mg \cos \theta.$$

- *A non-relativistic charged particle embedded in an electromagnetic field*

$$L(\boldsymbol{r}, \dot{\boldsymbol{r}}, t) = \frac{m\dot{\boldsymbol{r}}^2}{2} + q_e \left(\dot{\boldsymbol{r}} \cdot \boldsymbol{A}(\boldsymbol{r}, t) - U(\boldsymbol{r}, t) \right) \leftrightarrow$$

$$H(\boldsymbol{r}, \boldsymbol{p}, t) = \frac{(\boldsymbol{p} - q_e \boldsymbol{A}(\boldsymbol{r}, t))^2}{2m} + q_e U(\boldsymbol{r}, t).$$

---

[3] Note that the expression $H = T + V$, although frequent, is not systematic.

Summary

- A relativistic charged particle embedded in an electromagnetic field

$$L(\boldsymbol{r},\dot{\boldsymbol{r}},t) = -mc^2\sqrt{1-\dot{\boldsymbol{r}}^2/c^2} + q_e\left(\dot{\boldsymbol{r}} \cdot \boldsymbol{A}(\boldsymbol{r},t) - U(\boldsymbol{r},t)\right) \leftrightarrow$$
$$(H(\boldsymbol{r},\boldsymbol{p},t) - q_e U(\boldsymbol{r},t))^2 - (\boldsymbol{p} - q_e \boldsymbol{A}(\boldsymbol{r},t))^2 c^2 = m^2 c^4.$$

4.3. Hamilton's Equations

The temporal evolution of a system is governed by **Hamilton's equations**

$$\dot{q}_i = \frac{\partial}{\partial p_i} H(q,p,t); \tag{4.3}$$

$$\dot{p}_i = -\frac{\partial}{\partial q_i} H(q,p,t). \tag{4.4}$$

They replace Lagrange's equations. In this case, they form a system of $2n$ coupled differential equations, but of first order only in (q,p). They can be considered as giving, at each point (q,p) of phase space, the velocity vector $(\dot{q},\dot{p})$ used for building all the trajectories $(q(t),p(t))$. One of these trajectories (which of course depends on the initial conditions) is called the **flow** of the system.

These equations can be completed by

$$\frac{dH(q(t),p(t),t)}{dt} = \frac{\partial H(q,p,t)}{\partial t} = -\frac{\partial L(q,\dot{q},t)}{\partial t}. \tag{4.5}$$

Consequently, if the Hamilton function does not depend explicitly on time, there exists a constant of the motion: the energy. In this case, the system is said to be **autonomous** or **conservative**. The energy is the value taken by the Hamiltonian along the considered trajectory. Of course it depends on the initial conditions:

$$H(q(t),p(t)) = E. \tag{4.6}$$

4.4. Liouville's Theorem

The flow has the remarkable property of being incompressible, a property that can be expressed as:

Every closed surface in phase space preserves its hypervolume at each time

This property constitutes a succinct formulation of Liouville's theorem. In the simplest case of a one-dimensional system where the phase space is two dimensional, the hypervolume is just the area and Liouville's theorem is frequently named "the theorem of conservation of the area".

Thanks to Liouville's theorem, one can deduce another important theorem known as "the revisiting theorem". It is due to Poincaré and can be formulated as:

For an autonomous system moving in a finite phase space, then in any domain of this phase space, however small, there exist at least two points belonging to the same trajectory.

4.5. Autonomous One-dimensional Systems

We work in a two-dimensional phase space. The constant of the motion, the energy defined by (4.6), immediately gives the flow determined by the relation $H(q,p) = E$. The set of all curves $p(q, E)$, which follow from this equality, for different values of the energy is called the **phase portrait** of the system. Depending on the explicit value of the energy, one generally distinguishes several regimes of behaviour, separated by special curves, known as **separatrices**.

There exist trajectories that are restricted to one fixed point (equilibrium point): $\dot{p} = -\partial_q H(q,p) = 0$; such points correspond to extrema of the potential.[4] Close to these points, there exist two types of behaviour:

- If we are dealing with a potential minimum, the phase portrait is composed of closed curves which are nested inside each other and which look more and more similar to ellipses as we approach the fixed point; we speak in this case of an **elliptic point** or stable node.

- If we are in the presence of a potential maximum, we are dealing with an unstable equilibrium point; the phase portrait looks more and more similar to a set of hyperbolae as we approach the fixed point; we speak in this case of a **hyperbolic point** or unstable node.

Very often, the Hamiltonian is just the sum of the kinetic energy and the potential energy. Thus, for an autonomous system, one has $T = E - V$ on the trajectories. The kinetic energy being always a positive quantity, a real motion can exist only for points satisfying the condition $V \leq E$. The points for which $V = E$ are called **turning points**. Most of the time, the motion can occur only between (or possibly outside) the turning points, where the velocity vanishes (and changes sign).

---

[4] We forget about possible saddle points.

Summary

It is possible to obtain the time as a function of the coordinate (inverse of the temporal evolution) and the energy through an integral:

$$\dot{q} = \partial_p H(q,p) \rightarrow t - t_0 = \int_{q_0}^{q} \frac{dq}{\partial_p H(q,p)} \quad (4.7)$$

where, after calculation of the partial derivative $\partial_p H$, we substitute for p its expression in terms of q and E according to (4.6).

4.6. Periodic One-dimensional Hamiltonian Systems

For all periodic phenomena, it is worthwhile to observe the system at times separated by the period T. Indeed this is the principle of stroboscopy. In our case, this principle implies specifying the coordinate and the momentum once at every period. Thus, in phase space, we consider the application which transforms the point (q_{n-1}, p_{n-1}) at time t_{n-1} to the point $(q_n(q_{n-1}, p_{n-1}), p_n(q_{n-1}, p_{n-1}))$ at time $t_n = t_{n-1} + T$. If the system returns to its initial state after r periods, one speaks of a fixed point of order r (r-fold fixed point). Close to a fixed point (of order 1 to simplify the notation[5]), a small deviation $(\varepsilon_{n-1}, \eta_{n-1})$ of the fixed point is transformed, a period later, by a deviation (ε_n, η_n) which depends linearly on the initial deviation. This introduces a matrix, with unit determinant, called the **propagator**:

$$\left. \begin{pmatrix} \varepsilon \\ \eta \end{pmatrix} \right|_n = K \left. \begin{pmatrix} \varepsilon \\ \eta \end{pmatrix} \right|_{n-1} \quad (4.8)$$

It is possible to prove the following properties.
- If the eigenvalues of the propagator are complex, or if – and this is equivalent – the absolute value of the trace of the matrix is less than two, the excursion around the fixed point will remain limited. In phase space, the sequential points will occur close to an ellipse. We are in the presence of an elliptic fixed point or a stable fixed point. We have depicted this situation, for three different initial conditions, in the Fig. 4.1 where the sequence of points is plotted after a stroboscopic inspection. The ellipse will be filled in after an infinite number of periods.

---

[5] For a fixed point of order r, it is always possible to recover this special case by considering the *rth* power of the original application.

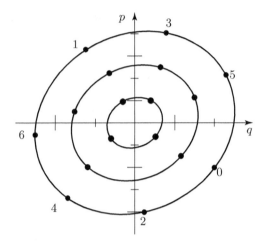

Fig. 4.1 Figure, in phase space, obtained from the positions of a system after periodic impulses labelled by the progressive numbers. This is the case of complex eigenvalues with unit modulus for the propagator

- If the eigenvalues of the propagator are real or, equivalently, if the absolute values of the trace are larger than two, the successive points lie on a hyperbola. Starting from a point located exactly on one of the asymptotes, the next point will lie closer to the fixed point according to a geometric progression. This is the convergent direction of the hyperbolic point. In contrast, starting from the other asymptote, the point will move away from the fixed point according to the inverse common ratio. A non singular initial deviation always leads to a departure from the fixed point. We say that we are faced with a **parametric resonance**. The fixed point is unstable. We have plotted such a situation in Fig. 4.2. We may remark that the stroboscopic order jumps from one branch to the other. This is a consequence of negative eigenvalues.

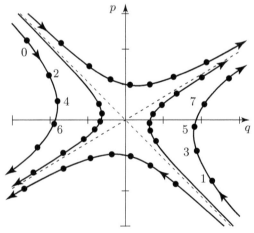

Fig. 4.2 Same situation as depicted in Fig. 4.1 in the case of a propagator having negative real eigenvalues

The transition between these two types of behaviour occurs for an absolute value of the trace equal to two. In this case, there exists a preferential direction. Close to the fixed point, the successive iterations regularly drift from the fixed point parallel to this straight line.

Problem Statements

4.1. Electric Charges Trapped in Conductors [Solution and Figure p. 188] ★★

An interesting problem dealing with an electrostatic image

A point-like electric charge q_e influences an infinite plane conducting wall which attracts it. The situation is identical to that of an opposite charge symmetrically placed with respect to the wall (electrostatic image). For a plane wall, the component of the motion parallel to the plate is uniform and straight and, thus, rather uninteresting.

We are concerned here with the perpendicular motion, along the $x'x$ axis; this axis intersects the plate at point O and the position M of the charge is specified by the generalized coordinate $x = \overline{OM}$. The particle, released without speed from point A such that $a = \overline{OA}$, accelerates and falls onto the plate whose role is to produce a reflection with an instantaneous inversion of the velocity.

1. Give the attractive force exerted by the plate on the charged particle. Deduce the corresponding potential.

2. Derive the Hamiltonian and give a constant of the motion. Plot the phase portrait. You will also consider particles coming from infinity ($a = \infty$).

3. Find the period of the oscillations that trap the particle. You will need to find the corresponding formula, consult a table of integrals (or a package such as Mathematica) or, lastly, ... use the relation

$$\int_0^a dx \left(\frac{1}{x} - \frac{1}{a}\right)^{-1/2} = \frac{\pi}{2} a^{3/2}.$$

4.2. Symmetry of the Trajectory
[Solution and Figure p. 190] ★

A simple geometrical property of the trajectory

In a central force field, one considers, in a plane, the trajectory of a particle with mass m (this could be a single particle subject to central forces or the virtual particle with reduced mass m in the two-body problem). This trajectory, expressed using polar coordinates $\rho(\phi)$, is assumed to be comprised between the two turning points $\rho_{\min}$ and $\rho_{\max}$ for the radial variable.

1. Show that the particle velocity is perpendicular to the radius vector at the turning points.
2. As usual, one defines the quantity $u(\phi) = 1/\rho(\phi)$. The reference axis for the polar angle can be chosen so as to coincide with the radius vector of a turning point; consequently $\phi = 0$ for that turning point. Show that, at this point, the condition of Question 1 implies $u'(0) = 0$.
3. Let $w(\phi) = u(-\phi)$ be the function obtained from u by a parity operation with respect to the axis. Show that the function $w(\phi)$ obeys the same Binet equation as the function $u(\phi)$.
4. Using the results of Questions 2 and 3, show that $w(\phi) = u(\phi)$, $\forall \phi$.
5. Deduce that the trajectory is symmetric with respect to the radius vectors of the turning points. Conclude that the whole trajectory can be obtained if it is known only between two consecutive turning points.
6. Show that these conclusions can be extended to the case of a differential equation of motion in u which is more general than Binet's equation provided that it does not contain derivatives of odd order.

4.3. Hamiltonian in a Rotating Frame
[Solution p. 192]   ★★

This problem is a natural consequence of Problem 2.8

A particle (or a set of particles) with mass m is studied in a frame rotating around the OZ axis, with angular velocity $\dot\phi = \omega$. We proved in Problem 2.8 that the potential of the inertial forces is

$$V = -\boldsymbol{\omega} \cdot \boldsymbol{L} - \frac{1}{2} m \left(\boldsymbol{\omega} \times \boldsymbol{r}\right)^2,$$

where $\boldsymbol{L}$ is the kinetic momentum around O.

1. Determine the generalized momenta taking as generalized coordinates either the Cartesian coordinates (X, Y, Z) or the cylindrical coordinates (ρ, ψ, Z), in this rotating frame. To simplify, one may choose the OZ axis along the rotation vector $\boldsymbol{\omega}$.
2. Perform the Legendre transform on the Lagrangian, assuming that there are no additional forces. Show that the resulting Hamiltonian is equal to

$H = H_0 - \boldsymbol{\omega} \cdot \boldsymbol{L}$, where H_0 is the Hamiltonian of a free particle in the rotating frame.

One should notice that the centrifugal term is hidden in the kinetic momentum. This expression must be used with care; it is not valid for an arbitrary choice of the coordinates.

4.4. Identical Hamiltonian Flows
[Solution and Figure p. 194] ★

Different Hamiltonians can lead to the same trajectories

Let $H(q, p)$ be an autonomous Hamiltonian and $(q(t), p(t))$ a corresponding trajectory in phase space, with energy E. We consider now a new Hamilton function $K(q, p)$, which is an arbitrary function of the original Hamiltonian $K(q, p) = F(H(q, p))$. We denote $(\tilde{q}(t), \tilde{p}(t))$ the corresponding trajectories with energy $F(E)$.

Show that these trajectories are the same as the previous ones, but governed by a different temporal law.

This implies that the flows of the Hamiltonians H and K are identical.

4.5. The Runge–Lenz Vector
[Solution and Figure p. 195] ★★

A very special vector

One considers the motion of a particle with mass m embedded in a central force field, the potential of which is given by $V(|\boldsymbol{r}|)$.

Generalities

1. Show that the Hamiltonian of the system is

$$H(\boldsymbol{r}, \boldsymbol{p}) = \frac{\boldsymbol{p}^2}{2m} + V(r).$$

2. Write down Hamilton's equations. What do they represent?

3. Demonstrate the following equality

$$\frac{d}{dt}\left(\frac{\boldsymbol{r}}{r}\right) = -\frac{1}{r^3}\left[\boldsymbol{r} \times (\boldsymbol{r} \times \dot{\boldsymbol{r}})\right].$$

4. In the case of a central force problem, it is well known that the kinetic momentum $\boldsymbol{\sigma} = \boldsymbol{r} \times \boldsymbol{p}$ is a constant of the motion.

Using Questions 2 and 3, deduce the relation

$$\frac{d}{dt}(\boldsymbol{p} \times \boldsymbol{\sigma}) = -mr^2 f(r)\frac{d}{dt}\left(\frac{\boldsymbol{r}}{r}\right),$$

where $f(r) = -dV(r)/dr$ is the value of the central force along the unit vector $\boldsymbol{r}/r$.

Kepler's problem

We deal now with the special case of the attractive Kepler problem for which: $V(r) = -K/r$ ($K > 0$).

1. Prove that the vector

$$\boldsymbol{C} = \boldsymbol{p} \times \boldsymbol{\sigma} - mK\frac{\boldsymbol{r}}{r},$$

called the Runge–Lenz vector, is a constant of the motion.

2. Show that this vector belongs to the plane of the orbit.

3. Choosing the reference axis of the polar coordinate ϕ along the vector $\boldsymbol{C}$, derive the relation $Cr\cos\phi = \sigma^2 - mKr$.

4. Relying on the previous question and on the general expression for the trajectory, show that $\boldsymbol{C}$ is directed along the radius vector corresponding to the perihelion, and that it is simply linked to the eccentricity e.

5. The problem of a particle in a central force field exhibits six constants of the motion (depending for example on the initial components of the position and velocity vectors). In fact, one of these pieces of information concerns the temporal law, for example the time of passage at the perihelion. There remain five independent constants concerning the orientation, the shape and the size of the ellipse. However, we may determine seven constants, namely the energy E, the three components of $\boldsymbol{\sigma}$ and the three components of $\boldsymbol{C}$. There must therefore exist two relationships between these quantities. Determine these relationships.

4.6. Quicker and More Ecologic than a Plane [Solution p. 198] ★ ★

An effective solution to conveyance problems

The Earth is considered as a sphere with radius $R_E = 6{,}371$ km and with uniform mass density. We set g, the acceleration due to gravity at ground level, equal to 9.81 ms^{-2}. Let imagine a straight tunnel between two arbitrary points A and B on the Earth's surface, which for instance connects Paris and Tokyo (see Fig. 4.3).

Problem Statements

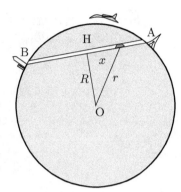

Fig. 4.3 Straight tunnel connecting Paris to Tokyo. The coach is located by its coordinate x along the tunnel, with the middle of the tunnel taken as the origin. The distance between the coach and the center of the Earth is denoted by r

Further imagine a coach which rolls without friction in this tunnel. Starting from Paris, it will accelerate due to the gravitational field then, after reaching the distance of least approach to the center of the Earth, it will decelerate.

Is its speed sufficient to reach Tokyo? If so in how much time? The effect due to the rotation of the Earth is neglected.

1. Calculate the gravity field $G(r)$ for a point inside the Earth located at a distance r from its center.
2. Calculate the potential $V(r)$ at the same point.
3. As a generalized coordinate x, you should take the distance between the coach and the middle of the tunnel. Give (up to a constant) the potential energy of the coach of mass m as a function of x. Write down the Hamiltonian of the system. Deduce Hamilton's equations and then the differential equation concerning $x(t)$.
4. Solve this equation for a departure from Paris. How much time does the journey take? Show that this time is independent of the choice for A and B.
5. Just for fun, calculate the length of the tunnel connecting Paris to Tokyo as well as the maximum speed reached in the middle. We give the latitudes and longitudes of both cities: Paris (48.52°N, 2.2°E), Tokyo (35.42°N, 139.46°E).

Hint: Be careful! The acceleration due to gravity depends on the distance to the center of the Earth where it vanishes. To find this dependence, you can use the analogy with the electric field inside a uniformly charged sphere. For the motion of the coach, you should find the traditional "harmonic oscillator".

4.7. Hamiltonian of a Charged Particle
[Solution p. 200] ★ ★ ★

A straightforward use of Lorentz's covariance

Let us consider a particle of mass m and electric charge q_e, placed in an electromagnetic field $(U, \boldsymbol{A})$. The aim of the problem is to check that the following Hamilton's functions lead to the equations of motion, namely that the time derivative of the linear momentum $\boldsymbol{\pi} = \boldsymbol{p} - q_e \boldsymbol{A}$ is equal to the Lorentz force $q_e \left[\boldsymbol{E} + \boldsymbol{v} \times \boldsymbol{B} \right]$. We consider two important cases.

1. The non-relativistic approximation for which the Hamiltonian H is

$$H = \frac{(\boldsymbol{p} - q_e \boldsymbol{A})^2}{2m} + q_e U.$$

2. The relativistic regime for which the Hamiltonian H is given by

$$(H - q_e U)^2 - (\boldsymbol{p} - q_e \boldsymbol{A})^2 c^2 = m^2 c^4.$$

3. *For the following questions a lecture course on relativity is necessary.*

It is more elegant to translate the Hamiltonian formalism into a covariant form. With this in mind, we use the following contravariant quadrivectors: $q^\mu = (ct, \boldsymbol{q}) = (ct, \boldsymbol{r})$; $p^\mu = (E/c, \boldsymbol{p})$; $A^\mu = (U/c, \boldsymbol{A})$. In this problem, the metric is that of special relativity, namely the Minkowski metric given by[6]

$$\begin{aligned} g^{\mu\nu} = g_{\mu\nu} &= g_{\mu\mu} \delta_{\mu\nu} \\ g^{ii} &= 1 \quad i = 1, 2, 3 \\ g^{00} &= -1. \end{aligned}$$

and the transformation of the corresponding contravariant and covariant quadrivectors

$$V_\mu = g_{\mu\mu} V^\mu; \qquad V^\mu = g^{\mu\mu} V_\mu.$$

Check that the equations of the motion are written

$$\frac{dq^\mu}{d\tau} = \frac{p^\mu - q_e A^\mu}{m}; \qquad \frac{dp^\mu}{d\tau} = \sum_\nu \frac{p^\nu - q_e A^\nu}{m} \frac{\partial A_\nu}{\partial q_\mu}.$$

where τ is the proper time defined by $d\tau^2 = \left(1 - \dot{\boldsymbol{q}}^2/c^2\right) dt^2$.

---

[6] Very often the metric is defined with the opposite sign. Of course, the physical results are independent of this choice. The choice employed here allows a more natural interpretation of the momentum signs if they are defined by the usual recipe relying on the Lagrangian.

To materialize the covariance of the formalism, we assume that the quadri-vector $q^\mu = (ct, \boldsymbol{q})$ depends on a continuous parameter ω, which will be identified with the proper time, and we set $q'^\mu = dq^\mu/d\omega$.

It is natural to define the conjugate momentum by the usual formula (take care of the indices; the derivative with respect to a contravariant component is indeed a covariant component): $p_\mu = \partial L/\partial q'^\mu$. However we must start from a correct Lagrangian, which must be a relativistic invariant and which leads to the correct equations of motion. One could define it in a standard way by

$$\int -mc\,ds - q_e(U\,dt - \boldsymbol{A} \cdot d\boldsymbol{q}) = \int L(q^\mu, q'^\mu)\,d\omega.$$

It is in fact easier to start from the expression, justified a posteriori,

$$L(q^\mu, q'^\mu) = \sum \frac{1}{2} m q'^\mu q'_\mu + q_e q'_\mu A^\mu$$

(the fans of Einstein's convention can remove the summation symbol).

4. Give the expression for the momenta. Show that if we interpret the parameter ω as the proper time τ, then p_μ represents the traditional momentum as given by Questions 2 and 3.

5. Calculate the Hamilton function $H(q^\mu, p^\mu)$. Find again the covariant Hamilton equation. This Hamiltonian is independent of the integration variable ω; its value must be constant. Fix this constant in order that the ω parameter is identified with the proper time τ.

4.8. The First Integral Invariant
[Solution and Figure p. 204] ★ ★

The simplest integral invariant of Cartan's theory[7]

For a one-dimensional system, the area limited by a strap which drifts following the flow remains invariant. We wish to generalize this case for a system with n degrees of freedom. Let us consider a point (q, p) in phase space and an infinitely close point $(q + \varepsilon, p + \eta)$. The flow corresponding to a time interval, dt, transports these points to (q', p') and $(q' + \varepsilon', p' + \eta')$ respectively.

---

[7] For Hamiltonian systems, Cartan proved that there exist n integral invariants. The total hypervolume, appearing in Liouville's theory, is one of them. In this problem, we present another rather simple one.

1. Show that

$$\varepsilon'_i = \varepsilon_i + dt \left(\sum_j \varepsilon_j \partial^2_{q_j p_i} H + \sum_j \eta_j \partial^2_{p_j p_i} H \right),$$

$$\eta'_i = \eta_i - dt \left(\sum_j \varepsilon_j \partial^2_{q_j q_i} H + \sum_j \eta_j \partial^2_{p_j q_i} H \right).$$

2. Similarly, take another infinitely close point $(q + \gamma, p + \delta)$ and expand, retaining only the first order terms in dt, the quantity

$$\sum_i (\varepsilon'_i \delta'_i - \eta'_i \gamma'_i).$$

Deduce that the sum of oriented areas projected on each (q_i, p_i) plane is an invariant:

$$\oint_\Gamma p \cdot dq = \text{const.}$$

4.9. What About Non-Autonomous Systems?
[Solution p. 206] ★

A possible use of time as a generalized coordinate

For simplicity, let us restrict ourselves to a non-autonomous one-dimensional system; its Hamiltonian is $H(q, p, t)$. We are free to chose the time t as a full generalized coordinate with a conjugate momentum denoted p_t. Now let us define, in an enlarged phase space, the function $\tilde{H}(q, t, p, p_t) = H(q, p, t) + p_t$ which now corresponds to a conservative but two-dimensional system.

1. Write the equations of the flow and interpret the flow parameter as well as the momentum p_t.
2. The pseudo-Hamiltonian $\tilde{H}(q, t, p, p_t)$ is autonomous; what is its value along the flow?

4.10. The Reverse Pendulum
[Solution and Figure p. 207] ★ ★ ★

An unusual manner to handle a pendulum! or how to stabilize a pendulum around its unstable equilibrium position

Can a simple pendulum with length l and mass m, remain close to its unstable equilibrium position?

The answer is yes if we compel the point of suspension A to follow a well chosen periodic motion $z(t)$. A sinusoidal motion makes the calculations very cumbersome. We will demonstrate the desired property for a periodic motion of A which, although not very realistic, makes the calculations easier. It consists in replacing an arc of a sinusoid by an arc of a parabola.

During a period $2T$, the driving motion is as follows. During the first half-period equal to T, the point of suspension follows the temporal law $z(t) = at(t-T)/2$. When it arrives precisely at its initial position $z = 0$, there is an instantaneous reversal of acceleration (the position and the velocity do not present a discontinuity) and during the second half-period the temporal law is $z(t) = a(t-T)(2T-t)/2$ which, at time $2T$, resets the point A in the same conditions as for the origin. The parameter a is chosen as a positive quantity. The cycle then repeats periodically (see Fig. 4.4). The angle θ between the pendulum direction and the **upward** vertical is chosen as the generalized coordinate.

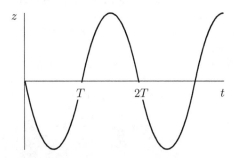

Fig. 4.4 Temporal law $z(t)$ imposed on the reverse pendulum

1. Using the Lagrangian or Hamiltonian formalism, find again the equation of motion $l\ddot{\theta} = (\ddot{z}+g)\sin\theta$.

2. We work in the vicinity of the **unstable equilibrium position**, that is for θ small. Given $\theta_0, \dot{\theta}_0$ for the point A at $z = 0$ and $t = 0$, calculate $\theta', \dot{\theta}'$ at the end of the first period of acceleration, that is for $z = 0$ after time T. It is interesting to introduce the proper angular frequency of the pendulum $\omega_0 = \sqrt{g/l}$ and the angular frequency in the presence of the acceleration $\omega = \sqrt{(a+g)/l} = \omega_0\sqrt{(a/g)+1}$.

3. Starting from these values $\theta', \dot{\theta}'$, calculate $\theta_1, \dot{\theta}_1$ after the second phase of the motion when $z = 0$ at time $2T$. Justify that stability can occur only if $a > g$; this is the case that will be assumed in what follows. We introduce the new angular frequency $\Omega = \sqrt{(a-g)/l} = \omega_0\sqrt{(a/g)-1}$.

4. Calculate the propagator matrix K, which brings $\theta_0, \dot{\theta}_0$ to $\theta_1, \dot{\theta}_1$ over a full period. Check the conservation of the area in phase space $(\theta, \dot{\theta})$. To study the stability, we rely on properties concerning the trace of the propagator (see Section 4.6). Give this trace.

5. Try to justify that the stability regions correspond to $\Omega T \cong (n + 1/2)\pi$ (n integer). Using a pocket calculator or a micro computer, investigate the possibility of a domain of stability as a function of a/g. In order to do this, you will proceed by successive approximations trying to impose the condition $|\text{Tr}(K)| = 2$.

4.11. The Paul Trap
[Solution and Figure p. 211] ★ ★ ★

A way to trap charged particles in a time dependent electromagnetic field

We already saw (see Problem 2.10) that it is possible to confine a particle, with mass m and electric charge q_e, in a region of space using a Penning trap. In this trap, it is the geometry of the electromagnetic field that guarantees the confinement along the symmetry axis. The transverse instability is compensated by a magnetic field which produces a drift along the equipotential curves.

The Paul trap provides an alternative to this type of trapping, with a similar geometry of the electrostatic field but without recourse to a magnetic field. As we will see, the confinement is achieved by the use of a periodic potential.[8] The form of the scalar potential (the vector potential is assumed to vanish, $\mathbf{A} = \mathbf{0}$) is given by (U_0 is a constant):

$$U(x, y, z, t) = \frac{1}{2} U_0 \left(x^2 + y^2 - 2z^2 \right) \cos(\omega t)$$

1. Give the three second order differential equations for the three functions $x(t)$, $y(t)$, $z(t)$. Use a non-relativistic treatment and the Hamiltonian formalism.

2. You will see that the three motions decouple. Each differential equation belongs to a general set of differential equations known as Mathieu's equations. These equations do not have analytical solutions and it is not obvious a priori that a force proportional to the distance, but which changes sign periodically, can give rise to a solution that is confined in space.

   To better understand how this could happen, but avoiding a complicated rigorous treatment, we simplify the time dependence of the potential assuming a "square wave shape", passing periodically from the constant value $U_0/2$ over a half-period $T/2 = \pi/\omega$, to the constant opposite value $-U_0/2$ over the other half-period.

---

[8] Note that such a potential satisfies Maxwell's equations only approximately, but better for the smaller angular frequency.

Give, under a "propagator matrix form", the relation between the position of the particle and its momentum projected along the three axes during the time when the potential is positive. Do the same for the other half-period. Check that the area in phase space is conserved in both cases.

3. Deduce the relation "propagator over a whole period" which links the components along the x axis for the position and the momentum after a full period.[9] Redo the same study for the position and momentum along the z axis.

4. We now follow the method proposed in Problem 4.10 for the reverse pendulum: this matrix can be diagonalized. If the eigenvalues are complex, the origin in phase space is an elliptic point and the charge remains confined. Using properties concerning the trace of a matrix, give the relationship between the various parameters in order to produce a confinement.

5. With the help of a graphical representation of an appropriate curve, and solving a transcendental equation, deduce an approximate numerical condition for the parameters in order to achieve confinement in the three directions.

Remark: For a variation exactly sinusoidal in time, and with the properties of the solutions of Mathieu's equations, the exact condition reads: $2q_e U_0/(m\omega^2) < 0.4539$.

4.12. Optical Hamilton's Equations
[Solution p. 214]   ★ ★ ★

Snell-Descartes law and optical systems

One considers a particular optical medium, for which the refraction index n depends only on the distance ρ to a given axis, the Oz axis for instance. You can imagine an optical fiber with an index gradient. Thus the lines parallel to Oz, at a distance ρ, are iso-index lines $n(\rho)$.

A light trajectory located in a plane containing the symmetry axis is described by an equation of the form $\rho(z)$. It is specified by Fermat's principle, the principle of least action which stipulates that the optical path between two given points A and B, $L_{\text{opt}} = \int_A^B n(\rho)\,ds$, is extremal along the light trajectory passing through these two points. As usual, we denote by $i(\rho)$ the angle between the trajectory and the normal to the iso-index lines.

---

[9] This way of investigating the system after only one period (stroboscopy) will be more intensively used in Chapter 8. The points in phase space for each period form what is known as a Poincaré section.

1. Express the optical path as an integral functional over the variable z. By analogy with classical mechanics, give the corresponding "Lagrangian" $L(\rho, \rho')$ for the light ray, taking $\rho, \rho' = d\rho/dz$ as generalized "coordinates" and "velocity", and z as the "time".
2. Give the expression of $\sin(i(\rho))$ and $\cos(i(\rho))$ as a function of ρ'.
3. Give the expression of the momentum p, conjugate to ρ. Ascribe a physical sense to this momentum.
4. Perform the Legendre transform to obtain the Hamiltonian.
5. Find a constant of the motion and recover the Snell-Descartes law.
6. Write down Hamilton's equations.
7. As an application of Question 5, one assumes that the index decreases with the distance to the symmetry axis. Under these conditions, it is possible to expand the index in powers of ρ and restrict oneself to lowest order, which gives: $n(\rho) = n_0 - a\rho^2$ (a is a positive constant, characteristic of the optical medium). We now consider the set of all possible light trajectories which originate from a single source located at the origin O of the Oz axis. Thus, for all trajectories, we have $\rho(0) = 0$. The trajectories differ by the emission angle with respect to symmetry axis, which is assumed to be small ($\rho'(0) \ll 1$). Differentiating the constant of the motion found in Question 5 and being consistent in the order of the truncated expansion, give the differential equation fulfilled by $\rho(z)$.
8. Show that all these trajectories intersect at a unique point I on the axis (this is the image of O by the optical system which can be virtual). Give the distance OI as a function of the characteristics of the system.

Note: The conservation of the area for a Hamiltonian system has a counterpart in geometrical optics. Let us take, for instance, a microscope and place an object of spatial extension dx_0 and extension in momentum dp_0 (due to the emission angle of the rays towards the optical set up). Leaving the microscope, the light rays of the object form an image with extension dx_i and with momentum dp_i on the retina. This image comes from the Hamiltonian flow of the object and the conservation of the area imposes $dx_i dp_i = dx_0 dp_0$. The extension of the object is fixed by our study and the momentum dp_i is limited by the physiology of the eye (ratio of the pupil diameter to the eye depth). Thus the ratio $dx_i/dp_0 = dx_0/dp_i$ is imposed on us. To obtain the largest image, one must manage to produce a dp_0 as large as possible. This last quantity depends on the incidence of the rays entering the microscope (which should therefore be as large as possible), but also on the optical index of the medium that we want to have also as large as possible.

4.13. Application to Billiard Balls
[Solution and Figure p. 216] ⋆ ⋆

A problem for addicts of the game of billiards

The behaviour of the ball on a billiard table and its quantum analog, a particle in a closed object, led to important progress in theoretical physics. For a two-dimensional planar billiard table, the problem is easy to describe. The ball maintains a uniform straight line motion until it strikes the cush, and rebounds with a velocity symmetric to the initial velocity with respect to the normal to the cush at the point of contact. The shape of the table is described by a mathematical equation; one can choose on the border a reference point. Then an arbitrary point of the cush is unambiguously specified by its curvilinear abscissa s. To describe the behaviour of the system, it is enough to know the abscissa s_n for the impact n and the angle of incidence i_n. In other words, in a two-dimensional phase space (s, i), the evolution of the system is represented by a sequence of points (s_n, i_n) which are deduced from the foregoing by an mathematical application.

How can we link the study of the billiard ball to the notions developed in this chapter? First it is a mechanical system with two degrees of freedom which specify its configuration. Inside the table, the Hamiltonian is just the kinetic energy; the ball is not subject to any force and the trajectory is straight. At the cush, the restoring force is infinite. Strictly speaking, a Hamilton function does not exist. Nevertheless, one can imagine a potential which, without discontinuity, exhibits a very rapid variation; thus the ball bounces without energy loss. The phase space has four dimensions. It is impossible to visualize it completely; a way to study it is to investigate it only for the bounces, because the couple (s_n, i_n) form a two-dimensional phase space.

In this problem, we will show that the momentum associated with the s_n coordinate is the sine of the incident angle $p_n = \sin(i_n)$. We will check that, in this space, the area is conserved for sequential applications. This method to reduce information (passing from four to two dimensions), restricting ourselves to only a part of phase space – in order to represent it graphically – is called the study of a Poincaré section. Nevertheless we will show that something remains from Liouville's theorem: the conservation of areas.

In order to carry out this investigation, one considers the straight line joining the points s_{n-1} to s_n with a reflection angle $-i_{n-1}$, namely a momentum $p_{n-1} = -\sin(i_{n-1})$ and an incident angle i_n corresponding to a momentum $p_n = \sin(i_n)$. Very close to a bounce, the shape of the cush is simulated by a straight segment, that can be considered indifferently as the tangent to this point or the segment joining this point to a closely adjacent point.

1. One considers first an infinitesimal variation of the starting abscissa δs_{n-1} with **constant incidence**. Calculate the variation of the curvilinear abscissa δs_n and the variation of the incidence δp_n at the next rebound.
2. One considers now an infinitesimal variation of the incident angle δp_{n-1} with **constant curvilinear abscissa**. Calculate the variation of the curvilinear abscissa δs_n and the variation of the incidence δp_n at the next rebound.
3. Superposing the previous elementary variations, write down the matrix which linearizes the application, that is such that

$$\begin{pmatrix} \delta s \\ \delta p \end{pmatrix}_n = M(s_n, p_n) \begin{pmatrix} \delta s \\ \delta p \end{pmatrix}_{n-1},$$

and show that its determinant is equal to unity; this corresponds, as we already saw, to the conservation of the area in phase space.

4.14. Parabolic Double Well
[Solution and Figure p. 219] ★ ★

A relatively common shape for a potential

This problem studies the motion of a particle of mass m along an axis $x'x$, in a symmetric double well represented by a very simple potential given by (see Fig. 4.5):

$$V(x) = \begin{cases} \infty & \text{if} \quad |x| > a \\ -V_0(x/a)^2 & \text{if} \quad |x| \leq a. \end{cases}$$

What is the Hamiltonian of the system for $|x| \leq a$? Is there a constant of the motion?

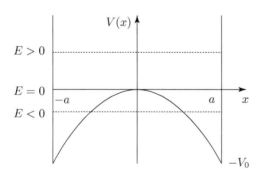

Fig. 4.5 Potential with a parabolic double well shape (full line). Three values for the mechanical energy are represented by dotted lines

There exist two distinct regimes.

First regime $E > 0$.

1. What are the turning points of the trajectory?
2. Plot the phase portrait of the system.
3. Integrate the equation of motion between two turning points, assuming that the particle starts from the lowest turning point at time $t = 0$.
4. What is the period of this motion.

Second regime $-V_0 \leq E < 0$.

Investigate again the previous questions for this case. What happens on the separatrix $E = 0$? In particular, what is the time necessary to travel from $x = -a$ to $x = 0$?

Note: This kind of potential, at least with a similar shape, is commonly used in quantum mechanics. In the case $E < 0$, it allows the system to pass from a bound state in the first well to a bound state in the second well due to the tunnel effect. The application to the ammonia molecule is classical.

4.15. Stability of Circular Trajectories in a Central Potential
[Solution p. 222]   ★ ★

Classical case of a power-law potential

Let us take the case of a very simple attractive potential[10] described by a power-law $V(\rho) = -\lambda/\rho^\alpha$ ($\lambda\alpha > 0$, α being an arbitrary real number). We restrict ourselves to the study of the trajectories in the plane of the motion.

The energy E, as well as the angular momentum σ of the particle (with reduced mass m in case of a two-body problem) are given. We are especially interested in the possible orbits, forgetting about the temporal law for describing them. With polar coordinates (ρ, ϕ), the equation of the trajectory $\rho(\phi)$ is given by an integral containing the potential $V(\rho)$.

Except in very specific cases, one does not have an analytical expression for the equation of the trajectory. Very often, it is more judicious to work with the function $u(\phi) = 1/\rho(\phi)$. We then obtain the equation

$$u'^2 + u^2 = \frac{2m}{\sigma^2}\left(E - V(1/u)\right),$$

---

[10] Power-law potentials are very important practically as a basis for some effective potentials.

which is nothing more than the integral form of Binet's equation. The equation can be recast in the form

$$E = \frac{\sigma^2}{2m} u'^2 + V_{eff}(u).$$

This is precisely the equation of a conservative one-dimensional system in which the variable ϕ replaces the variable t.

1. Give the expression of the effective potential $V_{eff}(u)$ and interpret each of the contributions.

2. Use the analogy of phase portraits to understand the circular orbits centered on the center of force? What is the radius of these orbits? Give the relationship between the total energy and this radius. In the total energy, what is the proportion of the kinetic energy and the potential energy (virial theorem).

   A small perturbation can modify one of these quantities and thus modify the circular trajectory. To explain what will happen, one can imagine linearizing the equations.

3. What condition must be verified by α in order that this circular orbit is stable? In this case, using an expansion of the potential $V_{eff}(u)$ truncated to second order in u, show that the new trajectory is comprised between a perihelion and an aphelion.

4. Working still in the harmonic approximation framework, calculate the angle $\Delta\phi$ between two successive passages through perihelia (or aphelia).

5. Again with the hypothesis of nearly circular motion, under which condition do there exist closed trajectories? Investigate the case for which $\alpha = n$ is an integer number. Check in particular that this is the case for the harmonic potential $\alpha = -2$ and the Coulomb potential $\alpha = 1$.

4.16. The Bead on the Hoop
[Solution and Figure p. 224] ★ ★

A problem that we already know well, but studied from another point of view and for which a broken symmetry is underlined, i.e., the presence of a bifurcation

We pursue the study of the bead on the hoop proposed in the Problem 1.4, but with a constant angular velocity ω for the rotation of the hoop.

1. Determine the Hamilton function. Show that it does not depend on time and thus gives rise to a constant of the motion, the energy E.

The system has one degree of freedom: the angle θ between the direction specifying the position of the bead and the upward vertical. The system is subject to an effective potential $V_\omega(\theta)$, which is the sum of the gravitational potential energy and the centrifugal potential energy. The aim of the problem is to study this system using the phase portrait technique.

2. Give the expression of the effective potential $V_\omega(\theta)$. Show that it is a periodic function symmetric with respect to $\theta = \pi$; consequently, one can restrict the study to the interval $[0, \pi]$.

Case of a rapid rotation $\omega > \sqrt{\dfrac{g}{R}} = \omega_0$

1. Show that the points $\theta = 0$ and $\theta = \pi$ are unstable equilibrium points and that there exist two stable equilibrium points for θ_s and $2\pi - \theta_s$. Calculate θ_s and the value of the potential at this point $E_m = V_\omega(\theta_s)$.

2. Study the behaviour of the system for an energy $E_m < E < -mgR$. Make the harmonic approximation in the vicinity of θ_s for an energy slightly larger than E_m. Give the corresponding angular frequency.

3. Study the behaviour of the system for an energy $-mgR < E < mgR$.

4. Study the behaviour of the system for an energy $E > mgR$.

5. Plot the phase portrait of the system in phase space, for all the previous cases. This portrait is similar for all the potentials of the type "double well".

Case of a slow rotation $0 < \omega < \omega_0$

1. Show that, in this situation, the point $\theta = 0$ is an unstable equilibrium point and that $\theta = \pi$ is a stable equilibrium point.

2. Study the behaviour of the system for an energy $-mgR < E < mgR$. Make the harmonic approximation in the vicinity of $\theta = \pi$ for an energy slightly larger than $-mgR$. Give the corresponding angular frequency.

3. Study the behaviour of the system for an energy $E > mgR$.

4. Plot the phase portrait of the system in phase space, for the two previous types of behaviour.

5. Plot the angles corresponding to stable equilibrium as a function of the angular velocity of the hoop. The splitting of the stable equilibrium position is called a bifurcation.

4.17. Trajectories in a Central Force Field
[Solution and Figure p. 228] ★★

Relativistic equations for the Coulomb problem

In the Problem 3.2, we wrote the differential equation which determines the trajectory of a charged particle in a central force field taking into account relativistic effects. We investigate in the present problem the first case, namely that of an electrostatic potential and more precisely the Coulomb potential, $V(\rho) = K/\rho$. The constant $K = q_e q'_e/(4\pi\varepsilon_0)$ may be positive (repulsive potential) or negative (attractive potential). The Kepler problem is just a particular case with a negative constant ($K = -Gmm'$). We will use as energy unit the rest mass energy of the particle and work with the reduced mass energy $\varepsilon = E/(mc^2)$, the dimensionless parameter connected to the angular momentum $\nu = [K/(\sigma c)]^2$, and the inverse of the radius vector in natural units $u = |K|/(mc^2\rho)$.

What are the types of different possible trajectories? To answer this question, the use of the phase portrait is a very practical tool. Indeed, we may consider the polar angle ϕ as a time, u as a coordinate and $u' = du/d\phi$ as a momentum.

1. Plot the phase portraits. Comment on the types of trajectories as a function of the attractive or repulsive character of the potential and of the value of the angular momentum σ.

2. As an application, you will estimate the relativistic corrections to the energy of a hydrogen atom considered as a bound state of a proton and an electron. The proton is supposed to remain at rest at the center of the force field; in this case, one has $K = -e^2/(4\pi\varepsilon_0)$. You could assume that the electron, with mass m, follows an orbit the radius of which equals the Bohr radius $a_0 = \hbar^2/(m|K|)$ and you should introduce the fine structure constant $\alpha = |K|/(\hbar c)$.

Problem Solutions

4.1. Electric Charges Trapped in Conductors [Statement p. 171]

1. The system formed by the charge q_e and the conducting plate is equivalent to that of the charge and a charge $-q_e$ symmetric with respect to the plate. This is an example of the principle of electrostatic images. For these two charges, the median is at a constant potential as it should be for a conductor.

We are concerned with the motion along the normal direction; choosing the origin O in the plane, this normal on which the particle moves is Ox, with unit vector $\boldsymbol{i}$. As generalized coordinate, we take the distance between the plane and the position of the charge: $x = \overline{OM}$. The force $\boldsymbol{F}$ exerted by the conductor on the charge can be reduced to the force exerted by the symmetric charge with opposite sign, but at the distance $2x$; it is thus attractive. From Coulomb's law and denoting $Q^2 = q_e^2/(4\pi\varepsilon_0)$ to simplify the notation, this force is:

$$\boldsymbol{F} = -\frac{Q^2}{4x^2}\boldsymbol{i}.$$

The particle potential energy is obtained after integration of the defining equation $F = -dV/dx$; this leads to the relation:

$$V(x) = -\frac{Q^2}{4x}.$$

2. The kinetic energy of the particle is $T = \frac{1}{2}m\dot{x}^2$, the Lagrangian $L = T - V$, the conjugate momentum $p = m\dot{x}$ and the corresponding Hamiltonian $H = \dot{x}p - L$, or:

$$H(x,p) = \frac{p^2}{2m} - \frac{Q^2}{4x}.$$

The Hamiltonian does not depend explicitly on time; its value is the energy E which remains constant along the trajectory. If the charge is released from point A ($\overline{OA} = a$), with a null speed, the value for the energy is $E = -Q^2/4a$. Thus

$$\frac{p^2}{2m} - \frac{Q^2}{4x} = -\frac{Q^2}{4a}.$$

This equality provides the relation $p(x)$, which is the phase portrait, depicted in the Fig. 4.6.

Explicitly:
$$p(x) = \pm\frac{Q}{2}\sqrt{2m}\sqrt{\frac{1}{x} - \frac{1}{a}}.$$

The particle starts from A with a null velocity ($p = 0$), then approaches the plane ($p < 0$) increasing its speed; it arrives on the plate with an infinite momentum $p = -\infty$. The contact with the plate produces an instantaneous change of the direction of the velocity and $p = \infty$. The particle then decelerates with a positive velocity ($p > 0$) to arrive at point A with a vanishing velocity ($p = 0$).

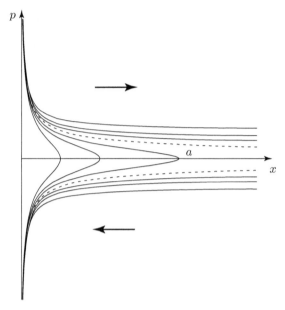

Fig. 4.6 Phase portrait for the motion of the charged particle trapped by the conductor. Just after contact with the conducting plate, the momentum instantaneously passes from $-\infty$ to ∞. There exist scattering trajectories for a positive energy (case $a < 0$) and periodic trapped trajectories for a negative energy ($a > 0$)

3. We are in presence of a periodic motion. The half-period $T/2$ is the time needed to travel from A to O, namely

$$\frac{T}{2} = \int_O^A dt = \int_O^A \frac{dx}{\dot{x}} = \int_O^A \frac{m\,dx}{p(x)}, \quad \text{or} \quad T = \frac{2\sqrt{2m}}{Q} \int_0^a \frac{dx}{\sqrt{1/x - 1/a}}.$$

With the proposed value for the integral, one obtains:

$$T = \pi \frac{\sqrt{2ma^3}}{Q}.$$

4.2. Symmetry of the Trajectory
[Statement p. 171]

1. In the plane of the trajectory, we always have the relation (it comes from Hamilton's equation on variable ρ) $\dot{\rho} = \sqrt{2\left(E - V_{\text{eff}}(\rho)\right)/m}$, with

$$V_{\text{eff}}(\rho) = V(\rho) + \frac{\sigma^2}{2m\rho^2}.$$

Problem Solutions

The turning points ρ_t are defined by $V_{\text{eff}}(\rho_t) = E$, which implies $\dot\rho = 0$ for this point. Furthermore, we have also $\boldsymbol{r} = \rho \boldsymbol{u}_\rho$, $\boldsymbol{v} = \dot{\boldsymbol{r}} = \dot\rho \boldsymbol{u}_\rho + \rho \dot\phi \boldsymbol{u}_\phi$. Because of the previous remark, at the turning point one has $\boldsymbol{v} = \rho_t \dot\phi_t \boldsymbol{u}_\phi$ and consequently:

$$\boldsymbol{r}.\boldsymbol{v} = 0.$$

At the turning point the velocity is perpendicular to the radius vector.

2. We begin with $d\rho/d\phi = \dot\rho/\dot\phi$, or using the definition of the angular momentum σ, $d\rho/d\phi = \rho' = m\rho^2 \dot\rho/\sigma$. Except in the case of a vanishing angular momentum, the condition concerning the turning point leads to $\rho' = 0$. Moreover, since $u = 1/\rho$, $u' = -u^2 \rho'$ and the condition for the turning point implies $u' = 0$. It is always possible to choose the polar angle such that $\phi = 0$ at the turning point. The previous condition is thus written

$$u'(0) = 0.$$

3. The equation for the trajectory is given by Binet's formula

$$u'' + u = -\frac{m}{\sigma^2} \frac{d(V(1/u))}{du} = \frac{m}{\sigma^2 u^2} V'(1/u)$$

where $V' = dV/d\rho$. This equation holds for an arbitrary ϕ value, in particular for $-\phi$:

$$u''(-\phi) + u(-\phi) = \frac{m}{\sigma^2 u(-\phi)^2} V'(1/u(-\phi)).$$

Let introduce the new function $w(\phi) = u(-\phi)$. It is easy to check that $w'(\phi) = -u'(-\phi)$ and $w''(\phi) = u''(-\phi)$. The previous equation can thus be written

$$w''(\phi) + w(\phi) = \frac{m}{\sigma^2 w(\phi)^2} V'(1/w(\phi)).$$

Therefore, $w(\phi)$ fulfills the same second order differential equation as $u(\phi)$.

4. At the turning point $\phi = 0$ so that $w(0) = u(-0) = u(0)$ and $w'(0) = -u'(-0) = -u'(0) = 0$. The function $w(\phi)$ satisfies the same differential equation with the same initial conditions as $u(\phi)$. It must therefore coincide with it

$$w(\phi) = u(\phi).$$

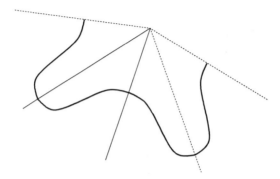

Fig. 4.7 The trajectory between the perihelion and the aphelion (dotted lines at the right hand side) is repeated symmetrically with respect to the aphelion axis for the portion between the aphelion and the next perihelion. This trajectory between two perihelia is then repeated by rotation around the center of force

5. The condition $u(-\phi) = u(\phi)$ is true $\forall \phi$; this proves that the trajectory is symmetric with respect to the radius vector of the turning point. As this turning point was chosen arbitrarily, this symmetry property is valid for every turning point. For instance, let us begin to follow the trajectory from an aphelion to the next perihelion. Then, we take the symmetric part of this portion with respect to the direction corresponding to the perihelion; we thus reach another aphelion (possibly the original one). We proceed this way from each aphelion to the next one to build the total trajectory step by step (see Fig. 4.7).

If the trajectory is a closed curve, this procedure needs a finite number of operations to obtain the whole trajectory; if not, it needs an infinite number.

6. All our previous conclusions relied on the fact that $w''(\phi) = u''(-\phi)$ and that the term $w'(\phi)$, which could invalidate the conclusion, was absent in Binet's equation. The same conclusions hold true for any differential equation in u, in which the odd order derivatives are absent. In particular, this is the case for some relativistic Binet's equations.

4.3. Hamiltonian in a Rotating Frame
[Statement p. 172]

1. We already proved that the potential due to inertial forces in a rotating frame reads $V = -\boldsymbol{\omega} \cdot \boldsymbol{L} - \frac{1}{2}m\left(\boldsymbol{\omega} \times \boldsymbol{r}\right)^2$. Choosing $\boldsymbol{\omega}$ along OZ: $\boldsymbol{\omega} = \dot{\phi}\hat{Z}$, and considering only the previous potential, the Lagrangian $\mathrm{L} = T - V$ (we denote by L the Lagrangian to avoid confusion with the angular momentum L) is easily obtained with Cartesian coordinates:

$$L(X,Y,Z,\dot{X},\dot{Y},\dot{Z}) =$$
$$\frac{m}{2}\left[\dot{X}^2 + \dot{Y}^2 + \dot{Z}^2 + 2\dot{\phi}(X\dot{Y} - \dot{X}Y) + \dot{\phi}^2(X^2 + Y^2)\right].$$

The generalized momenta are obtained by the usual recipe: $P = \partial_{\dot{Q}}L$. With the previous expression for the Lagrangian, the corresponding momenta are:

$$P_X = m\left(\dot{X} - \dot{\phi}Y\right); \qquad P_Y = m\left(\dot{Y} + \dot{\phi}X\right); \qquad P_Z = m\dot{Z}.$$

With cylindrical coordinates (ρ, ψ, Z) (we denote by ψ the polar angle in the rotating frame to avoid confusion with ϕ, the angle that specifies the rotating frame with respect to a Galilean frame), the previous Lagrangian reads:

$$L(\rho, \psi, Z, \dot{\rho}, \dot{\psi}, \dot{Z}) = \frac{1}{2}m\left[\dot{\rho}^2 + \rho^2\dot{\psi}^2 + \dot{Z}^2 + 2\rho^2\dot{\phi}\dot{\psi} + \rho^2\dot{\phi}^2\right].$$

Using this expression, we obtain the corresponding momenta:

$$P_\rho = m\dot{\rho}; \qquad P_\psi = m\rho^2(\dot{\psi} + \dot{\phi}); \qquad P_Z = m\dot{Z}.$$

The generalized momentum vector is simply the linear momentum vector in the original frame $\boldsymbol{P} = P_X\hat{X} + P_Y\hat{Y} = m\dot{x}\hat{x} + m\dot{y}\hat{y}$.

2. The Hamiltonian is obtained by a Legendre transform of the derivatives. Using Cartesian coordinates: $H = \dot{X}P_X + \dot{Y}P_Y + \dot{Z}P_Z - L$. From the expressions given in the first question, it can be shown that the Hamiltonian has the following form:

$$H = \frac{P_X^2}{2m} + \frac{P_Y^2}{2m} + \frac{P_Z^2}{2m} - \dot{\phi}(XP_Y - YP_X).$$

One notices that this expression can be recast in a more compact form

$$H = H_0 - \dot{\phi}L_Z = H_0 - \boldsymbol{\omega} \cdot \boldsymbol{L},$$

where H_0 is the Hamiltonian for a free particle.

With cylindrical coordinates: $H = \dot{\rho}P_\rho + \dot{\psi}P_\psi + \dot{Z}P_Z - L$. From the expression given in the first question, the Hamiltonian can be put into the form:

$$H = \frac{P_\rho^2}{2m} + \frac{P_\psi^2}{2m\rho^2} + \frac{P_Z^2}{2m} - \dot{\phi}P_\psi.$$

In this case $P_\psi = L_Z$ and we recover the previous expression

$$H = H_0 - \boldsymbol{\omega} \cdot \boldsymbol{L}.$$

4.4. Identical Hamiltonian Flows
[Statement p. 173]

The system is autonomous and described by the Hamiltonian $H(q,p)$. Let us define the functions $F_1(q,p) = \partial H/\partial p$ and $F_2(q,p) = -\partial H/\partial q$. The equations of motion are given by Hamilton's equations:

$$\dot{q} = F_1(q,p);$$
$$\dot{p} = F_2(q,p).$$

Let us denote by $\bar{q}(t)$ and $\bar{p}(t)$ the real trajectory of the system which is the solution of these equations. Along this trajectory, the Hamiltonian maintains a constant value equal to the energy, $H(\bar{q}(t), \bar{p}(t)) = E$.

Consider now a function $F(y)$, with the derivative $F'(y) = dF/dy$ and consider another system described by the new Hamiltonian $K(q,p) = F(H(q,p))$. The solutions for this new system are $\tilde{q}(t), \tilde{p}(t)$ which fulfil the new equations of motion:

$$\dot{\tilde{q}} = \frac{\partial K}{\partial p} = F'(H) F_1; \qquad \dot{\tilde{p}} = -\frac{\partial K}{\partial q} = F'(H) F_2.$$

Since $F'(H) = F'(E)$ is just a constant, this means that the velocities of both systems are collinear $\dot{\tilde{q}}(t) = F'(E)\dot{\bar{q}}(t)$, $\dot{\tilde{p}}(t) = F'(E)\dot{\bar{p}}(t)$ at all times. If we "forget" about the time, and restrict ourselves to the trajectory in phase space, we have

$$\frac{d\tilde{p}}{d\tilde{q}} = \frac{\dot{\tilde{p}}}{\dot{\tilde{q}}} = \frac{\dot{\bar{p}}}{\dot{\bar{q}}} = \frac{d\bar{p}}{d\bar{q}}$$

for all times, i.e. at every point of the trajectory. If the system possesses more than one degree of freedom, this conclusion holds for each of the degrees of freedom.

In phase space, we have a strict equality of the tangents at each point. The trajectory which is simply the envelop of these tangents is therefore the same for both systems.

The situation is more easily understood if one takes a glance at Fig. 4.8. The trajectories of both systems in phase space (the grey plane) are identical. This means that the two systems differ by their temporal evolutions. Indeed, let us define a new "time" $\tilde{t} = t F'(E)$. We see that $d(\bar{q}(\tilde{t}))/dt = F'(E)\dot{\bar{q}}(t) = \dot{\tilde{q}}(t)$ with an analogous property for the momentum. This equality being true at every time, we conclude that $\tilde{q}(t) = \bar{q}(\tilde{t})$ again with a similar relation for the momentum. This proves our claim that

$$\tilde{q}(t) = \bar{q}(t F'(E)); \qquad \tilde{p}(t) = \bar{p}(t F'(E)).$$

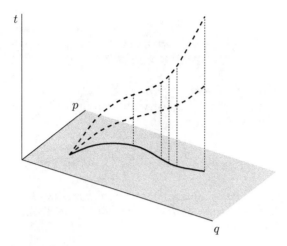

Fig. 4.8 In the extended phase space, the Hamiltonian K describes the upper dashed curve whereas the Hamiltonian H describes the lower dashed curve. Both Hamiltonians give rise to the same trajectories in phase space (full line)

4.5. The Runge–Lenz Vector
[Statement p. 173]

1. The Lagrangian for the system reads: $L(\boldsymbol{r}, \dot{\boldsymbol{r}}) = \frac{1}{2}m\dot{r}^2 - V(r)$. One deduces the momentum $\boldsymbol{p} = \partial_{\dot{\boldsymbol{r}}} L = m\dot{\boldsymbol{r}}$, then, after the Legendre transform, the Hamiltonian $H = \boldsymbol{p} \cdot \dot{\boldsymbol{r}} - L$, or:

$$H(\boldsymbol{r},\boldsymbol{p}) = \frac{\boldsymbol{p}^2}{2m} + V(r).$$

2. The first Hamilton equation gives

$$\dot{\boldsymbol{r}} = \frac{\partial H}{\partial \boldsymbol{p}} = \frac{\boldsymbol{p}}{m}$$

which is nothing more than the definition of the momentum already mentioned.

The second Hamilton equation gives:

$$\dot{\boldsymbol{p}} = -\frac{\partial H}{\partial \boldsymbol{r}} = -\boldsymbol{\nabla} V = \boldsymbol{f}.$$

Since $\dot{\boldsymbol{p}} = m\boldsymbol{a}$, it represents the fundamental equation of dynamics.

3. Simply differentiating, one sees that

$$\frac{d}{dt}\left(\frac{\boldsymbol{r}}{r}\right) = \frac{\dot{\boldsymbol{r}}}{r} - \frac{\boldsymbol{r}\dot{r}}{r^2} = -\frac{1}{r^3}\left[\boldsymbol{r}r\dot{r} - \dot{r}r^2\right].$$

Noting that $r\dot{r} = \boldsymbol{r}\cdot\dot{\boldsymbol{r}}$ and $r^2 = \boldsymbol{r}\cdot\boldsymbol{r}$, the expression between brackets can be identified with the double vector product $\boldsymbol{r}\times(\boldsymbol{r}\times\dot{\boldsymbol{r}})$. We thus obtain the required identity:

$$\frac{d}{dt}\left(\frac{\boldsymbol{r}}{r}\right) = -\frac{1}{r^3}[\boldsymbol{r}\times(\boldsymbol{r}\times\dot{\boldsymbol{r}})].$$

4. Because of the central character of the potential, only the radial component of the force $f(r) = -dV/dr$ exists. The second Hamilton equation simply reduces to $\dot{\boldsymbol{p}} = f(r)\boldsymbol{r}/r$. Consequently,

$$\dot{\boldsymbol{p}}\times\boldsymbol{\sigma} = f(r)\left(\frac{\boldsymbol{r}}{r}\right)\times(\boldsymbol{r}\times m\dot{\boldsymbol{r}}) = \frac{mf(r)}{r}[\boldsymbol{r}\times(\boldsymbol{r}\times\dot{\boldsymbol{r}})].$$

With the help of the identity derived in the previous question, this expression can be rewritten as: $\dot{\boldsymbol{p}}\times\boldsymbol{\sigma} = -mf(r)r^2 d(\boldsymbol{r}/r)/dt$. Moreover, the angular momentum is a constant of the motion so that $\dot{\boldsymbol{\sigma}} = 0$; this property allows us to write $\dot{\boldsymbol{p}}\times\boldsymbol{\sigma} = d(\boldsymbol{p}\times\boldsymbol{\sigma})/dt$. In summary, we obtain the following important formula, valid whatever the form of the potential:

$$\frac{d}{dt}(\boldsymbol{p}\times\boldsymbol{\sigma}) = -mf(r)r^2\frac{d}{dt}\left(\frac{\boldsymbol{r}}{r}\right).$$

5. Let us now consider the special case of Kepler's potential $V(r) = -K/r$, which leads to $-mf(r)r^2 = mK$. The equality of the last question can be recast in a very pleasing form $d[\boldsymbol{p}\times\boldsymbol{\sigma} - mK\boldsymbol{r}/r]/dt = 0$. This implies that the vector between brackets is a constant of the motion:

$$\boldsymbol{p}\times\boldsymbol{\sigma} - mK\frac{\boldsymbol{r}}{r} = \boldsymbol{C}.$$

This vector is called the Runge–Lenz vector. It is represented in Fig. 4.9.

6. Let us calculate $\boldsymbol{C}\cdot\boldsymbol{\sigma} = [\boldsymbol{p},\boldsymbol{\sigma},\boldsymbol{\sigma}] - mK(\boldsymbol{r}\cdot\boldsymbol{\sigma})/r$. The first term is a mixed product which vanishes. Using the definition for the angular momentum, one has further $\boldsymbol{r}\cdot\boldsymbol{\sigma} = [\boldsymbol{r},\boldsymbol{r},\boldsymbol{p}]$, which is again a vanishing mixed product. Consequently $\boldsymbol{C}\cdot\boldsymbol{\sigma} = 0$. This relation implies that $\boldsymbol{C}$ is in the plane perpendicular to $\boldsymbol{\sigma}$ which is precisely the plane containing the orbit. Thus the Runge–Lenz vector is contained in the plane of the orbit, i.e.

$$\boldsymbol{C}\cdot\boldsymbol{\sigma} = 0.$$

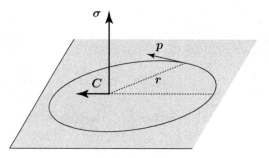

Fig. 4.9 The Runge–Lenz vector C is represented in the plane of the elliptic trajectory. It is directed along the perihelion. The radius vector and the linear momentum are also represented

7. In the plane of the orbit, let us take as the reference axis for the polar angle the direction of the Runge–Lenz vector and let us calculate

$$C \cdot r = Cr \cos \phi = [r, p, \sigma] - mK\frac{r^2}{r}.$$

The last term is simply $-mKr$ whereas the first one can be rewritten as $[r, p, \sigma] = (r \times p) \cdot \sigma = \sigma^2$.

Grouping these conclusions, we obtain the relation:

$$Cr \cos \phi = \sigma^2 - mKr.$$

8. This last equality can be written in the alternative form

$$r = \frac{\sigma^2}{mK}\left[1 + \frac{C}{mK}\cos\phi\right]^{-1}.$$

Comparing this expression to the equation of the trajectory in polar coordinates $r = (\sigma^2/mK)[1 + e\cos\phi]^{-1}$, we can make two conclusions. First the minimum radius is obtained for $\phi = 0$, that is along C; in other words, the Runge–Lenz vector is directed along the perihelion. Second, we have a very simple relationship between the modulus of the Runge–Lenz vector C and the orbit eccentricity e:

$$C = mKe.$$

9. One has to find two relations between C, E, and σ. The first one was already found in passing; it is:

$$C \cdot \sigma = 0.$$

This is a scalar equation which provides one relation. The second relation is obtained by considering the modulus C as a function of the eccentricity (see previous question): $C^2 = m^2 K^2 e^2$.

Furthermore, the eccentricity of the orbit is a function of the energy and angular momentum through the formula $e^2 = 1 + (2E\sigma^2)/(mK^2)$. This equality is used to obtain the second relation:

$$C^2 = m^2 K^2 + 2mE\sigma^2.$$

4.6. Quicker and More Ecologic than a Plane [Statement and Figure p. 174]

1. For a uniform mass distribution ρ, Gauss's theorem[11] tells us that the gravitational field is identical with that of a single mass located at the center and contained in the control sphere. In other word, the gravitational field at a distance r from the center is radial and takes the value

$$G(r) = \frac{GM(r)}{r^2},$$

where $M(r)$ is the mass contained in the sphere of radius r. In particular, just at the surface, it must be identified with the traditional gravitational field $g = GM_E/R_E^2$.

Inside the Earth

$$M(r) = \frac{4\pi}{3}\rho r^3, \quad \text{and thus} \quad G(r) = \frac{4\pi}{3}\rho G r$$

or, after replacement of the mass density in terms of the Earth's mass and the gravitational value at ground level,

$$G(r) = gr/R_E.$$

2. The gravitational potential is obtained from $dV_p/dr = G(r)$, which implies the relation: $V_p(r) = \frac{g}{2R_E} r^2$. The mechanical potential acting on an object with mass m is simply $V(r) = mV_p(r)$, or:

$$V(r) = \frac{mg}{2R_E} r^2.$$

3. Let O be the center of the Earth and H the middle of the tunnel AB. The triangle AOB being isosceles, H is also the extremity of the perpendicular height from O. Let us set $OH = R$. At a given time, the coach at point M in the tunnel is specified by its abscissa $x = \overline{HM}$ which is chosen as

---

[11] The gravitational interaction law being identical to Coulomb's law, Gauss's theorem can be applied in the same conditions.

a generalized coordinate. Obviously, one has $r = OM = \sqrt{R^2 + x^2}$, a relation that allows us to express the potential as a function of x:

$$V(x) = \frac{mgR^2}{2R_E} + \frac{mgx^2}{2R_E}.$$

The term $mgR^2/(2R_E)$ is an uninteresting constant. Let us consider only $V(x) = mgx^2/(2R_E)$; this potential describes a harmonic oscillator. The kinetic energy of the coach is $T = \frac{1}{2}m\dot{x}^2$ and the Lagrangian $L = \frac{1}{2}m\dot{x}^2 - V(x)$. The conjugate momentum is given by $p = \partial_{\dot{x}}L = m\dot{x}$, so that $\dot{x} = p/m$. The Hamiltonian is the Legendre transform of the Lagrangian $H = \dot{x}p - L$, namely:

$$H(x,p) = \frac{p^2}{2m} + \frac{mg}{2R_E}x^2.$$

The first Hamilton equation gives $\dot{x} = \partial_p H = p/m$, a relation encountered previously, and the second one $\dot{p} = -\partial_x H$, or $m\ddot{x} = -mgx/R_E$ which be rewritten, introducing the angular frequency $\omega = \sqrt{g/R_E}$ as:

$$\ddot{x} + \omega^2 x = 0.$$

4. For a movement starting at the extremity A of the tunnel, with a null velocity, the initial conditions are $x(0) = x_0 = HA$, $\dot{x}(0) = 0$. The previous differential equation can be integrated to give $x(t) = x_0 \cos(\omega t)$. This is a sinusoidal motion with period $T = 2\pi/\omega$ between the two ends of the tunnel. After a half-period, the coach reaches the other extremity of the tunnel. Thus the duration of the journey is simply $\tau = T/2 = \pi/\omega$. Explicitly:

$$\tau = \pi\sqrt{\frac{R_E}{g}}.$$

With the proposed data, one finds $\tau = 2{,}532\,\text{s} = 42\,\text{min}\ 12\,\text{s}$, independently of the tunnel length. A journey Paris–Tokyo in so short a time makes you wonder!

5. To calculate the tunnel length, the most convenient way is to work with the Cartesian coordinates of the two cities and to use the traditional formula for the distance between two points in a three-dimensional space. The length of the tunnel Paris-Tokyo is 8,838 km. In the middle of the tunnel the coach reaches the impressive speed of 5.48 km/s, or 19,740 km/h and passes 1,781 km under the Earth's surface. At such a depth, the heat is probably very oppressive; but it is true that we do not stay a long time!

4.7. Hamiltonian of a Charged Particle
[Statement p. 176]

1. Let us begin with the non-relativistic Hamilton function:

$$H(\boldsymbol{r},\boldsymbol{p},t) = \frac{1}{2m}\left(\boldsymbol{p} - q_e\boldsymbol{A}(\boldsymbol{r},t)\right)^2 + q_e U(\boldsymbol{r},t)$$

The first Hamilton equation gives $\dot{\boldsymbol{r}} = \partial_{\boldsymbol{p}} H = \left(\boldsymbol{p} - q_e \boldsymbol{A}(\boldsymbol{r},t)\right)/m$ or $\boldsymbol{\pi} = m\boldsymbol{v} = \boldsymbol{p} - q_e \boldsymbol{A}$. The second Hamilton equation gives $\dot{\boldsymbol{p}} = -\partial_{\boldsymbol{r}} H$.

We focus on the first component. Differentiate the first equation with respect to time to obtain

$$\dot{\pi}_x = \dot{p}_x - q_e \dot{A}_x = \dot{p}_x - q_e \boldsymbol{v}\cdot\boldsymbol{\nabla} A_x - q_e \partial_t A_x.$$

From the second Hamilton equation

$$\dot{p}_x = -\partial_x H = -q_e \partial_x U + q_e(\partial_x \boldsymbol{A})\cdot\frac{\boldsymbol{p}-q_e\boldsymbol{A}}{m} = -q_e \partial_x U + q_e(\partial_x \boldsymbol{A})\cdot\boldsymbol{v}.$$

In consequence,

$$\dot{\pi}_x = q_e\left[-\partial_x U - \partial_t A_x + (\partial_x \boldsymbol{A})\cdot\boldsymbol{v} - \boldsymbol{v}\cdot\boldsymbol{\nabla} A_x\right].$$

We note the appearance of the term $E_x = -\partial_x U - \partial_t A_x$. As for the term $(\partial_x \boldsymbol{A})\cdot\boldsymbol{v} - \boldsymbol{v}\cdot\boldsymbol{\nabla} A_x$, a simple calculation proves that it is equal to $v_y\left(\boldsymbol{\nabla}\times A\right)_z - v_z\left(\boldsymbol{\nabla}\times A\right)_y$ or $\left[\boldsymbol{v}\times\boldsymbol{B}\right]_x$. In summary, we obtain the result: $\dot{\pi}_x = q_e\left[E_x + (\boldsymbol{v}\times\boldsymbol{B})_x\right]$. After a cyclic permutation, one obtains a similar formula for the two other components, so that one can write in vectorial form:

$$\frac{d\boldsymbol{\pi}}{dt} = q_e\left[\boldsymbol{E} + \boldsymbol{v}\times\boldsymbol{B}\right].$$

This is the expected expression for Newton's equation of a charged particle submitted to the Lorentz force.

2. Let us consider now the relativistic expression for the Hamilton function

$$\left(H(\boldsymbol{r},\boldsymbol{p},t) - q_e U(\boldsymbol{r},t)\right)^2 - \left(\boldsymbol{p} - q_e\boldsymbol{A}(\boldsymbol{r},t)\right)^2 c^2 = m^2 c^4,$$

or, equivalently

$$H(\boldsymbol{r},\boldsymbol{p},t) = q_e U(\boldsymbol{r},t) + \sqrt{\left(\boldsymbol{p} - q_e\boldsymbol{A}(\boldsymbol{r},t)\right)^2 c^2 + m^2 c^4}.$$

The first Hamilton equation gives:

$$\dot{\boldsymbol{r}} = \partial_{\boldsymbol{p}} H = c^2(\boldsymbol{p} - q_e\boldsymbol{A})/\sqrt{(\boldsymbol{p} - q_e\boldsymbol{A})^2 c^2 + m^2 c^4}.$$

A simple derivation shows that
$$\sqrt{(\boldsymbol{p} - q_e \boldsymbol{A})^2 c^2 + m^2 c^4} = H - q_e U = mc^2/\sqrt{1 - v^2/c^2}.$$
Consequently, the first Hamilton equation can be recast as
$$\boldsymbol{\pi} = m\boldsymbol{v}/\sqrt{1 - v^2/c^2} = \boldsymbol{p} - q_e \boldsymbol{A}.$$
The second Hamilton equation gives the relation $\dot{\boldsymbol{p}} = -\partial_r H$. Here again, let us focus on the x component. As in the previous question
$$\dot{\pi}_x = \dot{p}_x - q_e \dot{A}_x = \dot{p}_x - q_e \boldsymbol{v} \cdot \boldsymbol{\nabla} A_x - q_e \partial_t A_x.$$
Moreover, the second equation gives
$$\begin{aligned}\dot{p}_x = -\partial_x H &= -q_e \partial_x U + q_e c^2 \frac{(\partial_x \boldsymbol{A}) \cdot (\boldsymbol{p} - q_e \boldsymbol{A})}{\sqrt{(\boldsymbol{p} - q_e \boldsymbol{A})^2 c^2 + m^2 c^4}} \\ &= -q_e \partial_x U + q_e (\partial_x \boldsymbol{A}) \cdot \boldsymbol{v}.\end{aligned}$$
The rest of the treatment is essentially the same as in the study developed in the previous question and, thus, we are led to the same final equation
$$\frac{d\boldsymbol{\pi}}{dt} = q_e \left[\boldsymbol{E} + \boldsymbol{v} \times \boldsymbol{B} \right].$$

3. In the previous question, we demonstrated the two relations (with the usual convention $\gamma = \left(1 - \dot{q}^2/c^2\right)^{-1/2}$). From now on, E denotes the energy.
$$E = \gamma mc^2 + q_e U; \qquad \boldsymbol{\pi} = \gamma m \dot{\boldsymbol{q}}.$$
Moreover the proper time τ is given by $d\tau = \sqrt{1 - \dot{q}^2/c^2}\, dt$, whence $\gamma = dt/d\tau$. The first equation, concerning the energy, can be written as $d(ct)/d\tau = (1/m)(E/c - q_e U/c)$. In other words, the equation providing the energy is the zeroth component of the required equation
$$\frac{dq^0}{d\tau} = \frac{p^0 - q_e A^0}{m}.$$
The second equation, concerning the momentum, can be expressed as
$$\frac{\boldsymbol{\pi}}{m} = \frac{\boldsymbol{p} - q_e \boldsymbol{A}}{m} = \frac{d\boldsymbol{q}}{d\tau}$$
which corresponds to the spatial part of the desired equation. In consequence, this equation holds in terms of quadrivectors:
$$\frac{dq^\mu}{d\tau} = \frac{p^\mu - q_e A^\mu}{m}.$$

The other equation is more cumbersome to derive; it relies on Hamilton's equations. In Question 2, we proved in passing that

$$\dot{p}_i = \dot{p}^i = -q_e \left(\partial_i U\right) + q_e \dot{\boldsymbol{q}} \cdot \left(\partial_i \boldsymbol{A}\right),$$

which can be rewritten, introducing the proper time, as $p'^i = -q_e \gamma \left(\partial_i U\right) + q_e \boldsymbol{q}' \cdot \left(\partial_i \boldsymbol{A}\right)$.

Let us work with the first term which is also

$$-q_e q'^0 \frac{\partial}{\partial x^i} \frac{U}{c} = q_e q'^0 \frac{\partial A_0}{\partial x^i}.$$

The sum of this term and the second one is expressed in the more compact form $q_e q'^\nu \partial(A_\nu)/\partial x_i$. Using the expression $q'^\nu = (p^\nu - q_e A^\nu)/m$, obtained just above, this equation can be recast as

$$\frac{dp^i}{d\tau} = \frac{q_e}{m}(p^\nu - q_e A^\nu)\frac{\partial A_\nu}{\partial x_i}$$

(with Einstein's summation convention) which is nothing more than the spatial part of the desired equation.

Let us begin now with $dH/dt = \partial H/\partial t$ and

$$H = q_e U + \sqrt{(\boldsymbol{p} - q_e \boldsymbol{A})^2 c^2 + m^2 c^4}.$$

We then obtain

$$\frac{dE}{d\tau} = q_e \frac{\partial U}{\partial \tau} - q_e c^2 \frac{(\boldsymbol{p} - q_e \boldsymbol{A}) \cdot (\partial \boldsymbol{A}/\partial \tau)}{\sqrt{(\boldsymbol{p} - q_e \boldsymbol{A})^2 c^2 + m^2 c^4}}.$$

Consider the first term which is also

$$q_e c \gamma \frac{\partial A^0}{\partial t} = \frac{q_e}{m}\gamma m c^2 \frac{\partial A^0}{\partial q^0}.$$

Owing to $\gamma m c^2 = E - q_e U$ and $(\partial A^0/\partial q^0) = (\partial A_0/\partial q_0)$, this term is rewritten as $(q_e/m)(E - q_e U)(\partial A_0/\partial q_0)$ or $(q_e c/m)(p^0 - q_e A^0)(\partial A_0/\partial q_0)$.

As for the second term, we use the fact that

$$\sqrt{(\boldsymbol{p} - q_e \boldsymbol{A})^2 c^2 + m^2 c^4} = \gamma m c^2$$

to transform it into the form

$$-\frac{q_e}{m}(\boldsymbol{p} - q_e \boldsymbol{A})\frac{\partial A^0}{\partial t} = \frac{q_e c}{m}(\boldsymbol{p} - q_e \boldsymbol{A})\frac{\partial \boldsymbol{A}}{\partial q_0}.$$

Summing both contributions and dividing by c, we finally obtain

$$\frac{dp^0}{d\tau} = \frac{q_e}{m}(p^\nu - q_e A^\nu)\frac{\partial A_\nu}{\partial q_0}$$

which is precisely the temporal part of the desired equation. Finally, the second equation can be expressed in terms of quadrivectors as:

$$\frac{dp^\mu}{d\tau} = q_e \sum_\nu \frac{p^\nu - q_e A^\nu}{m}\frac{\partial A_\nu}{\partial q_\mu}.$$

4. In order to calculate the derivatives, it is simpler to take the Lagrangian in the form

$$L = \sum_\nu \left[\frac{1}{2}mg_{\nu\nu}q'^\nu q'^\nu + q_e g_{\nu\nu}q'^\nu A^\nu\right].$$

After differentiation, one has $\partial L/\partial q'^\mu = mg_{\mu\mu}q'^\mu + q_e g_{\mu\mu}A^\mu$, or, introducing the covariant components

$$p_\mu = m\frac{dq_\mu}{d\omega} + q_e A_\mu.$$

If we rewrite this equation in the form $dq_\mu/d\omega = (p_\mu - q_e A_\mu)/m$, it can be identified with the first equation of Question 3 (in terms of covariant components) provided that the parameter ω is chosen as the proper time τ.

5. The Hamiltonian is defined as the Legendre transform of the Lagrangian

$$H = \sum_\nu p_\nu q'^\nu - L$$

or, with the proposed definition of the Lagrangian

$$H = \sum_\nu (mq'_\nu + q_e A_\nu) q'^\nu - \sum_\nu \left(\frac{m}{2}q'_\nu q'^\nu + q_e A_\nu q'^\nu\right) = \sum_\nu \left(\frac{m}{2}q'_\nu q'^\nu\right).$$

Substituting velocities by momenta, we obtain the final form of Hamilton's function:

$$H(q^\mu, p^\mu) = \sum_\mu \frac{(p_\mu - q_e A_\mu)(p^\mu - q_e A^\mu)}{2m}.$$

6. The first Lagrange equation is written in a covariant form as $q'^\mu = \partial H/\partial p_\mu = (p^\mu - q_e A^\mu)/m$ which is analogous to the first equation of Question 3.

The second Hamilton equation is written in the same manner

$$p'^\mu = -\frac{\partial H}{\partial q_\mu} = \sum_\nu \frac{g^{\nu\nu}}{m}(p_\nu - q_e A_\nu)q_e \frac{\partial A_\nu}{\partial q_\mu} = \sum_\nu \frac{q_e}{m}(p^\nu - q_e A^\nu)\frac{\partial A_\nu}{\partial q_\mu},$$

which is analogous to the second equation of Question 3. In summary:

$$\frac{dq^\mu}{d\omega} = \frac{p^\mu - q_e A^\mu}{m};$$
$$\frac{dp^\mu}{d\omega} = q_e \sum_\nu \frac{p^\nu - q_e A^\nu}{m}\frac{\partial A_\nu}{\partial q_\mu}.$$

Using the first Hamilton equation, the Hamiltonian on the trajectory can be written, in an alternative form, as

$$\sum_\mu \frac{1}{2}mq'^\mu q'_\mu = -\frac{1}{2}m\frac{c^2 dt^2 - d\mathbf{q}^2}{d\omega^2} = -\frac{1}{2}mc^2 \frac{d\tau^2}{d\omega^2}.$$

If the parameter ω is identified with the proper time τ, then $d\tau/d\omega = 1$ and the constant value of the Hamiltonian along the trajectory is simply:

$$H = -\frac{1}{2}mc^2.$$

4.8. The First Integral Invariant
[Statement p. 177]

1. In the $2n$-dimensional phase space, the system is at the point $A(q,p)$ at time t. It drifts with the flow and reaches the point $A'(q',p')$ at time $t + dt$. Obviously, one has $q' = q + \dot{q}\,dt = q + \partial_p H(A)\,dt$ and $p' = p + \dot{p}\,dt = p - \partial_q H(A)\,dt$. Remember that these equations are valid for any components (q_i, p_i).

Let $B(q + \varepsilon, p + \eta)$ be a point very close to A at time t. It drifts with the flow and reaches the point $B'(q' + \varepsilon', p' + \eta')$ at time $t + dt$. Similarly, one has $q'_i + \varepsilon'_i = q_i + \varepsilon_i + \partial_{p_i} H(B)\,dt$ and $p'_i + \eta'_i = p_i + \eta_i - \partial_{q_i} H(B)\,dt$.

Furthermore, one has also

$$\partial_{p_i} H(B) = \partial_{p_i} H(q + \varepsilon, p + \eta)$$
$$= \partial_{p_i} H(A) + \sum_j \left[\varepsilon_j \partial^2_{q_j p_i} H(A) + \eta_j \partial^2_{p_j p_i} H(A)\right]$$

and similarly

$$\partial_{q_i} H(B) = \partial_{q_i} H(q + \varepsilon, p + \eta)$$
$$= \partial_{q_i} H(A) + \sum_j \left[\varepsilon_j \partial^2_{q_j q_i} H(A) + \eta_j \partial^2_{p_j q_i} H(A)\right].$$

Taking into account Hamilton's equations $q_i' = q_i + \partial_{p_i} H(A)\, dt$ and $p_i' = p_i - \partial_{q_i} H(A)\, dt$, we can summarize the situation by the following expressions:

$$\varepsilon_i' = \varepsilon_i + dt \sum_j \left[\varepsilon_j \partial^2_{q_j p_i} H + \eta_j \partial^2_{p_j p_i} H\right];$$

$$\eta_i' = \eta_i - dt \sum_j \left[\varepsilon_j \partial^2_{q_j q_i} H + \eta_j \partial^2_{p_j q_i} H\right].$$

2. Let us choose now another point $C(q+\gamma, p+\delta)$ also very close to point A at time t. It drifts with the flow and reaches the point $C'(q'+\gamma', p'+\delta')$ at time $t + dt$.

Obviously we have relations concerning γ, δ which are completely identical to those concerning ε, η. From the projection a, b, c of points A, B, C and the projection a', b', c' of points A', B', C' onto a particular plane (q_i, p_i), one can build a small parallelogram the area of which, $\boldsymbol{ab} \times \boldsymbol{ac}$, drifts with the flow to become $\boldsymbol{a'b'} \times \boldsymbol{a'c'}$. The corresponding situation is depicted in Fig. 4.10.

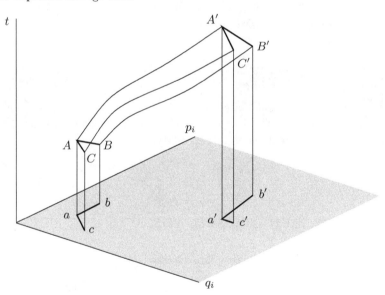

Fig. 4.10 Evolution, in the Hamiltonian flow, of three close points. In a particular plane (p_i, q_i), we also represent the projections of the original points and of the points transformed by the flow

The oriented area of this parallelogram takes the value $\varepsilon_i \delta_i - \eta_i \gamma_i$ and, after the drift, becomes $\varepsilon_i' \delta_i' - \eta_i' \gamma_i'$. Substituting in the latter expression the relations of Question 1, one notices that this particular area is not an invariant. On the contrary, if one sums over all the planes, an elementary algebraic calculation with judicious changes in the indices convinces us that, to first order in dt,

$$\sum_i (\varepsilon'_i \delta'_i - \eta'_i \gamma'_i) = \sum_i (\varepsilon_i \delta_i - \eta_i \gamma_i).$$

Indeed there is invariance for the **sum of projections of oriented areas** on every plane

$$\sum_i dA'_i = \sum_i dA_i.$$

In the presence of a finite domain, we can install a mesh of elementary small disjointed domains for which the previous equation is valid. Since the total area of the domain is the sum of the areas of each domain which forms the partition, the invariance property for an elementary domain can be extended to the whole domain:

$$\sum_i A'_i = \sum_i A_i.$$

Furthermore, the expression of the area can be also written as

$$A_i = \oint_{\Gamma_i} p_i dq_i,$$

where the integral is calculated on the border Γ_i of the projection of the domain onto the plane (q_i, p_i) so that, finally, the conservation of the projected areas is written as

$$\sum_i \oint_{\Gamma_i} p_i dq_i = \text{const},$$

which is recast in the symbolic simple form

$$\oint_\Gamma p \cdot dq = \text{const}.$$

4.9. What About Non-Autonomous Systems?
[Statement p. 178]

1. Since the system is not autonomous its Hamiltonian depends on time: $H(q, p, t)$. We therefore consider the time t as a full generalized coordinate so that there exists a conjugate variable: p_t. Let us define now a new

Hamilton function by $\tilde{H}(q,t,p,p_t) = H(q,p,t) + p_t$ and denote by τ the flow variable. This new Hamiltonian does not depend on τ; it represents a conservative system. Now, let us write down the corresponding Hamilton equations:

$$dq/d\tau = \partial_p \tilde{H} = \partial_p H$$
$$dt/d\tau = \partial_{p_t} \tilde{H} = 1$$
$$dp/d\tau = -\partial_q \tilde{H} = -\partial_q H$$
$$dp_t/d\tau = -\partial_t \tilde{H} = -\partial_t H.$$

The second equation shows clearly that the flow parameter can be identified with the time $\tau = t$. The first one and the third one represent the Hamilton equations for the original Hamiltonian. Introducing the energy function E via its traditional formula, the last equation reads $\dot{p}_t = -\partial_t H = -dE/dt$; this proves that one can interpret the variable, p_t, as the negative of the energy:

$$p_t = -E.$$

2. As the system described by $\tilde{H}$ is time independent, there exists a constant of the motion which is the Hamiltonian itself: $\tilde{H} = const$. Furthermore, the original Hamiltonian represents the energy (which is not a constant of the motion) $H = E$; thus $\tilde{H} = H + p_t$ and, owing to the previous results, one can write $\tilde{H} = E - E = 0$. Thus the constant of the motion is simply null:

$$\tilde{H} = 0.$$

4.10. The Reverse Pendulum
[Statement and Figure p. 178]

1. In the Galilean frame XOZ (OZ is the upward vertical) for the oscillation of the pendulum, the coordinates of the pendulum are: $X = l\sin\theta$, $Z = z + l\cos\theta$. The corresponding kinetic energy $T = \frac{1}{2}m(\dot{X}^2 + \dot{Z}^2)$ is deduced:

$$T = \frac{1}{2}m(l^2\dot{\theta}^2 - 2l\dot{z}\dot{\theta}\sin\theta + \dot{z}^2).$$

The potential energy is $V = mgZ = mg(z + l\cos\theta)$ and the Lagrangian $L = T - V$. The Lagrange equation provides the required relation:

$$l\ddot{\theta} = (\ddot{z} + g)\sin\theta.$$

This result is obvious. Indeed, one can work in the non-Galilean frame of the pendulum and add to the gravitational potential the inertial potential. The system seems to be subject to an apparent gravity $\ddot{z} + g$ directed downwards.

With the temporal law imposed from outside the system, the acceleration is constant $\ddot{z} = \pm a$. When the acceleration is directed upward $\ddot{z} = a$ (first half-period), the apparent force of gravity is larger than the natural gravity and the upper equilibrium position ($\theta = 0$) is even more unstable. When the acceleration is directed downward $\ddot{z} = -a$ and is larger than the natural gravity. One has an apparent force of gravity directed upward so that the upper equilibrium position becomes stable. Stability occurs only if the condition $a > g$ is satisfied. In this case, the condition for stability requires a more refined calculation.

First case: the acceleration is directed upward ($\ddot{z} = a$)

2. We consider the first half-period, from $t = 0$ to $t = T$. Let us set $\omega^2 = (a+g)/l = \omega_0^2(a/g+1)$. For a motion with a small amplitude ($\sin\theta \cong \theta$), the Lagrange equation is written as $\ddot{\theta} = \omega^2 \theta$. It exhibits an exponential behaviour characteristic of an unstable state and it is easily integrated, taking into account the initial conditions:

$$\theta(t) = \theta_0 \cosh(\omega t) + (\dot{\theta}_0/\omega)\sinh(\omega t).$$

This allows us to obtain the propagator matrix over the first half-period:

$$\begin{pmatrix}\theta\\ \dot{\theta}\end{pmatrix}_T = \begin{pmatrix}\cosh(\omega t) & \sinh(\omega t)/\omega\\ \omega\sinh(\omega t) & \cosh(\omega t)\end{pmatrix}\begin{pmatrix}\theta\\ \dot{\theta}\end{pmatrix}_0.$$

It can be easily checked that the determinant of this matrix is unity, a property that implies the conservation of the area in phase space (Liouville's theorem).

3. At the moment of change of the law, there is neither a change of position nor of velocity and, consequently, no force so that the initial conditions for this second study are precisely the final conditions of the previous case.

Second case: the acceleration is directed downward ($\ddot{z} = -a$)

We are now in the second half-period from $t = T$ to $t = 2T$. We set $\Omega^2 = (a-g)/l = \omega_0^2(a/g - 1)$. The Lagrange equation is written this time as $\ddot{\theta} = -\Omega^2\theta$ which exhibits an oscillating behaviour characteristic of a stable state. This equation is easily integrated and provides the desired relation using the initial condition.

This relation can be expressed in a form involving the propagator matrix:

$$\begin{pmatrix}\theta\\ \dot\theta\end{pmatrix}_1 = \begin{pmatrix}\theta\\ \dot\theta\end{pmatrix}_{2T} = \begin{pmatrix}\cos(\Omega t) & \sin(\Omega t)/\Omega\\ -\Omega\sin(\Omega t) & \cos(\Omega t)\end{pmatrix}\begin{pmatrix}\theta\\ \dot\theta\end{pmatrix}_T.$$

In this case also, the conservation of the area can be verified.

4. Using the two previous questions, one sees that the transformation over a full period can be written in the matrix form:

$$\begin{pmatrix}\theta\\ \dot\theta\end{pmatrix}_1 = K\begin{pmatrix}\theta\\ \dot\theta\end{pmatrix}_0.$$

The propagator K is simply the product of the two previous matrices. Of course, its determinant is unity which implies conservation of the area. Explicitly, we have:

$$K = \begin{pmatrix} \cosh(\omega T)\cos(\Omega T) + \dfrac{\omega}{\Omega}\sinh(\omega T)\sin(\Omega T) & \dfrac{1}{\omega}\sinh(\omega T)\cos(\Omega T) + \dfrac{1}{\Omega}\cosh(\omega T)\sin(\Omega T) \\ \omega\sinh(\omega T)\cos(\Omega T) - \Omega\cosh(\omega T)\sin(\Omega T) & \cosh(\omega T)\cos(\Omega T) - \dfrac{\Omega}{\omega}\sinh(\omega T)\sin(\Omega T) \end{pmatrix}.$$

5. In order to specify the stability conditions, one must rely on the eigenvalues of the propagator. If both eigenvalues are real, there is instability. Otherwise there is stability. The condition for stability corresponds to the relation $|\text{Tr}(K)| < 2$. It is thus necessary to calculate the trace of the matrix that can be written in the form:

$$\text{Tr}(K) = 2\cosh(\omega T)\cos(\Omega T) + (\omega/\Omega - \Omega/\omega)\sinh(\omega T)\sin(\Omega T).$$

Let us set $2u = (\omega/\Omega - \Omega/\omega) = \omega_0^2/(\omega\Omega)$. This quantity is always positive and tends to 0 when $a/g \to \infty$. It is interesting to investigate the regions of stability as a function of two independent dimensionless parameters which we choose as $\omega_0 T$ for abscissa and a/g for ordinate.

When $T = 0$, one has $\text{Tr}(K) = 2$ and as soon as T increases, the second term in the trace increases very rapidly, a situation that precludes the stability of the system. At least for large values of a/g, a way to attain stability is to impose a null value for the first term; this condition implies $\Omega T = (n + 1/2)\pi$.

One can go further seeking, in the plane defined by the parameters, the curves such that $\text{Tr}(K) = \pm 2$. The stability region is comprised between them.

Let examine for instance the condition $\mathrm{Tr}(K) = 2$ which can be rewritten as

$$\cos(\Omega T) + u \tanh(\omega T) \sin(\Omega T) = \frac{1}{\cosh(\omega T)}.$$

Introducing an angle $\alpha = \arctan\left(u \tanh(\omega T)\right)$ the previous equation becomes $\cos(\Omega T - \alpha) = \cos(\alpha)/\cosh(\omega T)$, or

$$\Omega T = (n + 1/2)\pi + \alpha + \arccos\left[\frac{\cos(\alpha)}{\cosh(\omega T)}\right].$$

This transcendental equation can be solved by successive iterations. Given a value for a/g, we start with a zero order approximation (region $n = 0$) $\Omega T^{(0)} = \pi/2$; we deduce $(\omega_0 T)^{(0)}$, calculate $\omega^{(0)}$, $\alpha^{(0)}$ then the new value for the first iteration $\Omega T^{(1)}$ with the help of the transcendental equation. We repeat this iterative procedure, improving, at each step, the value ΩT up to a value that does not change with successive iterations. We thus obtain the value $\omega_0 T$ associated with the initial value of a/g.

We then change the value of a/g and repeat the iterative procedure, which converges very rapidly in practice. In this way, we obtain, in the plane of the parameters, the curve $\omega_0 T$ as a function of a/g which fulfills the condition $\mathrm{Tr}(K) = 2$. The other regions are built starting instead from the zeroth order iteration $\Omega T^{(0)} = (n + 1/2)\pi$, $n = 1, 2, \ldots$

The result is presented in the Fig. 4.11.

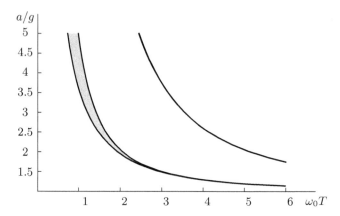

Fig. 4.11 Arnold tongue corresponding to the stability region for a reverse pendulum: the first region is comprised between the two adjacent curves on the left, and another zone on the right is restricted to within the thickness of the line

Problem Solutions

4.11. The Paul Trap
[Statement p. 180]

1. In the absence of a vector potential, the Hamilton function is written: $H(\boldsymbol{r},\boldsymbol{p},t) = \boldsymbol{p}^2/(2m) + q_e U(\boldsymbol{r},t)$. The corresponding Hamilton equations are: $\dot{\boldsymbol{r}} = \partial_p H = \boldsymbol{p}/m$ and $\dot{\boldsymbol{p}} = -\partial_r H = -q_e \nabla U$. Explicitly, we have for each component:

$$p_x = m\dot{x}, \quad p_y = m\dot{y}, \quad p_z = m\dot{z}$$

and
$$\dot{p}_x = -q_e U_0\, x\, \cos(\omega t), \quad \dot{p}_y = -q_e U_0\, y\, \cos(\omega t),$$
$$\dot{p}_z = 2 q_e U_0\, z\, \cos(\omega t).$$

After differentiation of the first set and use of the second one, we obtain the equations of motion for the particle:

$$m\ddot{x} = -q_e U_0\, x\, \cos(\omega t);$$
$$m\ddot{y} = -q_e U_0\, y\, \cos(\omega t);$$
$$m\ddot{z} = 2 q_e U_0\, z\, \cos(\omega t).$$

2. The equations are decoupled, but they are differential equations of Mathieu's type, and therefore not so easy to handle. With the proposed simplification, one substitutes, for the first half-period, $U_0 \cos(\omega t)$ by U_0. Let us set $\tilde{\omega}^2 = (q_e U_0)/m$ and $\Omega^2 = (2 q_e U_0)/m = 2\tilde{\omega}^2$; the equations of motion are then simplified and can be written as

$$\ddot{x} + \tilde{\omega}^2 x = 0;$$
$$\ddot{y} + \tilde{\omega}^2 y = 0;$$
$$\ddot{z} - \Omega^2 z = 0.$$

First, let us study the equation concerning the x variable; the general solution is $x(t) = A\cos(\tilde{\omega}t) + B\sin(\tilde{\omega}t)$. One has also $p_x = m\dot{x}$. The initial conditions for the position and the momentum lead to the determination of the unknown constants A and B. Thus $x(t) = x(0)\cos(\tilde{\omega}t) + p_x(0)/(m\tilde{\omega})\sin(\tilde{\omega}t)$ and $p_x(t) = p_x(0)\cos(\tilde{\omega}t) - m\tilde{\omega}\, x(0)\sin(\tilde{\omega}t)$. To obtain the expression of the position and momentum after a half-period, it is sufficient to substitute $t = T/2 = \pi/\omega$ in the previous expressions.

The phase $\varphi = \pi\tilde{\omega}/\omega$ appears naturally. It is then possible to write the variables relative to the half-period in terms of those relative to the time origin in the matrix form:

$$\begin{pmatrix} x \\ p_x \end{pmatrix}_{T/2} = \begin{pmatrix} \cos\varphi & \dfrac{1}{m\tilde{\omega}}\sin\varphi \\ -m\tilde{\omega}\sin\varphi & \cos\varphi \end{pmatrix} \begin{pmatrix} x \\ p_x \end{pmatrix}_0.$$

Of course, we have a similar equation for the y component.

Let us now study the equation concerning the z variable. The general solution is $z(t) = A \cosh(\Omega t) + B \sinh(\Omega t)$. A calculation analogous to the previous one leads to: $z(t) = z(0) \cosh(\Omega t) + p_z(0)/(m\Omega) \sinh(\Omega t)$ and $p_z(t) = p_z(0) \cosh(\Omega t) + m\Omega z(0) \sinh(\Omega t)$. In order to obtain the variable relative to the half-period, it is sufficient to substitute $t = T/2$ in the previous expressions. The phase $\Phi = \pi\Omega/\omega = \varphi\sqrt{2}$ naturally appears. We have the following matrix relation:

$$\begin{pmatrix} z \\ p_z \end{pmatrix}_{T/2} = \begin{pmatrix} \cosh \Phi & \dfrac{1}{m\Omega} \sinh \Phi \\ m\Omega \sinh \Phi & \cosh \Phi \end{pmatrix} \begin{pmatrix} z \\ p_z \end{pmatrix}_0.$$

In both cases, one easily checks that the determinant of the propagator matrix is unity, so that there is conservation of the area in phase space.

For the other half-period, one substitutes $U_0 \cos(\omega t)$ by $-U_0$; the equations of motion in this case are expressed as:

$$\ddot{x} - \tilde{\omega}^2 x = 0;$$
$$\ddot{y} - \tilde{\omega}^2 y = 0;$$
$$\ddot{z} + \Omega^2 z = 0.$$

A treatment completely analogous to the previous one (with the obvious translation by $T/2$ for the time origin) leads to the propagator matrix relative to the x variable:

$$\begin{pmatrix} x \\ p_x \end{pmatrix}_T = \begin{pmatrix} \cosh \varphi & \dfrac{1}{m\tilde{\omega}} \sinh \varphi \\ m\tilde{\omega} \sinh \varphi & \cosh \varphi \end{pmatrix} \begin{pmatrix} x \\ p_x \end{pmatrix}_{T/2}$$

and to a similar relation for the y variable. As for the propagator matrix relative to the z variable, it is:

$$\begin{pmatrix} z \\ p_z \end{pmatrix}_T = \begin{pmatrix} \cos \Phi & \dfrac{1}{m\Omega} \sin \Phi \\ -m\Omega \sin \Phi & \cos \Phi \end{pmatrix} \begin{pmatrix} z \\ p_z \end{pmatrix}_{T/2}.$$

Here too, the determinant of the propagator matrix is unity, and there is conservation of the area in phase space.

3. In order to obtain the propagator matrix over a complete period, we have only to calculate the product of propagator matrices relative to each of the half-periods. The calculation is straightforward.

One finds for the x component (with a similar relation for the y component):

$$\begin{pmatrix} x \\ p_x \end{pmatrix}_T =$$

$$\begin{pmatrix} \cosh\varphi\cos\varphi - \sinh\varphi\sin\varphi & \dfrac{\cosh\varphi\sin\varphi + \sinh\varphi\cos\varphi}{m\tilde{\omega}} \\ m\tilde{\omega}(-\cosh\varphi\sin\varphi + \sinh\varphi\cos\varphi) & \cosh\varphi\cos\varphi + \sinh\varphi\sin\varphi \end{pmatrix} \begin{pmatrix} x \\ p_x \end{pmatrix}_0$$

and for the z component

$$\begin{pmatrix} z \\ p_z \end{pmatrix}_T =$$

$$\begin{pmatrix} \cosh\Phi\cos\Phi + \sinh\Phi\sin\Phi & \dfrac{\cosh\Phi\sin\Phi + \sinh\Phi\cos\Phi}{m\Omega} \\ m\Omega(-\cosh\Phi\sin\Phi + \sinh\Phi\cos\Phi) & \cosh\Phi\cos\Phi - \sinh\Phi\sin\Phi \end{pmatrix} \begin{pmatrix} x \\ p_x \end{pmatrix}_0$$

4. To achieve stability (elliptic point in phase space), an arbitrary propagator matrix must have complex eigenvalues which is equivalent, in the case of the Hamiltonian under consideration (conservation of area), to the condition $|\text{Tr}(K)| < 2$. In the case of the x component, $\text{Tr}(K) = 2\cosh\varphi\cos\varphi$ and the stability condition becomes: $|\cosh\varphi\cos\varphi| < 1$ (with an equivalent relation for the y variable).

For the z component, $\text{Tr}(K) = 2\cosh\Phi\cos\Phi$ and the stability condition becomes: $|\cosh\Phi\cos\Phi| < 1$. In order for the orbit of the particle to remain confined, stability must hold obviously for the three axes. This implies the simultaneous inequalities:

$$|\cosh\varphi\,\cos\varphi| < 1$$
$$|\cosh(\varphi\sqrt{2})\,\cos(\varphi\sqrt{2})| < 1.$$

5. One can plot the curve $|\cosh\varphi\,\cos\varphi|$ as a function of φ. It is represented in the Fig. 4.12. Between 0 (value 1) and $\pi/2$ (value 0), it is always less than 1 and the condition is fulfilled. Beyond $\pi/2$, the curve begins to increase. It takes the value 1 for a particular value φ_0, solution of the transcendental equation $|\cosh\varphi_0\,\cos\varphi_0| = 1$. Numerically, one finds $\varphi_0 \cong 1.8751$. To summarize, one has a confined solution, if we fulfill the simultaneous conditions $0 < \varphi < \varphi_0$ and $0 < \varphi < \varphi_0/\sqrt{2}$. In fact, these two conditions are equivalent to the second one only. Substituting for φ by its value in terms of the physical parameters, we obtain the condition:

$$\frac{2q_e U_0}{m\omega^2} < 0.3562$$

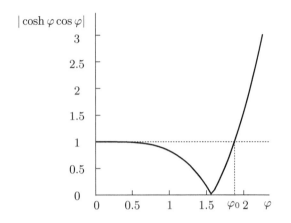

Fig. 4.12 Curve $|\cosh\varphi \cos\varphi|$ and value 1 which provides the special value $\phi_0 = 1.8751$ discussed in the text

This result does not differ significantly from a more rigorous calculation based on Mathieu's equations, for which the previous numerical value is 0.4539.

4.12. Optical Hamilton's Equations
[Statement p. 181]

1. The iso-index surfaces are cylinders with axes parallel to Oz; the trajectory of a ray between point A and point B is described by an equation $\rho(z)$ determined by the condition that the optical path $L_{\mathrm{opt}} = \int_A^B n(\rho)\,ds$ is an extremum. We are thus concerned with a classical variational problem, the optical path being a functional of the trajectory.

    One assumes that point B lies in the plane defined by the Oz axis and point A. In this case, the problem is two-dimensional. The length element is given as usual by $ds^2 = dz^2 + d\rho^2$, which implies $ds = \sqrt{1+\rho'^2}\,dz$. The optical path can thus be written in a form that is traditional for the action formalism, the z variable playing the role of the time. In this case, the "Lagrangian" is:

$$L(\rho, \rho') = n(\rho)\sqrt{1+\rho'^2}.$$

2. As usual, the angle of incidence i is defined as the angle between the direction of the ray and the normal to the iso-index surface (here the direction perpendicular to the Oz axis). It is easily seen that $\sin i = dz/ds$ and $\cos i = d\rho/ds$. Owing to the expression ds given above, we have:

$$\sin i(\rho) = \frac{1}{\sqrt{1+\rho'^2}}; \qquad \cos i(\rho) = \frac{\rho'}{\sqrt{1+\rho'^2}}.$$

3. The conjugate momentum to the ρ variable is defined by $p = \partial_{\rho'} L$. With the expression for the Lagrangian obtained in the first question, it is easy to obtain:

$$p = n(\rho) \cos i(\rho) = \frac{n(\rho)\rho'}{\sqrt{1+\rho'^2}}.$$

4. The Hamilton function is the Legendre transform of the Lagrangian: $H = p\rho' - L$. After a simple calculation based on the result of the first question and elimination of ρ' in terms of p with the relation given in Question 3

$$H(\rho, p) = -\sqrt{n(\rho)^2 - p^2}.$$

5. The Hamilton function does not depend explicitly on z. We know that, in this case, it remains constant along the trajectory. This property implies that $n(\rho)^2 - p^2 = \text{const}$, or, following Question 3, $n(\rho)^2 \sin^2 i(\rho) = \text{const}$, that is:

$$n(\rho) \sin i(\rho) = \text{const}$$

which is nothing more than the well known Snell–Descartes law.

6. With our conventions, Hamilton's equations are written as: $\rho' = \partial_p H$ and $p' = -\partial_\rho H$. Using the expression of the Hamiltonian given in Question 4, one deduces Hamilton's equations:

$$\rho' = \frac{p}{\sqrt{n(\rho)^2 - p^2}};$$

$$p' = \frac{n(\rho) n'(\rho)}{\sqrt{n(\rho)^2 - p^2}}.$$

7. Substituting $n(\rho) = n_0 - a\rho^2$ and replacing the sine by its expression given in Question 2 in the constant of the motion deduced in Question 5, we arrive at the differential equation providing the trajectory $n_0 - a\rho^2 = C\sqrt{1+\rho'^2}$, where C is the integration constant, function of the initial conditions. More precisely, $C = n_0/\sqrt{1+\rho_0'^2}$. Since the ray deviates only slightly from the axis of revolution $\rho_0' \ll 1$, one can consider that $C \cong n_0$.

Let us differentiate the differential equation and simplify by ρ'; we obtain: $\rho'' + 2a\rho\sqrt{1+\rho'^2}/C = 0 = \rho'' + 2a\rho(n_0 - a\rho^2)/C^2$. Still with the condition that the ray remains close to the axis of revolution, we can set $\rho^3 \ll \rho$ and, with the value $C \cong n_0$ already obtained, the final form of the differential equation is:

$$\rho'' + \frac{2a}{n_0}\rho = 0.$$

8. With the initial condition $\rho(0) = 0$, the solution of this equation is simply: $\rho(z) = A\sin(z\sqrt{2a/n_0})$. At the point I, with abscissa $z_I = OI$, where the trajectory intersects the optical axis (for the first time), one must have $\rho(z_I) = 0$, which leads to the value $z_I = OI$ with:

$$OI = \pi\sqrt{\frac{n_0}{2a}}.$$

This value is independent of the constant A, which governs the inclination with respect to the axis of the ray emitted by the source. The consequence is that all rays emitted from O (provided they remain close to the optical axis) intersect the optical axis at a single point I, known as the image of O in the optical system.

4.13. Application to Billiard Balls
[Statement p. 183]

Sign convention: one chooses a sense of rotation (for example the trigonometric sense) along the cush and the curvilinear abscissae are measured according to this sense with respect to an arbitrary origin. For the measure of angles of incidence, one can choose the angle between the normal and the direction of the incident trajectory.

In the following discussion, we neglect the curvature of the cush between two close points and consider that the difference between the curvilinear abscissae for these two points can be identified with the segment connecting them.

Between two consecutive rebounds, the ball is not subject to any force and thus follows a straight line.

After the impact, the angle of reflection is the negative of the angle of incidence (with our convention) since the trajectory is symmetric with respect to the normal.

1. Let A_{n-1} and A_n be the points on the cush where the impacts labelled $n-1$ and n occur. Furthermore let the respective curvilinear abscissae and incidence angles be (s_{n-1}, i_{n-1}) and (s_n, i_n).

    Assume that the ball arrives, at the impact $n-1$, from a point B_{n-1}, infinitely close to A_{n-1} and with the same angle of incidence i_{n-1}, but a slightly different curvilinear abscissa $A_{n-1}B_{n-1} = \delta s_{n-1}$. The next impact will occur at a point B_n infinitely close to A_n: $A_n B_n = -\delta s_n$ (pay attention to the sign convention), with the same incident angle, i_n, (the trajectories and the normal are parallel). The sines of the incidence angles are of course equal so that $\delta p_n = 0$.

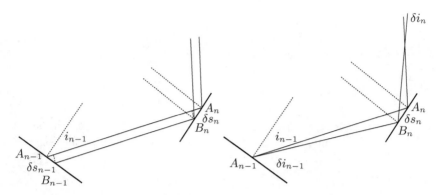

Fig. 4.13 Trajectories (full lines) of the billiard ball between the impacts $n-1$ and n and the rebound after the impact n. The normals to the cush are denoted by dotted lines. On the left, we represent two trajectories with the same angle of incidence but separated by a small curvilinear abscissa. On the right, we represent two trajectories originating at the same point but with slightly different angles of incidence

Let C be the perpendicular projection of B_{n-1} on $A_{n-1}A_n$ and D be the perpendicular projection of B_n on $A_{n-1}A_n$. Because of the previous remarks, one has $CB_{n-1} = DB_n = \Delta$. Furthermore, in the triangle $CB_{n-1}A_{n-1}$ the relation (see Fig. 4.13)

$$\Delta = A_{n-1}B_{n-1}\cos(i_{n-1}) = \delta s_{n-1}\cos(i_{n-1})$$

holds. Similarly in the triangle DA_nB_n, one has the relation $\Delta = \delta s_n \cos(i_n)$. Equating both expressions for Δ and taking care to use the correct sign, we arrive at the required relation:

$$\delta s_n = -\delta s_{n-1} \cos(i_{n-1})/\cos(i_n).$$

Thus, with a constant angle of incidence, we have the following equalities:

$$\delta s_n = -\delta s_{n-1} \frac{\cos(i_{n-1})}{\cos(i_n)};$$
$$\delta p_n = 0.$$

2. One works now with a constant angle of incidence, and the new impact $n-1$ thus takes place at A_{n-1} but with an angle of incidence $i_{n-1}+\delta i_{n-1}$. The new trajectory intersects the cush at B_n, close to A_n with an angle of incidence $i_n+\delta i_n$. Consider the triangle $A_{n-1}A_nB_n$ and set $A_{n-1}A_n = L$. The sine relation for triangles gives

$$\frac{A_nB_n}{\sin(\delta i_{n-1})} = \frac{A_{n-1}A_n}{\sin(\pi/2 - i_n - \delta i_n)}, \quad \text{or} \quad A_nB_n = L\frac{\sin(\delta i_{n-1})}{\cos(i_n + \delta i_n)}.$$

To first order, we retain only $A_n B_n = L\delta i_{n-1}/\cos(i_n)$. Furthermore $\delta p_{n-1} = \delta[\sin(i_{n-1})] = \cos(i_{n-1})\,\delta i_{n-1}$. Owing to the relation $A_n B_n = -\delta s_n$, the two previous equalities lead to

$$\delta s_n = -L\,\frac{\delta p_{n-1}}{\cos(i_{n-1})\cos(i_n)}.$$

Lastly, because the normals are parallel for impacts n, one sees that $\delta i_n = -\delta i_{n-1}$. This equality, translated in terms of variables p, implies the relation $\delta p_n = -\delta p_{n-1}\,(\cos(i_n)/\cos(i_{n-1}))$. Thus, with constant curvilinear abscissa, we have the following equalities:

$$\delta s_n = -\frac{L}{\cos(i_{n-1})\cos(i_n)}\delta p_{n-1};$$

$$\delta p_n = -\frac{\cos(i_n)}{\cos(i_{n-1})}\delta p_{n-1}.$$

3. Generally, one can write $s_n = F(s_{n-1}, p_{n-1})$ and $p_n = G(s_{n-1}, p_{n-1})$. Let us take the differential of these functions and employ the results of the previous questions to make the identifications:

$$\left(\frac{\partial F}{\partial s_{n-1}}\right)_{p_{n-1}} = -\frac{\cos(i_{n-1})}{\cos(i_n)}; \quad \left(\frac{\partial G}{\partial s_{n-1}}\right)_{p_{n-1}} = 0;$$

$$\left(\frac{\partial F}{\partial p_{n-1}}\right)_{s_{n-1}} = -\frac{L}{\cos(i_{n-1})\cos(i_n)}; \quad \left(\frac{\partial G}{\partial p_{n-1}}\right)_{s_{n-1}} = -\frac{\cos(i_n)}{\cos(i_{n-1})}.$$

These quantities allow us to write the infinitesimal variations of the generalized coordinates in a matrix form:

$$\begin{pmatrix}\delta s\\ \delta p\end{pmatrix}_n = M \begin{pmatrix}\delta s\\ \delta p\end{pmatrix}_{n-1}$$

with the matrix M given by:

$$M = \begin{pmatrix} -\dfrac{\cos(i_{n-1})}{\cos(i_n)} & -\dfrac{L}{\cos(i_n)\cos(i_{n-1})} \\ 0 & -\dfrac{\cos(i_n)}{\cos(i_{n-1})} \end{pmatrix}.$$

One can check immediately that $\det(M) = 1$, which implies the conservation of the area in phase space (provided of course that we choose s_n and p_n as generalized coordinates).

4.14. Parabolic Double Well
[Statement and Figure p. 184]

1. The potential does not depend on time and the kinetic energy is quadratic in the velocity so that the Hamiltonian is $H = T + V$. The kinetic energy is expressed as $p^2/(2m)$ and, for $|x| < a$ the potential is given by $V(x) = -V_0(x/a)^2$. Consequently:

$$H(x,p) = \frac{p^2}{2m} - V_0\left(\frac{x}{a}\right)^2.$$

The value taken by the Hamilton function remains constant all along the trajectory and can be identified with the energy E. One immediately sees that if $E < -V_0$, no motion is allowed because the kinetic energy must retain a positive value. Thus we have two possible situations: $E > 0$ and $-V_0 < E < 0$ for which the number of turning points is different.

First case: $E > 0$

2. Let us set $E = rV_0$ ($r > 0$ is a dimensionless real number) and $y = x/a$. The turning points are defined by $E = V(x)$. In our case the situation is somewhat special because the potential looks rather like a distribution with a passage from a negative value to infinity over a null distance (see Fig. 4.5). One can, however, agree that the potential takes precisely the value E at this distance. Furthermore the particle will never explore the region $|x| > a$. Consequently we have only two turning points:

$$x_0 = \pm a.$$

Between these two points, the motion is allowed.

3. With our conventions, the equation for the energy is written

$$rV_0 = \frac{p^2}{2m} - V_0 y^2,$$

that is: $$p(y) = \pm\sqrt{2mV_0}\sqrt{y^2 + r}.$$

This equation is nothing more than the phase portrait. The particle starts from $x = -a$ with a positive velocity; its velocity decreases to reach a minimum at $x = 0$, then it increases again. Attaining $x = a$, the velocity instantaneously changes its sign and becomes negative. The particle comes back in the opposite direction with a symmetrical motion. The phase portrait is illustrated in the Fig. 4.14.

4. Because of the symmetry of the problem, we restrict our study to the journey from $-a$ to a only. In this portion, $p = m\dot{x} = ma\dot{y}$ is positive.

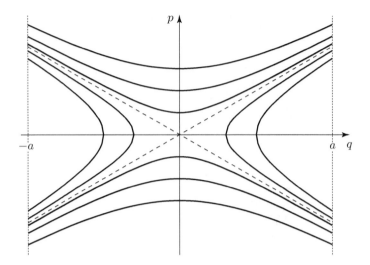

Fig. 4.14 Phase portrait for the parabolic double well. The curves must be understood as passing from a positive (negative) value for the momentum to a negative (positive) value when touching the wall. The dashed lines separate the two regimes (separatrices occurring for $E = 0$) and are also the asymptotes of the families of hyperbolae. The hyperbolae of the upper and lower parts correspond to the case $E > 0$ and those at the sides to the case $E < 0$

The relation proved in Question 3 provides the differential equation leading to $y(t)$, i.e.:

$$\dot{y} = \sqrt{\frac{2V_0}{ma^2}}\sqrt{y^2 + r}.$$

In this equation the variables can be separated and the integration can be performed without difficulty. The change of variable $y = \sqrt{r}\,u$ can be of some help. Bearing in mind that $x(t) = ay(t)$, we obtain the temporal law:

$$x(t) = a\sqrt{\frac{E}{V_0}}\sinh\left[\sqrt{\frac{2V_0}{ma^2}}t - \sinh^{-1}\sqrt{\frac{V_0}{E}}\right].$$

5. The motion is periodic with period T. For the one way journey, for which the temporal equation is given in the previous question, one has only a half-period $T/2$. Thus the condition $x(T/2) = a$ allows the determination of T:

$$T = 4\sqrt{\frac{ma^2}{2V_0}}\sinh^{-1}\sqrt{\frac{V_0}{E}}.$$

Problem Solutions

Second case: $-V_0 < E < 0$

6. In addition to the previous turning points, which persist for the same reason, there exist two further turning points for which $E = -V_0 y^2$. This time, we set $|E| = rV_0$. One finds $y_0 = \pm\sqrt{r}$, or $x_0 = \pm\sqrt{r}a$.

$$x_0 = \pm a \quad ; \quad x_0 = \pm\sqrt{\frac{|E|}{V_0}}a.$$

There exist two distinct regions corresponding to an allowed motion : between $-a$ and $-\sqrt{r}a$ or between $\sqrt{r}a$ and a. We will focuse only on this latter motion, the other one being described by a completely analogous treatment.

7. The equation still provides the phase portrait which, in this case, is written:

$$p(y) = \pm\sqrt{2mV_0}\sqrt{y^2 - r}.$$

Between $\sqrt{r}a$ (where the velocity vanishes) and a one must take the positive value for the previous expression; at $x = a$ the velocity instantaneously changes its sign and the particle comes back in the opposite direction with a negative velocity which vanishes again at $x = \sqrt{r}a$. The corresponding phase portrait is again illustrated in the Fig. 4.14.

8. In this case, the differential equation for the motion is written: $\dot{y} = \sqrt{2V_0/(ma^2)}\sqrt{y^2 - r}$, which can be integrated in the same way to give:

$$x(t) = a\sqrt{\frac{|E|}{V_0}} \cosh\left[\sqrt{\frac{2V_0}{ma^2}}t\right].$$

9. We are still dealing with a periodic motion. The half-period $T/2$ is obtained using the relation $x(T/2) = a$. A very simple calculation gives the result:

$$T = 2\sqrt{\frac{ma^2}{2V_0}} \cosh^{-1}\sqrt{\frac{V_0}{|E|}}.$$

When $E > 0$, the phase portrait consist of portions of hyperbolae, the asymptotes of which are the separatrices corresponding to $E = 0$. When $E < 0$, one still has portions of hyperbolae (which reduce to a single point for $-V_0 = E$), with the same asymptotes.

In any case, one can check that $T \to \infty$ when $E \to 0$.

4.15. Stability of Circular Trajectories in a Central Potential
[Statement p. 185]

1. Let us start from Binet's equation (in its integral form):

$$u'^2 + u^2 = \frac{2m}{\sigma^2}(E - V(1/u)).$$

With the proposed potential $V(1/u) = -\lambda u^\alpha$ (remember that $u(\phi) = 1/\rho(\phi)$), this equation can be recast as:

$$E = \frac{\sigma^2}{2m} u'^2 + V_{\text{eff}}(u); \quad V_{\text{eff}}(u) = \frac{\sigma^2}{2m} u^2 - \lambda u^\alpha.$$

The above equation can be put into the form $u' = u'(E, u)$, which is an equation analogous to that of the phase portrait $p = p(E, q)$, in which the ϕ variable is substituted for the time.

Notice that the first term of the effective potential is simply the potential corresponding to the centrifugal force. Indeed, the inertial centrifugal force is $f = m\omega^2 \rho$, which can be rewritten, with $\sigma = m\omega\rho^2$,

$$f = \frac{\sigma^2}{m\rho^3} = -\frac{d}{d\rho}\left[\frac{\sigma^2}{2m\rho^2}\right].$$

One recognizes at once that this force arises from the potential $\sigma^2/(2m\rho^2) = \sigma^2 u^2/(2m)$.

The other term of the effective potential corresponds simply to the potential associated with the real forces acting on the system.

2. One must still fulfill the condition $u'^2 > 0$, or $V_{\text{eff}}(u) < E$. The values of u for which $V_{\text{eff}}(u)$ exhibits an extremum correspond to equilibrium positions: it is easy to find the values u_c corresponding to $V'_{\text{eff}}(u_c) = 0$. A solution always exists: $u_c = [\sigma^2/(m\lambda\alpha)]^{1/(\alpha-2)}$. Calculating

$$V''_{\text{eff}}(u_c) = \frac{\sigma^2(2-\alpha)}{m},$$

one sees that the extremum is indeed a minimum (stable orbit) if $\alpha < 2$, and a maximum (unstable orbit) in the opposite case. The solution $u(\phi) = u_c$ corresponds to a circular orbit, centered at the center of force, provided that the energy is chosen so as to fulfill $u'(\phi) = 0$. The radius of the circle is given by $\rho_c = 1/u_c$ that is:

$$\rho_c = \left(\frac{m\lambda\alpha}{\sigma^2}\right)^{\frac{1}{\alpha-2}}.$$

Problem Solutions

The corresponding total mechanical energy E_c is equal to $V_{\text{eff}}(u_c) = E_c$. For the given potential, this equation leads to:

$$E_c = \frac{\lambda(\alpha-2)}{2}\left(\frac{\sigma^2}{m\lambda\alpha}\right)^{\frac{\alpha}{\alpha-2}}.$$

This energy is positive if $\alpha < 0$ or $\alpha > 2$; it is negative if $0 < \alpha < 2$. The potential energy on the circle is simply $V = -\lambda/\rho_c^\alpha$. Furthermore, Newton's equation implies $mv_c^2/\rho_c = |F| = |-dV/d\rho| = \lambda\alpha/\rho_c^{\alpha+1}$, leading to $T = \frac{1}{2}mv_c^2 = \lambda\alpha/(2\rho_c^\alpha)$, which implies $T = -\alpha V/2$. Since we have also $T + V = E_c$, we obtain finally:

$$\frac{V}{E_c} = -\frac{2}{\alpha-2}; \quad \frac{T}{E_c} = \frac{\alpha}{\alpha-2}.$$

These formulae express the virial theorem for this particular potential.

In its general version, the virial theorem stipulates the relation $\langle T \rangle = -\frac{1}{2}\langle \boldsymbol{F}\cdot\boldsymbol{\rho}\rangle$ where $\langle\rangle$ means an average value over a time sufficiently long. For the circular orbits under consideration, these quantities do not vary in time and their average value coincides simply with the value taken on the circle. With the proposed power-law potential, it is easy to check that $\boldsymbol{F}\cdot\boldsymbol{\rho} = \alpha V$ so that we recover directly the relation already given $T = -\alpha V/2$ (For the special case of a Coulomb potential one finds $T = -V/2$ and for a harmonic potential the relation $T = V$).

3. In the neighbourhood of the circular orbit, the energy differs only slightly from E_c so that we can write: $E = E_c + \Delta E$ and the radius $u(\phi)$ (which now depends on ϕ) differs slightly from u_c: $u(\phi) = v(\phi) + u_c$. The effective potential can be expanded up to second order in v (the first order term vanishes due to the extremum condition). Substituting the corresponding expressions into Binet's equation, one finds $v'^2 + (2-\alpha)v^2 = 2m\Delta E/\sigma^2$, or, after differentiation, $v'' + (2-\alpha)v = 0$.

If $\alpha > 2$, the solution of this differential equation is exponentially increasing and we are faced with an unstable circular orbit; on the contrary if $\alpha < 2$ the solution presents an oscillating behaviour $v(\phi) = \cos(\sqrt{2-\alpha}\phi)$. The circular orbit is stable and, in this case, the real orbit is comprised between a perihelion $\rho_{\min}$ and an aphelion $\rho_{\max}$:

$$\rho_{\min} = \frac{1}{u_c + A}; \quad \rho_{\max} = \frac{1}{u_c - A}.$$

4. ρ (or u) recovers its value together with $v(\phi)$, when $\phi \to \phi + 2\pi/\sqrt{2-\alpha}$; between two perihelia or aphelia the angle ϕ varies by

$$\Delta\phi = \frac{2\pi}{\sqrt{2-\alpha}}.$$

5. The perturbed orbit is closed if there is for sure coincidence between two perihelia: this means that $q\Delta\phi$ corresponds to an integer number of revolutions, that is $p\,2\pi$. The condition for the existence of a closed orbit is thus $\Delta\phi = (p/q)\,2\pi$. This equality implies a condition on α:

$$\alpha = 2 - \frac{q^2}{p^2}$$

where p and q are arbitrary positive integers.

If $\alpha = n$ is an integer number, the condition for a closed orbit is that $\sqrt{2-n}$ equals a rational number. One can check that this is indeed the case for $n = 1$ (Coulomb problem: $p = q = 1 \Rightarrow$ which requires a complete revolution to pass from one perihelion to the next one), and for $n = -2$ (harmonic problem: $p = 1, q = 2 \Rightarrow$ which requires only half a revolution to recover the next perihelion).

4.16. The Bead on the Hoop [Statement p. 186]

1. The Lagrange function of the system has already been determined (see Problem 1.4); it is expressed as

$$L(\theta, \dot\theta) = \frac{1}{2}mR^2(\dot\theta^2 + \omega^2 \sin^2\theta) - mgR\cos\theta.$$

The momentum is deduced at once $p = \partial_{\dot\theta} L = mR^2\dot\theta$, and thereby Hamilton's function: $H = p\dot\theta - L$, or explicitly

$$H(\theta, p) = \frac{p^2}{2mR^2} + \frac{1}{2}mR\left(2g\cos\theta - R\omega^2 \sin^2\theta\right).$$

2. Identifying the previous expression with the definition $H = T + V_\omega(\theta)$, one can derive the effective potential:

$$V_\omega(\theta) = \frac{1}{2}mR\left(2g\cos\theta - R\omega^2 \sin^2\theta\right).$$

This potential contains a term due to gravity and a term corresponding to the centrifugal force.

It is easy to check the properties: $V_\omega(2\pi + \theta) = V_\omega(\theta)$ (the function is periodic) and $V_\omega(\pi + \theta) = V_\omega(\pi - \theta)$ (the function is symmetric with respect to the vertical oriented downwards $\theta = \pi$). Consequently, it is sufficient to restrict our study to the interval $[0, \pi]$.

Furthermore, the system is autonomous so that the value taken by Hamilton's function on the trajectory is a constant identified with the energy:

$$E = \frac{p^2}{2mR^2} + V_\omega(\theta).$$

Let us calculate $dV_\omega/d\theta = -mR\sin\theta\,(g + R\omega^2\cos\theta)$ and introduce the proper angular frequency $\omega_0 = \sqrt{g/R}$ to write:

$$\frac{dV_\omega}{d\theta} = -mR^2\sin\theta\,(\omega_0^2 + \omega^2\cos\theta).$$

The derivative always vanishes for $\theta = 0$ and $\theta = \pi$ and, possibly, for the value $\cos\theta_s = -\omega_0^2/\omega^2$. The motion is possible only if the condition $V_\omega(\theta) < E$ is satisfied.

Case of a rapid rotation: $\omega > \omega_0$

3. In the case where the condition $|\omega_0^2/\omega^2| < 1$ is fulfilled, the value θ_s is indeed possible:

$$\theta_s = \arccos(-\omega_0^2/\omega^2).$$

One obtains the energy for this angle:

$$E_m = V_\omega(\theta_s) = -\frac{1}{2}mR^2\omega^2\left(1 + \frac{\omega_0^4}{\omega^4}\right).$$

The potential decreases from mgR for the value $\theta = 0$ to E_m for $\theta = \theta_s$ in a first step, then increases to $-mgR$ for $\theta = \pi$.

Consequently $\theta = \theta_s$ (and also $\theta = \pi + \theta_s$) corresponds to a minimum for the potential so that this angle corresponds to stable equilibrium, whereas $\theta = 0$ and $\theta = \pi$ correspond to maxima for the potential and thus are unstable equilibrium positions.

4. If $E_m < E < -mgR$, there exist two turning points; the motion is confined between these two points. Assume a motion around the equilibrium position θ_s and set $\theta = \theta_s + \varepsilon$. An expansion up to second order in ε for Hamilton's function provides

$$H = E_m + \frac{p^2}{2mR^2} + \frac{1}{2}mR^2\omega^2\left(1 - \frac{\omega_0^4}{\omega^4}\right)\varepsilon^2.$$

One recognizes the Hamiltonian of a harmonic oscillator. Hamilton's equations lead to the equation of motion $\ddot\varepsilon + \tilde\omega^2\varepsilon = 0$, with

$$\tilde\omega^2 = \omega^2\left(1 - \frac{\omega_0^4}{\omega^4}\right).$$

We are in presence of a sinusoidal motion with angular frequency:

$$\tilde\omega = \omega\sqrt{1 - \frac{\omega_0^4}{\omega^4}} = \omega\sin\theta_s.$$

5. Let us examine now the condition $-mgR < E < mgR$. There exists a turning point θ_1 comprised between 0 et θ_s and another one corresponding to the symmetric point $\pi + \theta_1$. The motion takes place between these these two turning points for which the velocity vanishes. From θ_1 to θ_s the speed increases, then from θ_s to π it decreases; from π to $2\pi - \theta_s$ it increases again, then from $2\pi - \theta_s$ to $2\pi - \theta_1$ it decreases up to θ_s where it vanishes. There the velocity changes its sign and the motion starts in the opposite direction and follows a symmetric behaviour. This type of motion is known as libration.

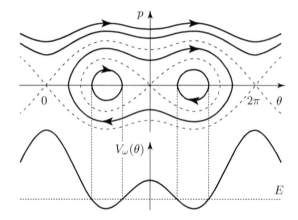

Fig. 4.15 Phase portrait of the bead for a rapid rotation of the hoop

6. If $E > mgR$, there is no turning point and the velocity retains a constant sign. We are dealing with a rotational motion (in one sense or the other). From the upper point to θ_s the speed increases, then it decreases from θ_s to π; from π to $2\pi - \theta_s$ it increases again, then from $2\pi - \theta_s$ to the upper point it decreases again but never vanishes.

7. The phase portrait is given by the equation:

$$p(\theta, E) = \pm\sqrt{2mR^2(E - V_\omega(\theta))}$$

which represents graphically the characteristics described above. It is plotted in Fig. 4.15. Each of the two curves obtained for the special values of the energy $E = -mgR$ and $E = mgR$, which delimit the two different regimes, is called a separatrix.

Case of a slow rotation: $0 < \omega < \omega_0$

8. In this case θ_s does not exist anymore. There are only two equilibrium positions: $\theta = 0$ for which the potential is maximal mgR (unstable equilibrium) and $\theta = \pi$ for which it is minimal $-mgR$ (stable equilibrium). Between these positions, the potential decreases (see Fig. 4.16).

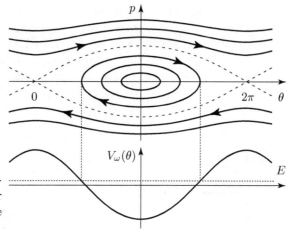

Fig. 4.16 Phase portrait of the bead for a slow rotation of the hoop

9. If $-mgR < E < mgR$, there exist two turning points θ_1 and $2\pi - \theta_1$ between which the motion can take place. Close to the equilibrium position, one sets $\theta = \pi + \varepsilon$ and one expands Hamilton's function up to second order in ε. A simple calculation, in complete analogy with that of Question 4, reveals a harmonic motion with the angular frequency:

$$\tilde{\omega} = \omega \sqrt{\frac{\omega_0^2}{\omega^2} - 1}$$

which again is referred to as libration.

10. When $E > mgR$, there exist no turning points and the velocity retains a constant sign. We are in the presence of a rotational motion. There is acceleration from the upper position to the lower position where the speed is maximum, then a deceleration from the lower position to the upper position where the speed is minimum.

11. The phase portrait results from the same equation as before, but it has a simpler structure. It is plotted in the Fig. 4.16. In this case, there exists only one separatrix, obtained for the value $E = mgR$.

12. The position $\theta = 0$ is in all cases an unstable equilibrium position. On the contrary, the position $\theta = \pi$ is a stable equilibrium position as long as the condition $\omega < \omega_0$ is satisfied but it becomes unstable as soon as $\omega > \omega_0$. For this particular value of the rotational frequency $\omega = \omega_0$, we find a pair of stable equilibrium positions for θ_s and $2\pi - \theta_s$ which move apart when the rotational speed increases and tend to the values $\pi/2$ and $3\pi/2$ for very large rotational speeds.

This phenomenon which consists in the switch from a stable equilibrium to an unstable one with the simultaneous appearance of a pair of stable equilibrium positions is known as a bifurcation.

The behaviour of the equilibrium points as a function of the rotational speed is illustrated in Fig. 4.17.

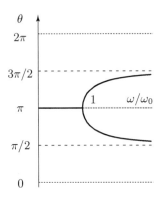

Fig. 4.17 Stable (full lines) and unstable (dotted lines) equilibrium angles as a function of the ratio between the proper angular frequency ω_0 and angular velocity of the hoop ω

4.17. Stability of Circular Trajectories in a Central Potential
[Statement p. 188]

1. Let us recall the dimensionless quantities that will be used: the energy $\varepsilon = E/(mc^2)$, an independent parameter linked to the angular momentum $\nu = [K/(\sigma c)]^2$ (larger values of ν are associated with smaller angular momenta or a weaker Coulomb force) and the inverse of the radius vector $u = |K|/(mc^2 \rho)$, a quantity that is always positive. The integral form of the relativistic Binet equation already found in Problem 3.2 is written in this special case with the natural units:

$$u'^2 + u^2 = \nu \left[(\varepsilon \pm u)^2 - 1 \right],$$

the positive sign being appropriate for an attractive potential and the negative sign for a repulsive potential. This equation can be recast in a more convenient form as:

$$u'^2 + (1-\nu)\left(u \mp \frac{\nu}{1-\nu}\varepsilon\right)^2 = \nu\left(\frac{\varepsilon^2}{1-\nu} - 1\right).$$

To this equation, one must add the following restrictions: $u > 0$ and a speed less than the speed of light, that is a relativistic factor $\gamma > 1$ which, owing to the expression for the energy $E = \gamma mc^2 + V$, leads to $u > 1 - \varepsilon$ in the case of attraction and $u < \varepsilon - 1$ in the case of repulsion.

First case $\nu < 1$
(large angular momentum or weak Coulomb force)

The phase portrait, plotted in the plane (u, u'), is an ellipse provided that the right hand member of the equation is positive. This is true if $\varepsilon > \varepsilon_c = \sqrt{1-\nu}$. ε_c is the lowest energy for a given angular momentum. In this case, one must have $u' = 0$ and $u = u_c = \nu/\sqrt{1-\nu}$. The trajectory is a circular orbit with radius $r_c = \sqrt{1-\nu}/\nu$. This condition for the energy follows the expression concerning the eccentricity of Kepler's orbits

$$e = \sqrt{1 + 2\frac{(E-mc^2)\sigma^2}{mK^2}} = \sqrt{1 + 2\frac{\varepsilon-1}{\nu}}$$

(don't forget that in the relativistic case the energy is the sum of the rest mass energy plus the classical energy). The eccentricity vanishes if $\varepsilon = 1 - \nu/2 \approx \sqrt{1-\nu}$. One can see that the need to work with a relativistic formalism increases with the value of ν: it decreases when the angular momentum increases.

For several values of the energy, the phase portrait consists of nested ellipses. In the case of an attractive potential, the phase portrait is represented in the left hand part of Fig. 4.18.

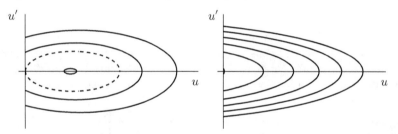

Fig. 4.18 Phase portrait in the case $\nu < 1$. At the left hand side, the potential is attractive; the ellipses are nested in order of decreasing energies. For $\varepsilon = 1$ we obtain the ellipse represented by a dashed line. For lower values of the energy, the ellipses are complete. They collapse to a single point for a circular orbit obtained for the energy $\varepsilon = \varepsilon_c$. On the right, the potential is repulsive; again the ellipses are nested in order of decreasing energies

For an energy $\varepsilon < 1$ the ellipse does not intersect the ordinate axis. The trajectory is comprised between an aphelion and a perihelion. In the Fig. 4.19, we represent an example of such an orbit. Let us notice that for $\varepsilon = 1$, one can have $u' = u = 0$; the ellipse is plotted with a dashed line. In this case the orbit can reach infinity but with a vanishing velocity.

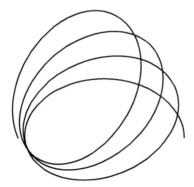

Fig. 4.19 Example of a trajectory corresponding to the left hand side of Fig. 4.18 for an energy $\varepsilon < 1$.

For energies $\varepsilon > 1$, the ellipse intersects the ordinate axis. The trajectory approaches the center of force at a minimum distance, possibly makes a revolution and goes off again to infinity without reaching a null velocity.

In the case of a repulsive potential, one always has $\varepsilon > 1$ and the ellipses are always truncated by the ordinate axis since the center is obtained for a negative value of u. The corresponding portrait is represented on the right hand side of Fig. 4.18. The ellipses are nested by order of decreasing energies. In all cases the trajectories extend to infinity and approach the center as the energy increases.

2. Let us apply these remarks to the hydrogen atom, with a circular orbit. We saw that, in this particular case, the relativistic energy is $\varepsilon = \sqrt{1-\nu}$ and the classical energy $\varepsilon = (1 - \nu/2)$. In this case $K = -e^2/(4\pi\varepsilon_0)$ and the potential energy is $V = -|K|/a_0$. The virial theorem for the Coulomb problem leads to $T = -\frac{1}{2}V$, whence the total energy

$$E = mc^2 + T + V = E = mc^2 + \frac{1}{2}V.$$

Lastly, using the value of the Bohr radius $a_0 = \hbar^2/(m|K|)$, we find

$$V = -mc^2 \frac{|K|^2}{\hbar^2 c^2} = -mc^2\alpha^2,$$

where we introduced the fine structure constant $\alpha = |K|/(\hbar c) = 1/137$. The classical value of the total energy is thus $E = mc^2(1 - \alpha^2/2)$; this relation provides the value $\nu = \alpha^2$. One deduces the relativistic value of the total energy:

$$E = mc^2\sqrt{1-\alpha^2}.$$

This result coincides precisely with the value obtained from a complete relativistic calculation performed in quantum mechanics.

Second case $\nu > 1$
(small angular momentum or large Coulomb force)

For the phase portrait, we are now dealing with hyperbolae, for each of which a branch is excluded because of the condition $\gamma > 1$. This portrait is represented in the Fig. 4.20.

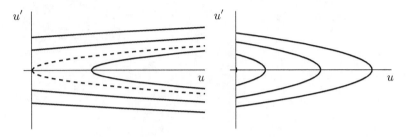

Fig. 4.20 Phase portrait in the case $\nu > 1$. On the left hand side, the potential is attractive; the hyperbolae are nested in order of decreasing energy (the curve represented with a dashed line corresponds to $\varepsilon = 1$). On the right hand side, the potential is repulsive; the hyperbolae are nested in order of decreasing energy but one always has the condition $\varepsilon > 1$

On the left hand side, the potential is attractive; all the orbits fall (or begin) from the center spirally. Those with a large energy $\varepsilon > 1$ extend to (or start from) infinity. Those with $\varepsilon = 1$ (dashed line) reach infinity with a vanishing velocity. Lastly, those with $\varepsilon < 1$ reach infinity, but move away from the center as the energy increases. An example of such a trajectory is illustrated in the Fig. 4.21.

Fig. 4.21 Example of trajectory corresponding to the left hand side of Fig. 4.20

On the right hand side of Fig. 4.20, the potential is repulsive. The hyperbolae are nested by order of decreasing energy, but we always have the condition $\varepsilon > 1$. All the orbits begin or end at infinity but none reaches the center of force. They come closest for the largest energies.

Chapter 5
Hamilton–Jacobi Formalism

Summary

This chapter will probably not be of great use for finding simple solutions to problems in mechanics. It is however of considerable historical interest and, furthermore is fundamental in that it establishes a bridge between mechanics, wave phenomena, optics and, above all, quantum mechanics.

The action considered here is still the integral of the Lagrangian. In Chapter 3, it was considered as a **functional of the path with fixed bounds**. In this chapter, the path is fixed once and for all as the physical trajectory; the action is regarded as a **function of space and time** relative to the end point of the trajectory. This function obeys a partial differential equation that can be solved only very rarely. If a solution exists, the mechanical problem is completely solved: one says that it is integrated.

5.1. The Action Function

Consider a trajectory[1] – that is a set of functions $\tilde{q}(\tilde{t}) = (\tilde{q}_1(\tilde{t}), \tilde{q}_2(\tilde{t}), \cdots, \tilde{q}_n(\tilde{t}))$ which obey Lagrange's equations – which originates at q_1 for time t_1 ($\tilde{q}(\tilde{t} = t_1) = q_1$) and terminates at q for time t ($\tilde{q}(\tilde{t} = t) = q$). The action function is defined as the integral of the Lagrangian along the trajectory:

$$S(q, t, q_1, t_1) = \int_{t_1}^{t} L(\tilde{q}(\tilde{t}), \dot{\tilde{q}}(\tilde{t}), \tilde{t}) \, d\tilde{t}. \tag{5.1}$$

---

[1] As we saw in Chapter 3, there may exist none or, in contrast, many of them.

Example – The free particle:
The Lagrangian is simply the kinetic energy and the trajectory is a straight line. From (5.1) the action is expressed as (see also Problem 5.1).

$$S(q,t,q_1,t_1) = \frac{m}{2}\frac{(q-q_1)^2}{(t-t_1)}.$$

In what follows, one assumes that q_1 and t_1 are fixed once and for all, and they will therefore often be omitted in the text. It is easy to prove the relations:

$$\frac{\partial S(q,t)}{\partial q} = p\,; \tag{5.2}$$

$$\frac{\partial S(q,t)}{\partial t} = -H(q,p,t). \tag{5.3}$$

Thus, the action function obeys a non linear partial differential equation of first order with $n+1$ variables (the coordinates and the time), known as the Hamilton–Jacobi equation:

$$\frac{\partial S(q,t)}{\partial t} = -H\left(q,\frac{\partial S(q,t)}{\partial q},t\right). \tag{5.4}$$

5.2. Reduced Action

By definition, for an **autonomous** system the energy is constant: $H(q,p) = E = p\cdot\dot{q} - L(q,\dot{q})$ and the action can be written as:

$$S(q,t,q_1,t_1) = -E(t-t_1) + \tilde{S}(q,q_1,E) \tag{5.5}$$

where we introduced the reduced action

$$\tilde{S}(q,q_1,E) = \int_{t_1}^{t} p\left(\tilde{q}(\tilde{t}),\dot{\tilde{q}}(\tilde{t})\right)\cdot\dot{\tilde{q}}(\tilde{t})d\tilde{t} = \int_{q_1}^{q} \tilde{p}(\tilde{q},E)\cdot d\tilde{q}. \tag{5.6}$$

Very important remark: q_1 and q being fixed, the orbit[2] can be covered with different temporal laws. This corresponds to different travel times and energies (for instance think of a free particle following a straight line with different speeds). Therefore, there exists a relationship between the energy and the travel time, which we wish to determine more precisely.

---

[2] By orbit, we mean the curve described in configuration space, independently of the manner it is covered.

Indeed, the orbit can be specified by a parameter τ. One obtains, up to a factor, all the generalized velocities as $\dot{q} = dq/dt = (dq/d\tau)\dot{\tau} = q'\dot{\tau}$ and for the momenta $p(q, q'\dot{\tau}) = \partial_{\dot{q}} L(q, \dot{q})$. The factor $\dot{\tau}$ is a function of the energy through the relation $H(q, p(q, q'\dot{\tau})) = E$. We thus have the relation $E(q, t)$ or $t(q, E)$. Consequently, the momentum $p(q, E)$ encountered in (5.6) is a function of the coordinates and the energy. This point will be discussed in more detail in the Problem 5.1.

The reduced action is no longer a function of time but of energy.
In contrast, the total action defined by (5.5), together with (5.6) is a function not of energy but of time.

Consider the case of a particle subject to a potential which is independent of the velocities. In the above approach by taking as parameter the curvilinear abscissa ds, we obtain:

$$\tilde{S}(q, q_1, E) = \int_{q_1}^{q} \sqrt{2m(E - V(q))} \, ds$$

The reduced action obeys the **characteristic (or secular) partial differential equation of Hamilton–Jacobi**:

$$H\left(q, \frac{\partial \tilde{S}(q)}{\partial q}\right) = E. \tag{5.7}$$

The elapsed time between the two extremities of the trajectory can be obtained from the formula:

$$T = t - t_1 = \frac{\partial \tilde{S}(q, q_1, E)}{\partial E}. \tag{5.8}$$

5.3. Maupertuis' Principle

For an autonomous system with more than one degree of freedom, one can solve the problem in two steps. One can first determine the orbit and then the speed along the orbit. The determination of the orbit can be achieved using Maupertuis' principle. To obtain the speed on the orbit is not really difficult: one calculates first the potential at the given point and then, by subtracting it from the mechanical energy, one obtains the kinetic energy hence the speed.

Maupertuis' principle stipulates that the real orbit between the points q_0 and q_1 corresponds to the minimization of the reduced action:

$$\tilde{S}(q, E) = \int_{q_0}^{q_1} \sum_i p_i(q, E)\, dq_i \tag{5.9}$$

In the case of a particle subject to a potential, one recognizes a kind of stationarity principle quite similar to that of Fermat's optical path with the index[3] $n(q, E) = \sqrt{2m\left(E - V(q)\right)}$ (see Problem 3.6).

5.4. Jacobi's Theorem

Every first order partial differential equation and in particular the Hamilton–Jacobi equation, possesses a solution which depends on an arbitrary function: this solution is the general integral. Nevertheless, for mechanical systems, it is not this general integral which is important, but a solution which depends on constants whose number equals the number of independent variables. This particular solution $S(q, t, \alpha)$ is called the complete integral of the Hamilton–Jacobi equation. It depends on an additive constant of no interest[4] and on n arbitrary constants labelled α for $(\alpha_1, \alpha_2, \ldots, \alpha_n)$.

Jacobi's theorem is expressed through the following equations:

$$\frac{\partial S(q, t, \alpha)}{\partial \alpha_i} = \beta_i \tag{5.10}$$

where $(\beta_1, \beta_2, \ldots, \beta_n)$ are n arbitrary constants. If we are fortunate enough to know the complete solution, the problem is completely solved; it is said to be integrated. Indeed, from Relation (5.10), it is possible (theoretically) by inversion to obtain the trajectory $q(t, \alpha, \beta)$, in which the $2n$ constants α and β are fixed by the initial conditions $q = q_1$ and $p_1 = \partial_q S(q, t, \alpha)|_{q=q_1, t=t_1}$.

5.5. Separation of Variables

Let assume that, in Hamilton's function, one is able to gather one coordinate (the last one for instance) and its corresponding momentum into a single group denoted $\alpha_n\left(q_n, \partial_{q_n} S(q, t)\right)$ which is independent of other degrees of freedom and time.

---

[3] This is the momentum modulus. Contrary to optics, it is for the largest velocity that the index is the largest.

[4] Since the action appears in the Hamilton–Jacobi equation only through its partial derivatives, adding a constant to the action yields a solution of the same equation. This constant does not change the momentum nor the equation of motion (5.2).

Summary

Then, one has the following results:
- the action can be split into a sum implying different coordinates

$$S(q, t) = S_n(q_n, \alpha_n) + S_{n-1}(q_1, \ldots, q_{n-1}, \alpha_n, t);$$

- furthermore α_n is one of the constants appearing in the complete Hamilton–Jacobi solution which allows us to determine $S_n(q_n, \alpha_n)$ through an integral obtained from the differential equation

$$\alpha_n \left(q_n, \frac{dS_n(q_n, \alpha_n)}{dq_n} \right) = \text{const}.$$

The variable q_n is said to be **separated**. This is always the case for an autonomous system since the time is separated and leads to the term $-Et$ in the action (see (5.5)). This is also the case for a cyclic coordinate q_n since in this case $p_n = \alpha_n$ and it is possible to exhibit a separable term $\alpha_n q_n$ in the action.

Once the variable q_n is separated, one substitutes in the Hamilton–Jacobi equation the corresponding group by a constant. A second coordinate (say the last but one) may also be separable in which case one introduces a new constant α_{n-1}.

If this procedure can be pursued for all the variables, the problem is said to be completely separable and it is integrable.[5] The action can be expressed in the form:

$$S(q, t, \alpha) = S_1(q_1, t, \alpha_1, \ldots, \alpha_n) + S_2(q_2, \alpha_2, \ldots, \alpha_n) + \cdots + S_n(q_n, \alpha_n). \tag{5.11}$$

A very important example: a particle in a central force field.

It is useful to work with spherical coordinates; in this case the Hamiltonian reads:

$$H = \frac{1}{2m} \left[p_r^2 + \frac{1}{r^2} \left(p_\theta^2 + \frac{p_\phi^2}{\sin^2 \theta} \right) \right] + V(r).$$

The ϕ coordinate appears only through its momentum[6] so that it is separable $p_\phi = \alpha_3$; this constant is nothing more than the projection of the angular momentum onto the OZ axis. The same thing happens for the θ coordinate which appears only through the group $p_\theta^2 + \alpha_3^2 / \sin^2 \theta$ which represents the square of the angular momentum.

---

[5] The last constant is automatically proportional to the energy.

[6] It is a cyclic coordinate.

The latter variable is also separated provided that the potential is time independent; the last constant α_1 corresponds simply to the energy. From the previous procedure, one obtains the expression of the action in a separated form:

$$S(q,t,\alpha) = -\alpha_1 t \pm \int \sqrt{2m\left(\alpha_1 - V(r) - \alpha_2^2/(2mr^2)\right)}\, dr$$
$$\pm \int \sqrt{\alpha_2^2 - \alpha_3^2/\sin^2(\theta)}\, d\theta + \alpha_3 \phi.$$

Application of Jacobi's theorem $\partial_{\alpha_i} S(q,t,\alpha) = \beta_i$ completely solves the problem. After differentiating with respect to α_3, and with a suitable choice of axes, one can set $\alpha_3 = 0$ and $\phi = \text{const}$. This implies a planar motion. In this plane, the angle θ represents the polar angle. Differentiating with respect to α_2, Jacobi's theorem provides:

$$\theta = \beta_2 \pm \int \frac{dr}{\sqrt{(2mr^4/\alpha_2^2)(E - V(r)) - r^2}}.$$

This is a function $\theta(r)$ which, after inversion, gives the orbit $r(\theta)$ in polar form.

Lastly Jacobi's theorem applied to the α_1 constant leads to the equation:

$$\beta_1 + t = \pm \int \sqrt{\frac{m}{2\left(E - V(r) - \alpha_2^2/2mr^2\right)}}\, dr$$

which provides the temporal evolution.

5.6. Huygens' Construction

We consider the most common situation: the system is autonomous and velocities and momenta are parallel.[7]

In configuration space, let us imagine a surface Σ_0 that we refer to as the initial wave front and which is taken as the reference origin for the actions (the action at every point of Σ_0 vanishes). From this surface, let us distribute over the same side "particles" with velocities that are perpendicular to it and whose modulus is fixed by the relation $H(q,p) = E$. Let these "particles" move on their own trajectories up to a point at which the action has a value given in advance.[8]

---

[7] One can get rid of this restriction by making the construction more cumbersome.

[8] Be careful: these "particles" do not take the same time to reach the iso-action surface.

Consequently, all the iso-action points will lie on a surface Σ which is constantly perpendicular to every trajectory. This iso-action surface moves with a speed E/p called the phase speed which moves faster as the "particles" move more slowly. One can refer to each trajectory as a ray and to each iso-action surface as a wave front. In contrast to the phase speed, the physical speed of the "particles" $\dot{q}$ is called the group speed.

As in optics, from a wave surface, one generates spheres with radius $r = \Delta L/n$ where ΔL is an infinitesimal variation of the optical path (the mechanical analog for this action is $n = \sqrt{2m(E-V)}$). The envelop of these spheres is a wave surface (with equal action), and one obtains the rays (trajectories) by building the curves normal to these surfaces.

Problem Statements

5.1. How to Manipulate the Action and the Reduced Action [Solution p. 252] ★ ★

To provide a better understanding of the basic notions concerning the action function and the reduced action

The purpose of this problem is to study in more detail the action for an autonomous system, defined as a time dependent function, and the reduced action for which the "good variable" is the energy.

We illustrate this point using three examples: the one-dimensional free particle, a particle in a constant gravitational field and the three-dimensional free particle.

Preliminary question:

Using the Relation (5.5) between the action and the reduced action, show that these two functions are Legendre transforms of one another, the transformation variables being the time and the energy. Equations (5.4) and (5.8) are useful. The trajectory begins at the point $(q_1 = 0, t_1 = 0)$ and ends at the point (q, t).

A – One-dimensional free particle

1. Recalling the expression for the Lagrangian, find the trajectory $\tilde{q}(\tilde{t})$ which passes through the fixed points and calculate $S(q, t)$ with the definition (5.1). Check that this quantity obeys the Hamilton–Jacobi equation.

2. Calculate the reduced action with the help of the integral $\int p\,dq$ (Formula (5.6)).

3. Solving the characteristic Jacobi equation, show that $\tilde{S}(q, E) = \pm\sqrt{2mE}q$. Deduce that the reduced action for a path between $(0,0)$ and (q,t) is $\tilde{S}(q, E) = \sqrt{2mE}\,|q|$. Find the relationship between the travel time and the energy.

4. Using the reduced action and expressing the energy as a function of the travel time, deduce once again the expression for the action obtained in the first question.

One now considers a particle with a high speed, governed by relativistic dynamics.

5. Give the expression of the action in this case.

6. Write down and solve the characteristic Jacobi equation and give the expression for the reduced action.

7. Deduce the travel time and find the action as a function of time.

B – Particle in a constant gravitational field (one dimension)

Here we consider a non-relativistic particle in a constant gravitational field arising from the potential $V(q) = mgq$.

1. Using the expression of the Lagrangian, find the trajectory $\tilde{q}(\tilde{t})$ which passes through the given points and calculate $S(q,t)$. Check that this quantity obeys the Hamilton–Jacobi equation. Deduce the expression of the energy on this trajectory. You will find that it is difficult to give the travel time in terms of the energy.

2. Calculate the reduced action with the help of the definition $\int p\,dq$. Show that the solution of the characteristic Jacobi equation is

$$\tilde{S}(q,E) = \pm \frac{2}{3g}\sqrt{2/m}\,(E - mgq)^{3/2}.$$

Give the relation between these results.

3. Making use of the anwers to the previous questions, deduce a simple expression for the travel time as a function of the energy.

C – Three-dimensional free particle

One considers a non-relativistic free particle with mass m, moving in a three-dimensional space. The Cartesian coordinates are denoted $(x,y,z) = \boldsymbol{r}$.

1. Give the expression of the action and of the reduced action.

2. Solve the characteristic equation to obtain the complete solution noting that the variables separate. Show that the result can be written as $\tilde{S}(\boldsymbol{r},\boldsymbol{\alpha}) = \alpha_x x + \alpha_y y + \alpha_z z = \boldsymbol{\alpha}\cdot\boldsymbol{r}$ where the three separation constants fulfill the relation $E = \boldsymbol{\alpha}^2/(2m)$. You will notice that by taking $\boldsymbol{\alpha}$ (which

is actually the momentum) parallel to r, one recovers the reduced action previously calculated.

3. Study the same questions using a relativistic treatment.

Note: the separation of the variables could also have been achieved using cylindrical or spherical coordinates.

5.2. Action for a One-dimensional Harmonic Oscillator [Solution p. 258] ★ ★

A simple and analytical example for the action function

One considers a particle with mass m moving on a straight line with origin at point O. The position of the particle is specified by the coordinate q. This particle is subject to a restoring force proportional to q (harmonic oscillator), which arises from the potential $V(q) = \frac{1}{2} m \omega^2 q^2$.

1. Using the definition of the action function, calculate this function on a trajectory starting at (q_1, t_1) and ending at (q, t).

2. Discuss the singularity in the denominator and check that the action function obtained obeys the Hamilton–Jacobi equation. Deduce the relationship between the energy and the travel time.

3. Find the reduced action between the same extremities and show that it satisfies the usual characteristic equation. Deduce the relationship between the energy and the travel time.

5.3. Motion on a Surface and Geodesic [Solution p. 260] ★ ★

Action for a free particle moving on a surface of arbitrary dimension

A point-like particle of mass m can move freely on an n dimensional surface. Its configuration is specified by generalized coordinates $q = (q_1, q_2, \ldots, q_n)$ in an n dimensional space. We may think, for example, of the two angles (θ, ϕ) that specify a point on a sphere or on a torus.

Let ds be an infinitesimal distance in the vicinity of a point q belonging to the surface. Using the definition of the (symmetric) metric tensor $g_{ij}(q)$, one can write

$$ds^2 = \sum_{i,j} g_{ij}(q) \, dq_i \, dq_j.$$

As a simple example we may cite a sphere of radius R: the dimension is 2 and $ds^2 = R^2 \left(d\theta^2 + \sin^2\theta\, d\phi^2 \right)$. Another less elementary example was proposed in Problem 3.15 concerning the precession of Mercury's orbit.

1. Give the expression of the Lagrangian and of the momenta. Give the constant of the motion (the energy) due to translational time invariance. Deduce that $\dot{s}$ is constant.

2. The particle is placed at point O at the time $t = 0$. Give the expression of the action function for a point A located at a distance $s(q)$ from O on the trajectory (unknown for the moment) for the time T. Give also the reduced action as a function of the point and of the energy.

3. Check your results in the case of a free particle moving on a straight line.

4. Conclusion: using Maupertuis' principle, it can be proved that the trajectory of the free particle on the surface is a geodesic on this surface.

5.4. Wave Surface for Free Fall
[Solution and Figure p. 261]   ★ ★

A simple example for wave fronts and trajectories

In the plane xOz, one is concerned by the free fall of a mass m in a uniform vertical gravitational field g. One chooses an arbitrary origin O, a horizontal Ox axis and a vertical Oz axis oriented in the sense opposed to the field.

1. Write down the characteristic Hamilton–Jacobi equation for the reduced action. Separating the variables, find a complete integral. It will be useful to introduce as a conserved quantity, in addition to the total energy E, the kinetic energy along the Ox axis: T_x (see also Problem 5.1 case B).

2. From this complete integral, determine the wave surfaces, defined as the surfaces with the same reduced action. Plot them with the help of a pocket calculator or a micro computer.

3. Calculate the momentum at an arbitrary point (x_1, z_1), for time $t = 0$. With these initial conditions, give $x(t)$ and $z(t)$. Deduce the equation of the trajectory.

4. Calculate at the point (x_1, z_1) the slope α_1 of the tangent to the trajectory. Calculate at the same point the slope β_1 of the tangent to the wave front passing through this point. Prove that the trajectory is orthogonal to the wave front at this point.

5.5. Peculiar Wave Fronts
[Solution and Figure p. 264]   ★ ★

Wave fronts and trajectories in a constant gravitational field

From a wave front, contained in the plane yOz, one wishes to build the successive wave fronts for a particle with mass m, in a constant gravitational field g, characterized by an energy $E = 0$, taken as null by convention. One chooses horizontal axes Ox and Oy and a vertical axis Oz oriented along the field direction (so that the potential energy is given by $-mgz$).

This can be achieved following Huygens' construction from the reference front which will be taken as the plane yOz, limited to the region $z > 0$.

To begin with, we need the trajectories that originate from the plane yOz with an initial velocity directed along Ox, and at the altitude $z_0 = h > 0$.

1. We recall that the vertical motion is uniformly accelerated and the horizontal motion is uniform. Give $z(t)$ and $x(t)$. Eliminate the time and use the condition $E = 0$ to find the equation of the trajectory beginning at $z_0 = h > 0$. Check that all the trajectories are tangent to the first bisector (this is a caustic).

2. The intersection of this trajectory with the wave front is characterized by the value $\tilde{S} = \int \sqrt{2m(E-V)}\, dl$ for the reduced action $\tilde{S}$ where the integration is performed over the trajectory. Compute this integral taking as integration variable the quantity $\tilde{x}$, which runs from 0 to x.

3. Solving the cubic equation of the previous question, express x and z as function of the h parameter for a given value $\tilde{S}$ of the action. These functions $x(h)$ and $z(h)$ provide a parametric representation of the wave front characterized by the value $\tilde{S}$ for the action. You could use the well known relation: $\sinh u = 4\sinh^3(u/3) + 3\sinh(u/3)$.

4. With the help of a pocket calculator or a micro-computer, plot the curves corresponding to the wave fronts and to the trajectories and check that they are orthogonal.

5.6. Electrostatic Lens [Solution p. 265]   ★ ★ ★

An electrostatic system that behaves like an optical system

Electrostatic lenses can be found in electronic microscopes, in particle accelerators and in many other devices. Their purpose is to focus the trajectory of charged particles (with charge q_e) which tend to spread out, in the same way that an ordinary lens focusses at the exit the light rays that are divergent at the entrance.

An electrostatic lens is composed of a set of conductors held at ad hoc electric potentials, this set up having a cylindrical symmetry.

The first obstacle to overcome is associated with electromagnetic problems. You will prove that the potential energy, at a distance r from the axis of revolution Oz and at a distance z along this axis is given approximately by

$$V(r,z) = q_e U(r,z) = q_e \left[U(0,z) - \frac{1}{4} r^2 \partial_{z^2}^2 U(0,z) \right].$$

Because of the revolution symmetry around the Oz axis, the electric potential is obviously of the form $U(r,z)$, with the natural choice of cylindrical coordinates. In the space left free by the conductors, this potential obeys Poisson's equation

$$\Delta U(r,z) = \frac{1}{r} \partial_r U(r,z) + \partial_{r^2}^2 U(r,z) + \partial_{z^2}^2 U(r,z) = 0.$$

Performing the Taylor expansion of the potential in the vicinity of the axis, and inserting in Poisson's equation, show that

$$\frac{1}{r} \partial_r U(0,z) + 2 \partial_{r^2}^2 U(0,z) + \partial_{z^2}^2 U(0,z) = 0.$$

1. Relying on physical arguments (Gauss's theorem for instance), justify that the radial electric field vanishes on the axis of revolution. Deduce the approximation $U(r,z) = U(0,z) - \frac{1}{4} r^2 \partial_{z^2}^2 U(0,z)$. The proposed form for the potential follows.

    Maupertuis' principle allows us to address the problem of trajectories in a very economical way.

    In the following, we set $V(r,z) = V(z) - \frac{1}{4} r^2 V''(z)$ and $T(z) = E - V(z)$, the kinetic energy of the charge if it is situated on the axis.

2. Using Maupertuis' principle, deduce the differential equation for $r(z)$ corresponding to the trajectory of the particle inside the conductor, provided it does not deviate too much from the axis. This means that we neglect all the terms of order greater than one for r or its derivative $r' = dr/dz$.

    This equation is linear and we speak of corpuscular optics. From a point located on the axis referred to as the object, let us emit identical charged particles with the same kinetic energy, but with different orientations for the velocities.

3. Show, relying on the linearity of the differential equation, that they leave the lens either parallel, or diverging from or converging to the same point (image).

Remark: Why do we build microscopes working with electrons rather than light ? Just as the light wave focuses onto special points given by geometrical optics, the wave associated with each electron focuses onto the same points, as we saw in this problem. However, the wave length of the De Broglie wave associated with the electron can be made much smaller than the wave length of light; the resolution power of the electronic microscope is thus much improved.

5.7. Maupertuis' Principle with an Electromagnetic Field
[Solution p. 268] ★ ★ ★

Maupertuis' principle is also useful in the presence of an electromagnetic field

How can we apply Maupertuis' principle if the Hamiltonian is not expressed in the form $T+V$? Consider a particle with charge q_e and mass m, specified by its Cartesian coordinates $(x,y,z) = \boldsymbol{r}$ and subject to the electromagnetic potential $(U(\boldsymbol{r},t), \boldsymbol{A}(\boldsymbol{r},t))$. We will restrict ourselves to a non-relativistic treatment.

1. Show that the reduced action is

$$\int \boldsymbol{p}.d\boldsymbol{r} = \int \left(\sqrt{2m(H - q_e U)}\, dl + q_e \boldsymbol{A}.d\boldsymbol{r} \right),$$

where dl is the length element along the trajectory.

2. Using Maupertuis' principle we study the motion of this particle in the plane xOy, subject only to a uniform magnetic field $\boldsymbol{B}$ perpendicular to this plane. We recall that we can choose a gauge for which the vector potential is given indifferently by $\boldsymbol{A}(-yB,0,0)$, or by $\boldsymbol{A}(0,xB,0)$ or by the half sum (have a glance at Problem 2.9).

    Demonstrate the analogy between Maupertuis' principle applied to this system and the isoperimetric problem. Deduce the equations of motion and check that the trajectory is a circle with a radius R to be determined.

3. One subjects the particle to a motion on a closed curve enclosing a magnetic field which vanishes on this contour. The particle experiences no force except the reaction force normal to the contour (see Fig. 5.1). For simplicity we take into account only the vector potential. Use as a generalized coordinate the curvilinear abscissa s along this curve, referred to an arbitrary origin. Give the expression of the Lagrangian (denote by $\boldsymbol{t}(s)$ the unit vector tangent to the curve at the point with coordinate s). Deduce the momentum, the Hamilton function and the reduced action. Show that $\dot{s}$ is constant but not p.

Show that this magnetic field (although invisible for the particle around its trajectory) nevertheless induces a variation of the reduced action, the value of which is $q_e \Phi$ over a complete revolution, where Φ is the magnetic flux crossing the loop.[9]

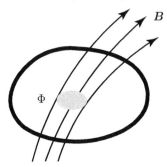

Fig. 5.1 –Charged particle moving on a closed contour (thick line) enclosing a magnetic flux tube

5.8. Separable Hamiltonian, Separable Action [Solution p. 270] ★

Separable Hamiltonian with several degrees of freedom

In the case of separation of variables, it can be shown that, with a judicious choice of the n coordinates, the action can be expressed as a sum of n separate terms, each one being a function of one coordinate only. This favorable situation allows us to find a complete action function and, using Jacobi's theorem, to completely solve the problem. The original Hamiltonian may not exhibit this property; the Hamiltonian of the particle subject to a central potential belongs to this category. In this problem, we will illustrate as an example the very simple general property corresponding to the converse, i.e., if the Hamiltonian is separable in the above sense, then the action is separable and the problem is solved.

Consider the study of the trajectory, in a uniform gravitational field g directed along the Oz axis, of a particle with mass m moving in the two-dimensional space of the xOz plane.

The Hamiltonian $H = p_x^2/(2m) + p_z^2/(2m) + mgz$ is manifestly separable.

1. Write down the characteristic Hamilton–Jacobi equation for the reduced action $\tilde{S}(x, z, E)$.

2. Seek "separable" solutions under the form $\tilde{S}(x, z, E) = S_x(x, E) + S_z(z, E)$. What happens in this case to the preceding equation?

---

[9] In the quantum mechanical framework, this additional term can be investigated experimentally. This is the Aharonov–Bohm effect.

3. Differentiating this equation with respect to x (or with respect to z), show that these functions must obey two separate characteristic Hamilton–Jacobi equations

$$\frac{1}{2m}\left(\frac{dS_x(x)}{dx}\right)^2 = e; \qquad \frac{1}{2m}\left(\frac{dS_z(x)}{dz}\right)^2 + mgz = E - e$$

where e is a (positive) constant.

4. Give the complete Jacobi integral for these two separate actions and thereby the total action.

5. Integrate the problem completely using Jacobi's theorem. Check in particular that the trajectory is the expected parabola.

5.9. Stark Effect
[Solution and Figures p. 271] ★ ★ ★

Parabolic coordinates can be used to separate variables in a problem which, initially, does not appear separable

Although rare, the cases for which the variables can be separated are important in physics. One of them, the so-called Stark effect, concerns an electric charge subject to a central electrostatic potential $-K/r$, on which one superposes a uniform electric field (for instance along the Oz axis) which arises from the potential kz.

In this case, it is astute to work with parabolic coordinates defined by

$$\xi = r + z; \qquad \eta = r - z; \qquad \phi$$

where $(\rho = \sqrt{x^2 + y^2}, \phi, z)$ are the standard cylindrical coordinates and $r = \sqrt{\rho^2 + z^2}$ is the modulus of the radius vector.

1. Show that the potential can be written as $(V_\xi(\xi) + V_\eta(\eta))/(\xi + \eta)$ and write, in explicit form, the two functions $V_\xi(\xi)$ and $V_\eta(\eta)$.

2. Using the expression of the kinetic energy in terms of parabolic coordinates, write down the Hamilton function (in this case, since the vector potential is null, this function is the sum of the kinetic energy and the potential energy given above).

3. Write down the Hamilton–Jacobi equation, using the fact that the ϕ coordinate is cyclic. Multiply the Hamilton–Jacobi equation by $\xi + \eta$ and remark that the variables can be separated. Split the initial problem into three one-dimensional problems, introducing three constants. Do not attempt to obtain an analytical form for the reduced action.

Just as for an autonomous one-dimensional system, there exist, for each separate variable, allowed regions where the square of the momentum is positive. For trajectories that do not deviate too much from the attractive center, the values of the constants (energy E, projection of the angular momentum σ, and separation energy β) are such that there exist two cusps.

4. Show that, if this is the case, these restrictions in the space (ξ, η) impose that the trajectory in the plane (ρ, z) is maintained in a region limited by two parabolae oriented upwards ($\eta = $ const) and two parabolae oriented downwards ($\xi = $ const).

We encourage the reader to make a numerical application with the following parameters $m = K = 1$ (this prejudices in no way a choice for the length and time units), $k = 0.4$, $\sigma = 0.45$, $\beta = 0.2$, $E = -1.3$.

In your answer, you will see how a trajectory (obtained by numerically solving Hamilton's equations with parabolic coordinates) fills the space.

5.10. Orbits of Earth's Satellites
[Solution and Figures p. 275] ★ ★ ★

Separation of variables in a problem that is not very simple: the motion of Earth's satellites. This is the classical "two-center problem", which is separable with the use of elliptic coordinates

The orbit of Earth's satellites is much more complex than the simple ellipse given by Kepler's laws. Among the numerous reasons for these complications, one is the fact that the Earth is not spherical, but flattened at the poles so that the potential to which the satellite is subjected does not exhibit the $1/r$ dependence. As a consequence, the motion is not planar and the orbit is not closed.

The purpose of this problem is to demonstrate that, with a better approximation and a judicious choice of coordinates, the problem can be completely solved by quadratures.

We showed in the Problem 2.11 dealing with the equinox precession (it is sufficient to assume in this problem that the Sun plays the role of the satellite) that, due to the Earth (mass M), a satellite of mass m is subject to a gravitational potential of the form:

$$V(r, u) = -GmM\left(\frac{1}{r} + \frac{(I - I_3)}{Mr^3}\frac{(3u^2 - 1)}{2}\right) + O(1/r^4)$$

The Earth is assumed to be an ellipsoid with cylindrical symmetry around the pole axis and with a uniform density.

I_3 is the moment of inertia with respect to the axis of revolution and I, the moment of inertia with respect to a perpendicular axis. r represents the distance of the satellite to the center of the Earth and u, the cosine of the angle between the vector radius of the satellite and the pole axis.

Let us assume for the moment that the Earth exhibits the shape of a rugby ball ($I > I_3$), the so-called prolate shape. One can approximately simulate the gravitational attraction exerted by the Earth as that due to two very close equal masses, located on the axis of revolution and separated by 2σ (quadrupolar approximation). In this simplified model, a better description of the potential would be thus to consider an expression of the form $1/(2r_1) + 1/(2r_2)$, where r_i is the distance between the satellite and the mass $i = 1$ or 2.

Let us study the potential corresponding to this model, which is illustrated in the Fig. 5.2.

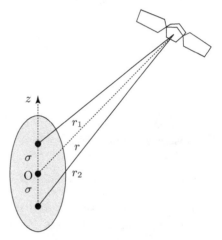

Fig. 5.2 The prolate "Earth" is schematized by two identical masses placed on the axis of revolution at a distance $\pm\sigma$ from the center. The satellite is subjected to a gravitational attraction from the two masses situated at the distances r_1 and r_2

1. The potential can be written as $V = -\dfrac{K}{2}\left(\dfrac{1}{|\boldsymbol{r} - \sigma\hat{z}|} + \dfrac{1}{|\boldsymbol{r} + \sigma\hat{z}|}\right)$
   where $\hat{z}$ is the unit vector along Oz.

   Find the expansion of the potential in terms of powers[10] of σ/r:

   $$\frac{1}{|\boldsymbol{r} - \sigma\hat{z}|} = \frac{1}{r} + \frac{\sigma u}{r^2} + \frac{(3u^2 - 1)\sigma^2}{2r^3} + O(\sigma^3/r^4)$$

---

[10] An expansion to all orders generates the Legendre polynomials.

if u is the angle between the direction of $\boldsymbol{r}$ and $\hat{\boldsymbol{z}}$. Give the expression of the potential V and show that it can be identified with the expression $V(r, u)$ given previously. Deduce the values of K and σ as a function of the characteristics of the Earth.

Remark: The previous relation does not imply that we can replace the gravitational problem due to an ellipsoid by that of two distant masses. This is true only up to order $1/r^3$. For higher order, the two multipole expansions differ. Nevertheless this simple model already provides a good approximate description of the phenomenon.

For this type of potential, the problem is separable if one works with the set of elliptic coordinates. This corresponds to the famous "two-center problem" which is treated in many textbooks.

However the true problem of interest is not quite this one. Indeed the Earth is not prolate but it is flat at the poles; this oblate shape $(I < I_3)$ is incompatible with the condition $\sigma^2 > 0$ that arises from this set of coordinates. Physicists however are resourceful people: why not consider σ as a purely imaginary number and maintain the use of elliptic coordinates.

In this framework, the potential still reads

$$-\frac{K}{2}\left(\frac{1}{r_1} + \frac{1}{r_2}\right),$$

but now with $r_{1,2} = \sqrt{\rho^2 + (z \mp i\sigma)^2}$. From the cylindrical coordinates (ρ, ϕ, z), one defines the elliptic coordinates (ξ, η, ϕ) by

$$\xi = \frac{r_1 + r_2}{2\sigma}, \qquad \eta = \frac{r_2 - r_1}{2i\sigma}.$$

Notice that, with such a treatment, the distances r_1 and r_2 are two complex conjugate numbers. In contrast, the elliptic coordinates ξ and η, as well as the potential, are real quantities. The contour curves of these coordinates are represented in the Fig. 5.3.

2. With this set of coordinates calculate the elementary length. A preliminary calculation of $r_1^2 - r_2^2$, $r_1^2 + r_2^2$, $r_1^2 r_2^2$ can help a little.

3. Deduce the kinetic energy. Express the potential in terms of these coordinates. Deduce the Lagrangian for the satellite. Then, perform the Legendre transform to obtain the Hamiltonian. Lastly, write down the characteristic Hamilton–Jacobi equation for this problem.

4. We will see in this question that, with this set of coordinates, the problem is separable.

   First notice that ϕ is a cyclic coordinate and that $p_\phi = \partial_\phi S = \alpha$ is a constant.

Multiply the equation by $\xi^2 + \eta^2$; the separation of the variables becomes evident if one uses the following equality:

$$\frac{1}{(1+\xi^2)(1-\eta^2)} = \frac{1}{(\xi^2+\eta^2)}\left(\frac{1}{1-\eta^2} - \frac{1}{1+\xi^2}\right).$$

Introducing a second constant β, separate the variables and give the expression of the complete action. Do not try to calculate the integrals.

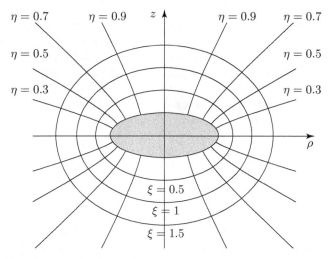

Fig. 5.3 Mesh of elliptic orthogonal coordinates ξ and η obtained for some particular values

Thanks to Jacobi's theorem, the problem is fully solved: it is enough to identify three new constants as the derivatives of the action with respect to the three constants α, β, E of the complete solution. The three initial coordinates and the three components of the initial momentum will fix the values of these six constants.

In this problem, one can adopt the same approach as that employed in Problem 5.9 to determine the allowed regions and their implications on the limitation for trajectories.

5.11. Phase and Group Velocities
[Solution p. 279] ★

Some aspects of phase and group velocities

Consider a particle of mass m, possibly subject to a potential $V(\mathbf{r})$.

It can be proved that the speed of the displacement of wave fronts for a trajectory with given energy E, also called phase velocity, is given by $v_\varphi = E/p$. It depends on the position of the particle on the trajectory. The group velocity is defined as the physical velocity of the particle $v_g = |\dot{r}|$. It depends also on the position on the trajectory. The aim of this problem is to study the relation between these two types of velocities for some particular cases.

1. In the case of a free particle $V(r) = 0$, what is the relationship between v_φ and v_g?

   In quantum mechanics, a free particle is described by a plane wave wave function: $\psi(r,t) = e^{i(k \cdot r - \omega t)}$. The angular frequency of the wave ω is linked to the particle energy through $E = \hbar\omega$ ($\hbar$= Planck constant / 2π). This is Einstein's postulate. The wave vector k is linked to the momentum through $p = \hbar k$. This is the De Broglie's postulate. Starting from the relation between E and p for the free particle, find the relation between ω and k for the corresponding plane wave. The phase velocity corresponds to the conservation of the phase, i.e. $v_\varphi = \omega/k$. The group velocity is only meaningful for wave packets formed by a superposition of plane waves. The group velocity corresponds to the speed of the maximum of the wave packet; it can be shown that $v_g = d\omega/dk$.

2. Calculate v_φ and v_g and give the relation between them. Conclusion.

3. In the case of a particle subject to an arbitrary potential, show that the product $v_\varphi v_g$ is constant. What is the value of this constant?

4. We consider now a relativistic example. Calculate v_φ and v_g, as well as $v_\varphi v_g$ for a free particle. Generalize to the broader case of a particle subject to a scalar potential, whose Hamiltonian is given by $H(r,p) = \sqrt{(mc^2 + V(r))^2 + p^2 c^2}$. Comment on this property.

Problem Solutions

5.1. How to Manipulate the Action and the Reduced Action [Statement p. 239]

Preliminary question

For an autonomous system and for a given orbit ($t_1 = 0$, $q_1 = 0$ is the starting point by convention), there is a relation between the energy E and the travel time t. The total action is linked to the reduced action by Equation (5.5), which we rewrite as: $S(q,t) = -Et - (-\tilde{S}(q,E))$ in which E must be understood as $E(t)$. Moreover, from (5.2), $\partial_t S = -H = -E$.

Problem Solutions

The Legendre transform of the action S is, by definition, $(-E)t - S(q,t)$, a quantity in which the time t must be understood as $t(E)$. This transform is precisely $-\tilde{S}(q, E)$.

One knows that the Legendre transform is involutive; this means that S is the Legendre transform of $-\tilde{S}$. Let us check. From Equation (5.8), one has $\partial_E(-\tilde{S}) = -t$ and the Legendre transform of $-\tilde{S}$ is by definition $E(-t) - (-\tilde{S}(q,E))$, so that the result is $S(q,t)$, as it should be.

A – One-dimensional free particle

1. The Lagrangian is $L(q,\dot q) = \frac{1}{2}m\dot q^2$, the Hamiltonian $H(q,p) = p^2/(2m)$ and the trajectory passing through the imposed points is obviously $\tilde q(\tilde t) = (q/t)\tilde t$. One deduces the general action $S(q,t) = \int_0^t \frac{1}{2}m(q^2/t^2)\,d\tilde t$ i.e.,

$$S(q,t) = \frac{1}{2}m\frac{q^2}{t}.$$

The partial derivatives $\partial_t S = -\frac{1}{2}mq^2/t^2$ and $\partial_q S = mq/t$ are calculated. Consequently

$$H(q, \partial_q S) = \frac{(\partial_q S)^2}{2m} = \frac{mq^2}{2t^2} = -\partial_t S$$

so that the Hamilton–Jacobi equation is satisfied:

$$H\left(q, \frac{\partial S}{\partial q}\right) = -\frac{\partial S}{\partial t}.$$

2. The Hamiltonian is equal to the energy $H = E$ and one derives $\tilde p(E) = \sqrt{2mE}$. The reduced action follows (see (5.6)) $\tilde S(q,E) = \int_0^q \sqrt{2mE}\,d\tilde q = \sqrt{2mE}\,q$. Thus

$$\tilde S(q,E) = \sqrt{2mE}\,q.$$

3. The characteristic Hamilton–Jacobi equation $H(q, \partial_q \tilde S) = E$ reads in this case $(\partial_q \tilde S)^2/(2m) = E$, or $\partial_q \tilde S = \pm\sqrt{2mE} = p$ which is integrated to give $\tilde S(q,E) = \pm\sqrt{2mE}\,q$. If $q > 0$, the velocity is positive, so that $p > 0$. One must retain the $+$ sign and the corresponding reduced action is $\tilde S(q,E) = \sqrt{2mE}\,q = \sqrt{2mE}\,|q|$. If $q < 0$, the velocity is negative $p < 0$. One retains the $-$ sign and $\tilde S(q,E) = -\sqrt{2mE}\,q = \sqrt{2mE}\,|q|$. Thus, in any case:

$$\tilde S(q,E) = \sqrt{2mE}\,|q|.$$

The travel time is given by (5.8) $t = \partial_E \tilde S$, i.e., $t = \sqrt{mq^2/(2E)}$ or $E = \frac{1}{2}m(q/t)^2$, as expected.

4. With this last value of the energy, we have $-Et = -\frac{1}{2}mq^2/t$ and the reduced action $\tilde{S} = \sqrt{2mE}\,|q| = mq^2/t$. Consequently the total action is $S = -Et + \tilde{S} = \frac{1}{2}mq^2/t$ which is identical with the answer to the first question.

5. For a relativistic particle, the Lagrangian is $L = -mc^2\sqrt{1 - \dot{q}^2/c^2}$, the momentum $p = \partial_{\dot{q}}L = m\dot{q}/\sqrt{1 - \dot{q}^2/c^2}$ and the Hamiltonian

$$H = \sqrt{p^2c^2 + m^2c^4}.$$

The Hamiltonian does not depend on time; on the trajectory its value remains constant and equal to the energy E. One deduces that the momentum remains also constant as does the velocity $\dot{q} = q/t$. The action is calculated from (5.1) and, since the Lagrangian is constant on the trajectory, one has $S(q,t) = Lt$, or after inserting the value of the velocity:

$$S(q,t) = -mc\sqrt{c^2t^2 - q^2}.$$

6. The characteristic Hamilton–Jacobi equation $H(q, \partial_q \tilde{S}) = E$ is written in this case: $E^2 = m^2c^4 + (\partial_q \tilde{S})^2 c^2$, or $\partial_q \tilde{S} = \pm\sqrt{E^2 - m^2c^4}/c$, which may be integrated to give $\tilde{S} = \pm\sqrt{E^2 - m^2c^4}\,q/c$. A remark analogous to the non-relativistic case concerning the choice for the sign leads to the expression (valid in all cases):

$$\tilde{S}(q, E) = \frac{1}{c}\sqrt{E^2 - m^2c^4}\,|q|\,.$$

7. The travel time is given by (5.8), that is $t = E|q|/\left(c\sqrt{E^2 - m^2c^4}\right)$ or, after inversion, $E = mc^2/\sqrt{1 - q^2/(c^2t^2)}$. With this latter value for the energy, the relation between time and energy is $-Et = -mc^3t^2/\sqrt{c^2t^2 - q^2}$ and the reduced action becomes

$$\tilde{S} = \frac{1}{c}\sqrt{E^2 - m^2c^4}\,|q| = \frac{mcq^2}{\sqrt{c^2t^2 - q^2}}.$$

Thus the total action reads $S = -Et + \tilde{S} = -mc\sqrt{c^2t^2 - q^2}$ as expected from Question 5.

B – Particle in a constant gravitational field

1. In this case, the Lagrangian is $L(q, \dot{q}) = \frac{1}{2}m\dot{q}^2 - mgq$, the momentum $p = m\dot{q}$ and the Hamiltonian

$$H(q, p) = \frac{p^2}{2m} + mgq.$$

It is easy to check that the temporal evolution on the trajectory passing through the imposed points is expressed as: $\tilde{q}(\tilde{t}) = -\frac{1}{2}g\tilde{t}^2 + v_0(q,t)\tilde{t}$ with $v_0(q,t) = \frac{1}{2}gt + q/t$. Now, the Lagrangian can be calculated on the trajectory, $L = mg^2\tilde{t}^2 - 2mgv_0\tilde{t} + mv_0^2/2$, and the general action is the integral of this expression between $\tilde{t} = 0$ and $\tilde{t} = t$. Finally, we obtain:

$$S(q,t) = m\left[-\frac{1}{24}g^2t^3 - \frac{1}{2}gtq + \frac{1}{2}\frac{q^2}{t}\right].$$

One deduces the partial derivatives $\partial_t S = -\frac{1}{2}m\left[\frac{1}{4}g^2t^2 + gq + q^2/t^2\right]$ and $\partial_q S = m\left[-\frac{1}{2}gt + q/t\right]$. The substitution in $H(q, \partial_q S) = (\partial_q S)^2/(2m) + mgq$ shows that the value obtained is simply $-\partial_t S$. The Hamilton–Jacobi equation is thus satisfied:

$$H\left(q, \frac{\partial S}{\partial q}\right) = -\frac{\partial S}{\partial t}.$$

Since the Hamiltonian does not depend on time, its value on the trajectory is constant and equal to the energy; consequently $E = -\frac{\partial S}{\partial t}$ that is $E(t) = \frac{1}{2}m\left[\frac{1}{4}g^2t^2 + gq + q^2/t^2\right] = \frac{1}{2}mv_0^2$. Thus,

$$E(t) = \frac{1}{2}m\left[\frac{1}{4}g^2t^2 + gq + \frac{q^2}{t^2}\right].$$

One sees that it is not easy to invert this relation to obtain $t(E)$, because one must solve a bi-squared equation.

2. From the expression for Hamilton's function, we derive the momentum as a function of the energy $\tilde{p} = \sqrt{2m(E - mg\tilde{q})}$ (choosing $\tilde{p} > 0$ is equivalent to stating that we pass through the extremity before the maximum is attained) and the reduced action is obtained from (5.6) or

$$\tilde{S}(q,E) = \int_0^q \sqrt{2m(E - mg\tilde{q})}\, d\tilde{q} :$$

$$\tilde{S}(q,E) = -\frac{2}{3g}\sqrt{\frac{2}{m}}(E - mgq)^{3/2} + \frac{m}{3g}\left(\frac{2E}{m}\right)^{3/2}.$$

Let us start from $\tilde{S}(q,E) = \pm\frac{2}{3g}\sqrt{2/m}\,(E - mgq)^{3/2}$. One calculates the partial derivative $\partial_q \tilde{S} = \mp\sqrt{2m(E - mgq)}$. Then, substituting in $H(q, \partial_q \tilde{S})$, one obtains $(\partial_q \tilde{S})^2/(2m) + mgq = E$. The characteristic Hamilton–Jacobi equation is satisfied. To remain consistent with the sign chosen for the momentum, one must take the $-$ sign and one sees that the two expressions for the reduced action differ only by the constant $(m/3g)(2E/m)^{3/2}$, which has no influence on the equation of motion.

3. The travel time is given by (5.8) $t = \partial_E \tilde{S}$. With the expression for the reduced action obtained above, the travel time as a function of the energy is easily found and exhibits the simple form:

$$t(E) = \frac{1}{g}\left[\sqrt{\frac{2E}{m}} - \sqrt{\frac{2(E-mgq)}{m}}\right].$$

With a little courage, one can check that this expression does indeed fulfill the bi-squared equation resulting from the inversion of $E(t)$.

C – Three-dimensional free particle

1. The Lagrangian reads $L(\dot{\bm{r}}) = \frac{1}{2}m\dot{r}^2$, the momentum $\bm{p} = m\dot{\bm{r}}$ and the Hamiltonian $H(\bm{p}) = \bm{p}^2/(2m)$. The temporal evolution on the trajectory, which satisfies the required conditions, is given by $\tilde{x}(\tilde{t}) = (x/t)\tilde{t}$, $\tilde{y}(\tilde{t}) = (y/t)\tilde{t}$, $\tilde{z}(\tilde{t}) = (z/t)\tilde{t}$. The Lagrangian on the trajectory is thus $L = \frac{1}{2}mr^2/t^2$ and the action, which is the corresponding integral between $\tilde{t} = 0$ and $\tilde{t} = t$ is obtained as:

$$S(\bm{r},t) = m\frac{r^2}{2t}.$$

On the trajectory, the momentum and the velocity are constant and the reduced action is given by (5.6): $\tilde{S}(\bm{r}, E) = \int_0^t \bm{p}\cdot\dot{\bm{r}}\,d\tilde{t} = \bm{p}\cdot\dot{\bm{r}}\,t = \bm{p}^2 t/m$ $= 2Et$. On the other hand, one has $E = \frac{1}{2}m(r/t)^2$, so that the time can be eliminated in favour of the energy to obtain:

$$\tilde{S}(\bm{r}, E) = \sqrt{2mEr^2}.$$

It is easy to check that this function obeys the characteristic Hamilton–Jacobi equation.

2. Let us start from the characteristic equation

$$\frac{1}{2m}\left[(\partial_x \tilde{S})^2 + (\partial_y \tilde{S})^2 + (\partial_z \tilde{S})^2\right] = E.$$

The z variable is cyclic and the reduced action is recast in the form $\tilde{S}(x, y, z, E) = \hat{S}(x, y, E) + \alpha_z z$. However the same argument applies to any of the variables so that the reduced action can be expressed in the form $\tilde{S}(x, y, z, E) = \alpha_x x + \alpha_y y + \alpha_z z$. The characteristic equation is satisfied if $\bm{\alpha}^2 = 2mE$. The action is written in the separable form:

$$\tilde{S}(\bm{r}, \bm{\alpha}) = \bm{\alpha}\cdot\bm{r}; \quad \bm{\alpha}^2 = 2mE.$$

Manifestly the constant vector $\bm{\alpha}$ coincides with the momentum.

In general, there exists no relation between the two reduced actions given in Questions 1 and 2. Nevertheless, one may remark that if the momentum is taken as parallel to the radius vector $\bm{\alpha} = \sqrt{2mE/r^2}\,\bm{r}$ the two expressions are equal. Jacobi's theorem $\partial_{\alpha} S = \beta$ applied to this complete solution gives directly the equations of motion $r(t) = (\alpha/m)\, t + \beta$.

3. We carry out the same study with the new Lagrangian

$$L(\dot{r}) = -mc^2 \sqrt{1 - \frac{\dot{r}^2}{c^2}},$$

momentum $p = m\dot{r}/\sqrt{1 - \dot{r}^2/c^2}$ and Hamiltonian $H(p) = \sqrt{p^2 c^2 + m^2 c^4}$. The momentum, as well as the velocity, is a constant vector and the equation of the trajectory is $\tilde{r}(\tilde{t}) = (r/t)\, \tilde{t}$. Since the Lagrangian is constant along the trajectory, the action is simply given by $S = Lt$, that is:

$$S(r,t) = -mc\sqrt{c^2 t^2 - r^2}.$$

It is not difficult to check that it fulfills the Hamilton–Jacobi equation. One can use a treatment similar to that employed in the one-dimensional case.

One still has

$$\tilde{S}(r, E) = \int_0^t \tilde{p} \cdot \dot{\tilde{r}}\, d\tilde{t} = \tilde{p} \cdot \dot{\tilde{r}}\, t = m \frac{\dot{r}^2 t}{\sqrt{1 - \dot{r}^2/c^2}} = \frac{E}{c^2} \frac{r^2}{t^2} t.$$

On the other hand $E^2 = m^2 c^6/(c^2 - r^2/t^2)$, so that the time can be eliminated in favour of the energy to find:

$$\tilde{S}(r, E) = \frac{1}{c}\sqrt{(E^2 - m^2 c^4)\, r^2}.$$

It is easy to check that this function fulfills the characteristic Hamilton–Jacobi equation.

4. The characteristic equation

$$\left[(\partial_x \tilde{S})^2 + (\partial_y \tilde{S})^2 + (\partial_z \tilde{S})^2\right] c^2 = E^2 - m^2 c^4$$

is separable exactly as for the non-relativistic case, so that the reduced action is given as $\tilde{S}(x, y, z, E) = \alpha_x x + \alpha_y y + \alpha_z z$. The characteristic equation is satisfied if $\alpha^2 c^2 = E^2 - m^2 c^4$. The action is expressed as:

$$\tilde{S}(r, \alpha) = \alpha \cdot r; \quad \alpha^2 c^2 = E^2 - m^2 c^4.$$

Manifestly in this case also, the constant vector α is simply the momentum.

5.2. Action for a One-Dimensional Harmonic Oscillator [Statement p. 241]

1. For the harmonic potential, Lagrange's equation leads to the well known differential equation $\ddot{\tilde{q}}+\omega^2\tilde{q}=0$ whose general solution is $\tilde{q}(\tilde{t}) = A\cos(\omega\tilde{t}) + B\sin(\omega\tilde{t})$.

   The integration constants A and B are determined by requiring that the trajectory passes through the points q_1, t_1 and q, t. A simple calculation gives, with $T = t - t_1$:

$$A = \frac{(q_1 \sin(\omega t) - q\sin(\omega t_1))}{\sin(\omega T)}; \quad B = \frac{(q\cos(\omega t_1) - q_1 \cos(\omega t))}{\sin(\omega T)}.$$

   With the definition $\tau = t + t_1$, these values leads to:

$$B^2 - A^2 = \frac{(q^2 \cos(2\omega t_1) + q_1^2 \cos(2\omega t) - 2qq_1 \cos(\omega\tau))}{\sin^2(\omega T)};$$

$$2AB = \frac{(-q^2 \sin(2\omega t_1) - q_1^2 \sin(2\omega t) + 2qq_1 \sin(\omega\tau))}{\sin^2(\omega T)}.$$

   The Lagrangian can now be obtained as $L = \frac{1}{2}m(\dot{\tilde{q}}^2 - \omega^2\tilde{q}^2)$ along the trajectory so that the corresponding action is $S = \int_{t_1}^{t} L(\dot{\tilde{q}}, \tilde{q})\, d\tilde{t}$, which gives: $S = \frac{1}{4}m\omega\left[(B^2 - A^2)\cos(\omega\tau) - 2AB\sin(\omega\tau)\right]$. Inserting the values of the constants obtained above, a simple calculation gives the final expression:

$$S(q, t, q_1, t_1) = \frac{m\omega}{2\sin(\omega T)}\left[-2qq_1 + (q^2 + q_1^2)\cos(\omega T)\right]$$

   which depends only on the time difference $T = t - t_1$.

2. If $\omega T = 2k\pi$, $S = m\omega(q - q_1)^2/(2\sin(\omega T))$ so that a difficulty occurs unless $q = q_1$.

   If $\omega T = (2k+1)\pi$, $S = -m\omega(q + q_1)^2/(2\sin(\omega T))$ and there is a difficulty unless if $q = -q_1$.

   These remarks are connected to the existence (or non-existence) of the trajectory passing through the required points (see Problem 3.4).

   From the general expression for the action, one deduces:

$$\partial_t S = \partial_T S = \frac{m\omega^2}{2\sin^2(\omega T)}\left[-(q^2 + q_1^2) + 2qq_1 \cos(\omega T)\right]$$

   and

$$\partial_q S = \frac{m\omega}{\sin(\omega T)}\left[-q_1 + q\cos(\omega T)\right].$$

   The Hamilton–Jacobi equation $\partial_t S = -H(q, \partial_q S)$ for the harmonic oscillator is expressed as $\partial_t S = -(\partial_q S)^2/(2m) - \frac{1}{2}m\omega^2 q^2$. Substituting for

Problem Solutions

the partial derivatives by their previous values, it is not difficult to show that the Hamilton–Jacobi equation is indeed verified:

$$\frac{\partial S}{\partial t} = -H\left(q, \frac{\partial S}{\partial q}\right).$$

Since $\partial_t S = -E$, we have the relationship between the energy E and the travel time T:

$$E = m\omega^2 \frac{[-(q^2 + q_1^2) + 2qq_1 \cos(\omega T)]}{2\sin^2(\omega T)}.$$

3. The reduced action is defined by $\tilde{S}(q, E) = \int_{q_1}^{q} \tilde{p}\, d\tilde{q}$. Moreover, the energy along the trajectory is $E = \tilde{p}^2/(2m) + \frac{1}{2}m\omega^2 \tilde{q}^2$, so that $\tilde{p}(q, q_1, E) = \sqrt{2mE - m^2\omega^2\tilde{q}^2}$. The calculation of the integral is performed with the help of the change of variable $\tilde{x} = \tilde{q}\omega\sqrt{m/(2E)}$ to give

$$\tilde{S} = \frac{2E}{\omega}\int_{x_1}^{x}\sqrt{1-\tilde{x}^2}\, d\tilde{x}.$$

Finally, one obtains:

$$\tilde{S}(q, q_1, E) = \frac{E}{\omega}\left[y\sqrt{1-y^2} + \arcsin y\right]_{y_1 = \omega q_1 \sqrt{m/(2E)}}^{y = \omega q \sqrt{m/(2E)}}.$$

From this equation, one deduces the partial derivative

$$\partial_q \tilde{S} = \sqrt{2mE - m^2\omega^2 q^2}.$$

We can now calculate $H(q, \partial_q \tilde{S}) = \frac{1}{2}m\omega^2 q^2 + (\partial_q \tilde{S})^2/(2m)$, or, with the value of $\partial_q \tilde{S}$ obtained above, $\frac{1}{2}m\omega^2 q^2 + E - \frac{1}{2}m\omega^2 q^2$:

$$E = H\left(q, \frac{\partial \tilde{S}}{\partial q}\right).$$

The characteristic equation is thus fulfilled.

The travel time is obtained from (5.8):

$$T = \partial_E \tilde{S} = \frac{1}{\omega}\left[y\sqrt{1-y^2} + \arcsin y - y_1\sqrt{1-y_1^2} - \arcsin y_1\right]$$

$$+ \frac{E}{\omega}\left[2\sqrt{1-y^2}(dy/dE) - 2\sqrt{1-y_1^2}(dy_1/dE)\right].$$

With $(dy/dE) = -y/(2E)$, $(dy_1/dE) = -y_1/(2E)$, this value can be simplified to give finally:

$$T = \frac{1}{\omega}[\arcsin y - \arcsin y_1]; \quad y = \omega q\sqrt{\frac{m}{2E}}; \quad y_1 = \omega q_1\sqrt{\frac{m}{2E}}.$$

5.3. Motion on a Surface and Geodesic
[Statement p. 241]

1. The Lagrangian is defined on the surface by

$$L = T = \frac{1}{2}mv^2 = \frac{1}{2}m\frac{ds^2}{dt^2},$$

or, using the expression of the length element,

$$L = \frac{1}{2}m \sum_{i,j} g_{ij}(q)\, \dot{q}_i\, \dot{q}_j.$$

One deduces the momentum $p_k = m \sum_i g_{ik}(q)\, \dot{q}_i$, and thereby the Hamiltonian

$$H = \sum_i p_i \dot{q}_i - L = m \sum_{i,j} g_{ij}(q)\, \dot{q}_i\, \dot{q}_j - L = 2L - L = L.$$

The Hamiltonian does not depend explicitly on time and its value on the trajectory is a constant, the energy E.

$$H = L = \frac{1}{2}m \sum_{i,j} g_{ij}(q)\, \dot{q}_i\, \dot{q}_j = E.$$

In the expression of the Hamiltonian, one must substitute the velocity by the momentum, a procedure which necessitates the inversion of the metric tensor.

One has $E = \frac{1}{2}m\dot{s}^2$, which implies that $\dot{s}$ is a constant quantity.

2. The trajectory originates from a point O, at time $t = 0$, and terminates at a point A, with curvilinear abscissa $s(q)$, at time t. The action is defined by

$$S = \int_0^t L(q, \dot{q})\, d\tilde{t} = \int_0^t E\, d\tilde{t} = Et.$$

Furthermore $E = \frac{1}{2}m(ds/dt)^2$, so that $ds/dt = \sqrt{2E/m}$ and, after integration, $s(q) = t\sqrt{2E/m}$, whence $E = ms(q)^2/(2t^2)$. Consequently the action is expressed as:

$$S(q,t) = \frac{1}{2}m\frac{s(q)^2}{t}.$$

On the other hand, one has $S(q,t) = \tilde{S}(q, E) - Et$, i.e., $\tilde{S}(q, E) = 2Et$. But in the reduced action $\tilde{S}$, time must be eliminated in favour of the energy or, as we already saw, $t = \sqrt{m/(2E)}\, s(q)$ which implies the expression of the reduced action:

$$\tilde{S}(q, E) = \sqrt{2mE}\, s(q).$$

Problem Solutions

3. In the case of a particle moving on a straight line, there is only one degree of freedom $s(q) = q$. One finds the action:

$$S(q,t) = \frac{1}{2}m\frac{q^2}{t}$$

as expected (see also Problem 5.1).

Concerning the reduced action, one has $\tilde{S}(q,E) = \sqrt{2mE}\,q$, so that the action $S(q,t) = \tilde{S}(q,E) - Et$ is given equivalently by:

$$S(q,t) = \sqrt{2mE}\,q - Et$$

expression that was already proved in Problem 5.1.

4. The Maupertuis principle stipulates that $\int p\,dq$ is stationary on the trajectory. With the expression of the momentum obtained previously, this integral can be rewritten as

$$m\sum_{i,j}\int_O^A g_{ij}(q)\dot{q}_j dq_i = m\sum_{i,j}\int_O^A g_{ij}(q)dq_j \frac{dq_i}{dt} = m\int_O^A \frac{ds^2}{dt} = \int_O^A \left[m\frac{ds}{dt}\right]ds.$$

Moreover, we remarked that $\frac{ds}{dt} = \sqrt{\frac{2E}{m}}$. It follows that $\sqrt{2mE}\int_O^A ds$ is stationary, or, equivalently, $\int_O^A ds$ is stationary.

This simply means that the trajectory follows, from the two points O and A, a geodesic on the surface.

5.4. Wave Surface for Free Fall
[Statement and Figure p. 242]

1. The Hamiltonian of the system reads

$$H(x,z,p_x,p_z) = \frac{p_x^2}{2m} + \frac{p_z^2}{2m} + mgz$$

and the associated characteristic equation is $H(x,z,\partial_x\tilde{S},\partial_z\tilde{S}) = E$, which is written in this case:

$$\frac{1}{2m}\left(\frac{\partial\tilde{S}}{\partial x}\right)^2 + \frac{1}{2m}\left(\frac{\partial\tilde{S}}{\partial z}\right)^2 + mgz = E.$$

One notices that the x variable is cyclic, with the consequence that the action can be written in the form $\tilde{S}(x,z) = S_z(z) + cx$. We note that $\partial_x\tilde{S} = p_x = c = const$. Denoting $T_x = p_x^2/(2m)$ the constant is written as $c = \sqrt{2mT_x}$. Moreover, $\partial_z\tilde{S} = dS_z/dz$ so that the characteristic equation reduces to $dS_z/dz = \pm\sqrt{2m(E - T_x - mgz)}$. The function $S_z(z, E, T_x)$ is obtained by integration:

$$S_z(z, E, T_x) = \mp 2\sqrt{2m}/(3mg)\,(E - T_x - mgz)^{3/2}.$$

Finally the reduced action is expressed as:

$$\tilde{S}(x, z, E, T_x) = \pm \frac{2}{3g}\sqrt{\frac{2}{m}}\,(E - T_x - mgz)^{3/2} + \sqrt{2mT_x}\,x.$$

2. The wave surfaces are defined as the surfaces with equal action $\tilde{S}(x, z, E, T_x) = \tilde{S}$. Substituting this value in the previous relation and isolating the z variable, one obtains (the $-$ sign is chosen in order for the momentum to be oriented in the increasing (x, z) direction):

$$z(x) = \frac{E - T_x}{mg} - \left[\frac{3}{2m\sqrt{2g}}\left(\sqrt{2mT_x}\,x - \tilde{S}\right)\right]^{2/3}.$$

3. The momentum is given by $p_x = \partial_x \tilde{S} = const$. On the other hand, $p_x = m\dot{x}$ which, after integration and use of the initial conditions, provides the horizontal temporal evolution:

$$x(t) = \frac{p_x}{m}t + x_1.$$

As for the vertical variable, one has $p_z = \partial_z \tilde{S} = \sqrt{2m(E - T_x - mgz)}$ and thus $p_{z_1} = \sqrt{2m(E - T_x - mgz_1)}$. Moreover, $p_z = m\dot{z}$. The resulting differential equation is separable, $\sqrt{2/m}\,dt = dz/\sqrt{(E - T_x - mgz)}$. After integration, one obtains $mgt = p_{z_1} - \sqrt{2m(E - T_x - mgz)}$. One eliminates the square root by squaring this equation and using the property $E = T_x + p_{z_1}^2/(2m) + mgz_1$. The vertical temporal evolution is then found to be:

$$z(t) = -\frac{1}{2}gt^2 + \frac{p_{z_1}}{m}t + z_1.$$

The travel time is extracted from the horizontal time law and inserted in the preceding (vertical) equation. After some algebra, the equation of the trajectory is obtained in the form:

$$z(x) = -\frac{mg}{4T_x}(x - x_1)^2 + \sqrt{\frac{E - T_x - mgz_1}{T_x}}(x - x_1) + z_1.$$

This is the equation of a parabola, as expected.

4. The slope of the trajectory at the point (x_1, z_1) is easily calculated:

$$\alpha_1 = \left(\frac{dz}{dx}\right)\bigg|_{x=x_1} = \sqrt{\frac{E - T_x - mgz_1}{T_x}},$$

where $z(x)$ is given in the previous question.

The slope of the wave front is more difficult to find but, relying on the value of the constant $\tilde{S}$ given by

$$\sqrt{2mT_x}\, x_1 - \tilde{S} = 2\sqrt{\frac{2}{m}}\, \frac{(E - T_x - mgz_1)^{3/2}}{3g},$$

one obtains (with a little patience) the value:

$$\beta_1 = \left(\frac{dz}{dx}\right)\bigg|_{x_1} = -\sqrt{\frac{T_x}{E - T_x - mgz_1}},$$

$z(x)$ being given in Question 2. The relation $\alpha_1 \beta_1 = -1$ proves that these two tangents are orthogonal. The point (x_1, z_1) is arbitrary; the latter property thus holds for any point and leads to the conclusion that the wave fronts are orthogonal to the trajectories.

The wave fronts (dotted lines) and the trajectories (full lines) are illustrated in Fig. 5.4.

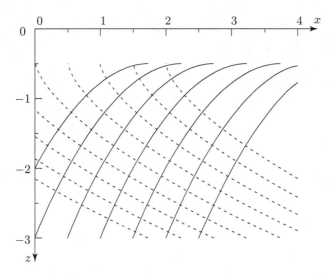

Fig. 5.4 Representation in the plane xOz of the iso-action wave front (dotted lines) and the trajectories (full lines) for a particle in a constant gravitational field

5.5. Peculiar Wave Fronts [Statement p. 243]

1. With our convention for the axes, the equations of motion are given by $\ddot{x} = 0$ and $\ddot{z} = g$. Using the initial conditions, they can be integrated easily to give $x(t) = v_0 t$ and $z(t) = \frac{1}{2}gt^2 + h$. The mechanical energy is constant and, at the starting point, it is equal to $E = 0 = \frac{1}{2}mv_0^2 - mgh$, so that $v_0 = \sqrt{2gh}$. Thus the temporal laws become

$$x(t) = \sqrt{2gh}\, t; \qquad z(t) = \frac{1}{2}gt^2 + h.$$

Elimination of time allows us to obtain the equation for the trajectory:

$$z(x) = \frac{x^2}{4h} + h.$$

At any point x_1, the slope of the tangent to the trajectory is $x_1/(2h)$ and the equation for this tangent is $z - z_1 = \frac{x_1}{2h}(x - x_1)$ with $z_1 = \frac{x_1^2}{4h} + h$. Let us choose the point $x_1 = 2h$; one sees easily that the equation of the tangent is simply $z = x$, that is the first bisector. This conclusion remains valid for any value of h, or for any trajectory. In other words, the first bisector is the envelop to all trajectories which is known as a caustic.

2. One seeks the wave front which takes the value $\tilde{S}$. It is obtained by integrating $\sqrt{2m(E-V)}$ over the trajectory. With $E = 0$, $V = -mgz$ and

$$dl = \sqrt{dx^2 + dz^2} = dx\sqrt{1 + z'^2},$$

the integral reduces to

$$\tilde{S} = m\sqrt{2g} \int_0^x \sqrt{z}\sqrt{1 + z'^2}\, d\tilde{x}.$$

Using the equation of the trajectory $z(\tilde{x}) = h + \tilde{x}^2/(4h)$, this integral becomes $\tilde{S} = m\sqrt{2gh} \int_0^x \left(1 + \frac{\tilde{x}^2}{4h^2}\right) d\tilde{x}$, which can be easily integrated to give:

$$\tilde{S} = \frac{m\sqrt{2g}}{12h^{3/2}} \left(x^3 + 12h^2 x\right).$$

3. Setting $x = 4h \sinh(u/3)$, a simple calculation shows that

$$\tilde{S} = \frac{4}{3} m\sqrt{2g}\, h^{3/2} \left(4\sinh^3(u/3) + 3\sinh(u/3)\right),$$

or, using the well known relation concerning the hyperbolic sines, $\tilde{S} = \frac{4}{3}m\sqrt{2g}\,h^{3/2}\sinh u$. The rest follows. For a given $\tilde{S}$ and for each value of h, one derives a quantity $u(h)$ from the previous relation. The abscissa of the wave front is given by $x(h) = 4h\sinh(u(h)/3)$ and the ordinate, which is on the same trajectory, by $z = h + x^2/(4h) = h\left(1 + 4\sinh^2(u/3)\right)$, or, using the relation $2\sinh^2(u/3) = \cosh(2u/3) - 1$, $z = h\left(2\cosh(2u/3) - 1\right)$. In summary, the equation of the wave fronts with a given value of $\tilde{S}$ are obtained parametrically from the following formulae:

$$u(h) = \sinh^{-1}\left(\frac{3\tilde{S}}{4m\sqrt{2gh^3}}\right); \qquad x(h) = 4h\sinh\left(\frac{u(h)}{3}\right);$$

$$z(h) = h\left[2\cosh\left(\frac{2u(h)}{3}\right) - 1\right].$$

The trajectories (full lines) and wave fronts (dotted lines) are illustrated in Fig. 5.5.

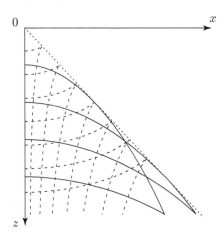

Fig. 5.5 Trajectories with null energy (full lines) and corresponding wave fronts (dashed lines) for a particle in a constant gravitational field. Before the trajectory touches the caustic (dotted line), it is orthogonal to the descending wave fronts. After meeting the caustic, it is orthogonal to the wave fronts that are curved upwards

5.6. Electrostatic Lens [Statement p. 243]

1. There is a symmetry of revolution around the Oz axis; the scalar electrostatic potential reads $U(r, z)$ and the mechanical potential exerted on the particle is $V(r, z) = q_e U(r, z)$.

In the space where the particles move, there is no charge source and the potential obeys Laplace's equation which, in cylindrical coordinates, is:

$$\Delta U = 0 = \partial_{r^2}^2 U + \frac{1}{r}\partial_r U + \partial_{z^2}^2 U.$$

We expand the potential in a Taylor series to obtain

$$U(r,z) = U(0,z) + r\,\partial_r U\big|_{r=0} + \frac{1}{2}r^2\,\partial_{r^2}^2 U\big|_{r=0};$$

$$\partial_r U = \partial_r U\big|_{r=0} + r\,\partial_{r^2}^2 U\big|_{r=0};$$

$$\partial_{r^2}^2 U = \partial_{r^2}^2 U\big|_{r=0};$$

$$\partial_{z^2}^2 U = \partial_{z^2}^2 U\big|_{r=0}.$$

Up to this order, the Laplacian becomes:

$$0 = 2\,\partial_{r^2}^2 U\big|_{r=0} + (1/r)\,\partial_r U\big|_{r=0} + \partial_{z^2}^2 U\big|_{r=0}$$

Consider a given value of z; the terms $\partial_{r^2}^2 U\big|_{r=0}$, $\partial_{z^2}^2 U\big|_{r=0}$ depend only on z and have a finite value. Let r tend to 0; these two terms maintain constant values. In order for the Laplacian to vanish, it is necessary that the other term also retains a constant value; this is possible only if $\partial_r U\big|_{r=0} = 0$. The same conclusion is obtained if one uses Gauss's theorem.

Laplace's equation now implies that $\partial_{r^2}^2 U\big|_{r=0} = -\frac{1}{2}\partial_{z^2}^2 U\big|_{r=0}$ and the truncated expansion for the potential reads simply

$$U(r,z) = U(0,z) + \frac{1}{2}r^2\,\partial_{r^2}^2 U\big|_{r=0} = U(0,z) - \frac{1}{4}r^2\partial_{z^2}^2 U\big|_{r=0}.$$

Finally, for small values of r, the mechanical potential is expressed as:

$$V(r,z) = q_e\left[U(0,z) - \frac{1}{4}r^2\frac{\partial^2 U(0,z)}{\partial z^2}\right].$$

The electrostatic field, which is the gradient of the potential, behaves linearly in r close to the axis and therefore vanishes on the axis itself.

2. The notation can be somewhat simplified if one sets

$$V(r,z) = V(z) - \frac{1}{4}r^2 V''(z) \text{ with } V(z) = q_e U(0,z)$$

and

$$V''(z) = q_e\,\partial_{z^2}^2 U\big|_{r=0} = q_e\left(d^2 U(0,z)/dz^2\right).$$

Let us apply the principle of Maupertuis to determine the trajectory: the quantity $\int \sqrt{2m\left(E - V(r,z)\right)}\,dl$ is stationary. In a given plane $dl^2 = dr^2 + dz^2$, or $dl = dz\sqrt{1 + r'^2(z)}$ where $r'(z) = dr/dz$. Thus, we must render stationary the quantity $\int F(r,r',z)\,dz$ with

$$F(r,r',z) = \sqrt{2m\left(E - V(r,z)\right)(1 + r'^2)}.$$

Setting $T(z) = E - V(z)$ the kinetic energy of the particle along the axis of revolution; we have $E - V(r, z) = T(z) + (r^2/4)V''(z)$.

We are now faced with a classical problem of variations, and the equation to be solved is $d(\partial_{r'} F)/dz = \partial_r F$.

To first order in r and r', one calculates

$$\frac{d(\partial_{r'} F)}{dz} \approx r''\sqrt{2mT} + \frac{2mr'T'}{2\sqrt{2mT}}; \qquad \partial_r F \approx \frac{mrV''}{2\sqrt{2mT}}.$$

Finally the resulting differential equation reads:

$$r''(z) + \frac{T'(z)}{2T(z)}r'(z) - \frac{V''(z)}{4T(z)}r(z) = 0.$$

3. This equation $r''(z) + A(z)r'(z) - B(z)r(z) = 0$ is linear. For particles with the same mass ejected from the axis with the same energy, the $A(z)$ and $B(z)$ functions are identical. This means that all the trajectories for these particles obey the same linear differential equation.

All particle trajectories start from the same point, O, on the axis, chosen as the origin and referred to as the source point in optics. This means that $r(0) = 0$ for all trajectories which differ only by the angle of emittance, that is by $r'(0)$. Because of the linearity of the differential equation, we have the proportionality of all trajectories. For any two trajectories, the ratio of the slopes remains constant all along the trajectory (and equal to the initial value).

If at point $z = s$, (extremity of the region where the differential equation is applicable), the particles are no longer subject to forces, they will continue to follow a straight line. Two cases must be considered.

- A particle leaves parallel to the axis: $r_1'(s) = 0$. Owing to the previous remark, this is true for all other particles. In other words, all the particles which are emitted from the same point with the same speed (but with a different angle of emittance) leave with parallel directions. One can say that the emission point is situated at the focus of the electromagnetic lens.

- If $r_1'(s) \neq 0$, the straight line followed by a particle has the equation $r_1(z) - r_1(s) = (z - s)r_1'(s)$; it intersects the axis at a point, z_1, such that $r_1(z_1) = 0$, that is $z_1 = s - r_1(s)/r_1'(s)$. The same thing is true for a second particle which cuts the axis at point $z_2 = s - r_2(s)/r_2'(s)$. Since $r_2(s)/r_2'(s) = r_1(s)/r_1'(s)$, it follows that $z_2 = z_1$. All the particles emitted from the same point with the same speed converge to the same point ($r_1'(s) < 0$) which is called a real image, or, alternatively, diverge from the same point ($r_1'(s) > 0$) which is referred to as a virtual image.

5.7. Maupertuis' Principle with an Electromagnetic Field
[Statement and Figure p. 245]

1. The reduced action is defined by $\tilde{S} = \int \boldsymbol{p} \cdot d\boldsymbol{r}$. In this expression, the generalized momentum $\boldsymbol{p}$ is connected to the linear momentum $\boldsymbol{\pi} = m\dot{\boldsymbol{r}}$ by the well known relation $\boldsymbol{p} = \boldsymbol{\pi} + q_e \boldsymbol{A}$, so that

$$\tilde{S} = \int \boldsymbol{\pi} \cdot d\boldsymbol{r} + \int q_e \boldsymbol{A} \cdot d\boldsymbol{r}.$$

Let us focus on the first term that can be written in the form $\int \boldsymbol{\pi} \cdot \dot{\boldsymbol{r}} \, dt = \int m v^2 \, dt = \int |\boldsymbol{\pi}| \, dl$. The non-relativistic Hamiltonian may be taken as $H = \pi^2/(2m) + q_e U$ and thus the latter integral can be written as $\int \sqrt{2m(H - q_e U)} \, dl$. Grouping both contributions, we obtain:

$$\tilde{S} = \int \sqrt{2m(H - q_e U)} \, dl + \int q_e \boldsymbol{A} \cdot d\boldsymbol{r}.$$

2. In the case under consideration, $U = 0$ and $H = E = $ const since the vector potential $\boldsymbol{A} = -\frac{1}{2} \boldsymbol{r} \times \boldsymbol{B}$ is time independent. The reduced action becomes

$$\tilde{S} = \int \left[\sqrt{2mE} \sqrt{dx^2 + dy^2} + (q_e B/2)(x \, dy - y \, dx) \right].$$

Maupertuis' principle implies the property of stationarity for this quantity; this problem is thus a variational problem quite similar to the isoperimetric problem. One first seeks a parametric representation of the trajectory $(x(\tau), y(\tau))$. The reduced action can be written in the traditional form $\tilde{S} = \int d\tau \, L(x, y, x', y')$, with

$$L(x, y, x', y') = \sqrt{2mE} \sqrt{x'^2 + y'^2} + (q_e B/2)(xy' - yx').$$

One obtains two Euler–Lagrange equations both of which can be integrated to give:

$$\frac{\sqrt{2mE}\, x'}{\sqrt{x'^2 + y'^2}} = q_e B y + C_1, \qquad \frac{\sqrt{2mE}\, y'}{\sqrt{x'^2 + y'^2}} = -q_e B x + C_2,$$

where C_1 and C_2 are two integration constants.

Multiply the first expression by y', the second by x' and substract to obtain $2xx' + 2yy' - 2x_0 x' - 2y_0 y' = 0$ with $x_0 = C_2/(q_e B)$ and $y_0 = -C_1/(q_e B)$. This latter equation can be integrated at once to give $(x - x_0)^2 + (y - y_0)^2 = \text{const} = R^2$. This is the equation of a circle with center (x_0, y_0) and radius R. The origin can be chosen precisely at the center so that $C_1 = 0$, $C_2 = 0$. Lastly, one can choose $\tau = t$ as the time and assume that $x(0) = R$, $y(0) = 0$. Substituting these values in the equations of motion, one sees that $x'(0) = 0$, $y'(0) < 0$ (the particles rotates clockwise) and that $\sqrt{2mE} = q_e B R$, which leads to the determination of the radius of the circle:

$$R = \frac{\sqrt{2mE}}{q_e B}.$$

3. The curvilinear abscissa s plays the role of a generalized coordinate. Of course, $\boldsymbol{v}(s) = \dot{s}\boldsymbol{t}(s)$, where $\boldsymbol{t}(s)$ is the unit vector tangent to the trajectory at the point labelled by s. The kinetic energy is $T = \frac{1}{2} m \dot{s}^2$ and the potential $V = -q_e \boldsymbol{v} \cdot \boldsymbol{A} = -q_e \dot{s} \boldsymbol{t} \cdot \boldsymbol{A}$. The Lagrangian is the difference between the kinetic energy and the potential energy, namely:

$$L(s, \dot{s}) = \frac{1}{2} m \dot{s}^2 + q_e \dot{s} \boldsymbol{t}(s) \cdot \boldsymbol{A}(s).$$

The momentum is given, as usual, by $p = \partial_{\dot{s}} L$, or, in this particular case,

$$p = m\dot{s} + q_e \, \boldsymbol{t}(s) \cdot \boldsymbol{A}(s),$$

so that one deduces $\dot{s} = (p - q_e \, \boldsymbol{t}(s) \cdot \boldsymbol{A}(s))/m$. Hamilton's function is obtained through the Legendre transform $H = p\dot{s} - L$, or, after substitution of $\dot{s}$ by p:

$$H(s, p) = \frac{(p - q_e \, \boldsymbol{t}(s) \cdot \boldsymbol{A}(s))^2}{2m}.$$

The Hamiltonian does not depend on time; it remains constant on the trajectory and takes a value E, which is the energy. Indeed, one notices that $H = \frac{1}{2} m \dot{s}^2$ and thus can be identified with the kinetic energy. One deduces that the speed along the trajectory remains constant $\dot{s} = \sqrt{2E/m}$, whereas the momentum, p, varies because of the presence of the term $\boldsymbol{t}(s) \cdot \boldsymbol{A}(s)$.

Let us focus on the variation of the reduced action over one revolution. It is expressed as

$$\Delta \tilde{S} = \oint p \, ds = \oint m\dot{s} \, ds + \oint q_e \, \boldsymbol{t}(s) \cdot \boldsymbol{A}(s) \, ds.$$

Since $\dot{s}$ remains constant over one revolution and since the curvilinear abscissa recovers its value, the first contribution vanishes. As for the second one, it can be rewritten, using Stokes' theorem, as

$$q_e \iint_\Sigma (\nabla \times \boldsymbol{A}) \cdot d\boldsymbol{\sigma} = q_e \iint_\Sigma \boldsymbol{B} \cdot d\boldsymbol{\sigma}.$$

The integral on the surface Σ enclosed by the curve is nothing more than the flux Φ of the magnetic field through this surface. In consequence, we arrive at the desired relation:

$$\Delta \tilde{S} = q_e \Phi.$$

5.8. Separable Hamiltonian, Separable Action [Statement p. 246]

1. Let $\tilde{S}(x, z, E)$ be the reduced action for the system. The characteristic Hamilton–Jacobi equation reads $H(x, z, \partial_x \tilde{S}, \partial_z \tilde{S}) = E$, or, with the form proposed for the Hamiltonian:

$$\frac{1}{2m}\left[\left(\frac{\partial \tilde{S}}{\partial x}\right)^2 + \left(\frac{\partial \tilde{S}}{\partial z}\right)^2\right] + mgz = E.$$

2. The Hamiltonian is manifestly separable and this property encourages us to search for a separable form of the reduced action: $\tilde{S}(x, z, E) = S_x(x, E) + S_z(z, E)$. Now, since $\partial_x \tilde{S} = dS_x/dx$ and $\partial_z \tilde{S} = dS_z/dz$, the characteristic equation involves total derivatives instead of partial derivatives:

$$\frac{1}{2m}\left[\left(\frac{dS_x}{dx}\right)^2 + \left(\frac{dS_z}{dz}\right)^2\right] + mgz = E.$$

3. Let us differentiate this equation with respect to x. This gives

$$\frac{1}{2m}\partial_x \left(\frac{dS_x}{dx}\right)^2 = 0, \quad \text{or} \quad \frac{1}{2m}\left(\frac{dS_x}{dx}\right)^2 = e,$$

where e is some constant. The characteristic equation then provides the other equation. In summary:

$$\frac{1}{2m}\left(\frac{dS_x}{dx}\right)^2 = e; \qquad \frac{1}{2m}\left(\frac{dS_z}{dz}\right)^2 + mgz = E - e.$$

4. The first equation is easily integrated to give $S_x(x,e) = \pm\sqrt{2me}\,x$. The second one gives

$$dS_z/dz = \pm\sqrt{2m(E - e - mgz)}.$$

The corresponding primitive is somewhat complicated but elementary:

$$S_z(z, E, e) = \pm(2/(3g))\sqrt{2/m}\,(E - e - mgz)^{3/2}.$$

Therefore, the complete reduced action is obtained in the separable form:

$$\tilde{S}(x, z, E, e) = \pm\sqrt{2me}\,x \pm \frac{2}{3g}\sqrt{\frac{2}{m}}\,(E - e - mgz)^{3/2}.$$

The total action is obtained from the classical formula $S = \tilde{S} - Et$, that is:

$$S(x, z, E, e, t) = \pm\sqrt{2me}\,x \pm \frac{2}{3g}\sqrt{\frac{2}{m}}\,(E - e - mgz)^{3/2} - Et.$$

5. Jacobi's theorem leads to a complete solution of the problem. One has first $\partial S/\partial E = \alpha = const$ which implies

$$z(t) = -\frac{1}{2}g(t+\alpha)^2 + \frac{E-e}{mg}.$$

This is the temporal law for the vertical motion.

On the other hand, one also has $\partial S/\partial e = \beta = const$ which leads to $x/\sqrt{e} \mp (2/(mg))\sqrt{E - e - mgz} = 2\beta/m = C$, which can be recast, after some algebra, in the form:

$$z(x) = \frac{1}{2}\left(\frac{x}{\sqrt{e}} - C\right)^2 + \frac{E-e}{mg}.$$

This is precisely the equation for the trajectory. It is a parabola, as expected.

These last two equations provide the complete solution to the problem.

5.9. Stark Effect [Statement p. 247]

1. The parabolic coordinates are defined by $\xi = r + z$, $\eta = r - z$ so that $r = \frac{1}{2}(\xi + \eta)$, $z = \frac{1}{2}(\xi - \eta)$. The potential of the system is

$$V = -\frac{K}{r} + kz = -\frac{2K}{\xi + \eta} + \frac{k(\xi - \eta)}{2}$$

which can be rewritten as
$$V = \frac{V_\xi(\xi) + V_\eta(\eta)}{\xi + \eta},$$
with the symmetric formulation:
$$V_\xi(\xi) = \frac{k}{2}\xi^2 - K \quad ; \quad V_\eta(\eta) = -\frac{k}{2}\eta^2 - K.$$

2. To calculate the kinetic energy, we start from its expression in cylindrical coordinates
$$T = \frac{1}{2}m\left(\dot{\rho}^2 + \rho^2\dot{\phi}^2 + \dot{z}^2\right).$$
We then switch to parabolic coordinates and use the relation $\xi\eta = r^2 - z^2 = \rho^2$. There is no major difficulty and one obtains:
$$T = \frac{1}{2}m\left[\frac{\eta^2\dot{\xi}^2 + \xi^2\dot{\eta}^2 + 2\xi\eta\dot{\xi}\dot{\eta}}{4\xi\eta} + \xi\eta\dot{\phi}^2 + \frac{1}{4}\left(\dot{\xi}^2 + \dot{\eta}^2 - 2\dot{\xi}\dot{\eta}\right)\right].$$
From the Lagrangian $L = T - V$, one deduces the momenta
$$p_\xi = \frac{\partial L}{\partial \dot{\xi}} = m(\xi+\eta)\frac{\dot{\xi}}{4\xi}, \quad p_\eta = \frac{\partial L}{\partial \dot{\eta}} = m(\xi+\eta)\frac{\dot{\eta}}{4\eta}, \quad p_\phi = \frac{\partial L}{\partial \dot{\phi}} = m\xi\eta\dot{\phi}.$$
Hamilton's function is calculated thanks to the usual formula
$$H = p_\xi \dot{\xi} + p_\eta \dot{\eta} + p_\phi \dot{\phi} - L$$
substituting velocities by momenta, which are given by the previous relations. One finally obtains:
$$H(\xi, \eta, \phi, p_\xi, p_\eta, p_\phi) = \frac{2}{m}\left(\frac{\xi p_\xi^2 + \eta p_\eta^2}{\xi + \eta} + \frac{p_\phi^2}{4\xi\eta}\right) + \frac{V_\xi(\xi) + V_\eta(\eta)}{\xi + \eta}.$$

3. The reduced action $\tilde{S}(\xi, \eta, \phi, E)$ is obtained from the characteristic Hamilton-Jacobi equation, which reads in this case:
$$\frac{2}{m}\left(\frac{\xi}{\xi+\eta}\left(\frac{\partial \tilde{S}}{\partial \xi}\right)^2 + \frac{\eta}{\xi+\eta}\left(\frac{\partial \tilde{S}}{\partial \eta}\right)^2 + \frac{1}{4\xi\eta}\left(\frac{\partial \tilde{S}}{\partial \phi}\right)^2\right)$$
$$+ \frac{V_\xi(\xi) + V_\eta(\eta)}{\xi + \eta} = E.$$
One may remark that the ϕ coordinate is cyclic and, of course, one sets:
$\tilde{S}(\xi, \eta, \phi, E, c) = \hat{S}(\xi, \eta, E, c) + c\phi.$

Substituting this equality into the characteristic equation and multiplying by $m(\xi+\eta)$, the latter equation can be rewritten in the form:

$$\left[2\xi\left(\frac{\partial \hat{S}}{\partial \xi}\right)^2 - mE\xi + \frac{c^2}{2\xi} + mV_\xi(\xi)\right]$$
$$+ \left[2\eta\left(\frac{\partial \hat{S}}{\partial \eta}\eta\right)^2 - mE\eta + \frac{c^2}{2\eta} + mV_\eta(\eta)\right] = 0.$$

This new equation is itself separable; it is natural to set: $\hat{S}(\xi,\eta,E,c) = S_\xi(\xi,E,c) + S_\eta(\eta,E,c)$. Substitution in the preceding equation leads to:

$$\left[2\xi\left(\frac{dS_\xi}{d\xi}\right)^2 - mE\xi + \frac{c^2}{2\xi} + mV_\xi(\xi)\right]$$
$$+ \left[2\eta\left(\frac{dS_\eta}{d\eta}\right)^2 - mE\eta + \frac{c^2}{2\eta} + mV_\eta(\eta)\right] = 0.$$

The partial derivatives are transformed into total derivatives; under these conditions the first bracket is a function of the variable ξ alone and the second one of the variable η alone. In order for the characteristic equation to be fulfilled for all values of the variables (ξ,η), it is necessary that the first bracket is a constant β and the second one a constant $-\beta$. We arrive at two differential equations:

$$\left[2\xi\left(\frac{dS_\xi}{d\xi}\right)^2 - mE\xi + \frac{c^2}{2\xi} + mV_\xi(\xi)\right] = \beta$$
$$\left[2\eta\left(\frac{dS_\eta}{d\eta}\right)^2 - mE\eta + \frac{c^2}{2\eta} + mV_\eta(\eta)\right] = -\beta$$

which lead to the determination of the functions S_ξ and S_η.

The problem is now completely solved since the reduced action is expressed as:

$$\tilde{S}(\xi,\eta,\phi,E,\beta,c) = S_\xi(\xi,E,\beta,c) + S_\eta(\eta,E,\beta,c) + c\phi.$$

4. To conform to the usual notation, let us call $\sigma = c$ the angular momentum ($p_\phi = \partial_\phi S$) and introduce the effective potentials

$$W_\xi(\xi) = -\beta/(2\xi) + \sigma^2/(4\xi^2) + mV_\xi(\xi)/(2\xi),$$
$$W_\eta(\eta) = \beta/(2\eta) + \sigma^2/(4\eta^2) + mV_\eta(\eta)/(2\eta).$$

They are represented in Fig. 5.6, with the parameters specified in the caption. The equations giving the actions are now written in a simpler form:

$$\left(\frac{dS_\xi}{d\xi}\right)^2 = \frac{mE}{2} - W_\xi(\xi) \quad \text{and} \quad \left(\frac{dS_\eta}{d\eta}\right)^2 = \frac{mE}{2} - W_\eta(\eta).$$

Consequently, one must have the following constraints

$$\frac{mE}{2} \geq W_\xi(\xi), \qquad \frac{mE}{2} \geq W_\eta(\eta).$$

The equalities in these expressions correspond to turning points; with the proposed parameters, one has explicitly

$$\xi_m = 0.094, \xi_M = 0.734 \quad \text{and} \quad \eta_m = 0.175, \eta_M = 0.496.$$

In the plane (ξ, η), the region corresponding to allowed trajectories is the rectangle

$$\xi_m \leq \xi \leq \xi_M, \qquad \eta_m \leq \eta \leq \eta_M.$$

Switching back to the original variables (ρ, z), the corresponding region is enclosed between two parabolae $(\eta = \eta_m, \eta = \eta_M)$ oriented upwards and two parabolae $(\xi = \xi_m, \xi = \xi_M)$ oriented downwards.

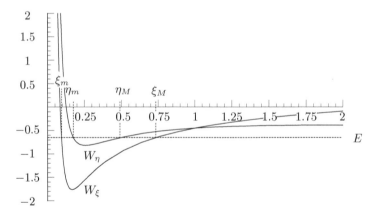

Fig. 5.6 Effective potentials $W_\xi(\xi)$ and $W_\eta(\eta)$ as a function of their arguments, for the following values of the parameters $m = K = 1$, $k = 0.4$, $\sigma = 0.45$, $\beta = 0.2$ The intersections of these curves with the straight line $mE/2$ give the turning points (ξ, η)

A trajectory, obtained by numerically solving Hamilton's equations, is represented in the Fig. 5.7. The allowed region for the exploration is indicated by dashed lines.

Problem Solutions

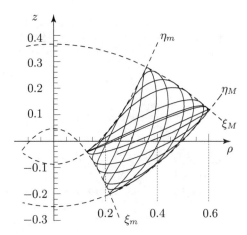

Fig. 5.7 Trajectory, in the plane (ρ, z), for a particle subjected to an attractive central electrostatic potential and a constant electric field along Oz

5.10. Orbits of Earth's Satellites
[Statement and Figures p. 248]

1. Let us begin with the obvious equality $|\boldsymbol{r} - \sigma\hat{z}|^2 = (\boldsymbol{r} - \sigma\hat{z})^2 = r^2 + \sigma^2 - 2\sigma r u$, where $u = \cos(\hat{r}, \hat{z})$ is the cosine between the radius vector and the Oz axis. It follows that

$$\frac{1}{|\boldsymbol{r} - \sigma\hat{z}|} = \frac{1}{r}\left(1 - \frac{2\sigma u}{r} + \frac{\sigma^2}{r^2}\right)^{-1/2}.$$

One works at a distance much larger than the distance between the two centers $\sigma/r \ll 1$; it is therefore justified to perform a truncated expansion of the square root, which relies on the well known formula $(1+\varepsilon)^{-1/2} = 1 - \varepsilon/2 + 3\varepsilon^2/8$. We arrive at the desired formula:

$$\frac{1}{|\boldsymbol{r} - \sigma\hat{z}|} = \frac{1}{r} + \frac{\sigma u}{r^2} + \frac{(3u^2 - 1)\sigma^2}{2r^3} + O(\sigma^3/r^4).$$

The expression for $1/|\boldsymbol{r} + \sigma\hat{z}|$ follows simply by changing σ into $-\sigma$.

The potential is deduced at once

$$V = -K\left[\frac{1}{r} + \frac{(3u^2 - 1)\sigma^2}{2r^3}\right].$$

This expression should be compared with the analog obtained from a revolution ellipsoid:

$$V = -GmM\left[\frac{1}{r} + \frac{(3u^2 - 1)(I - I_3)}{2Mr^3}\right].$$

We deduce that the potential due to the ellipsoid can be replaced by the two center potential provided that the following identification is made:

$$K = GmM; \quad \sigma = \sqrt{\frac{I - I_3}{M}}$$

For a prolate system, $I > I_3$ and the parameter σ should be taken as a real quantity.

2. In reality the Earth is oblate $I < I_3$. Let us set $\sigma = \sqrt{(I_3 - I)/M}$ and work with $i\sigma$. Define $r_1 = \sqrt{\rho^2 + (z - i\sigma)^2}$ and $r_2 = \sqrt{\rho^2 + (z + i\sigma)^2}$. These lengths are complex conjugate numbers but the potential

$$V = -\frac{1}{2}K\left(\frac{1}{r_1} + \frac{1}{r_2}\right)$$

is a real quantity. The same property is true for the two dimensionless elliptic coordinates $\xi = (r_1 + r_2)/(2\sigma)$ and $\eta = (r_2 - r_1)/(2i\sigma)$; one maintains the polar angle ϕ as the third coordinate.

It is easy to check the identities $r_1^2 + r_2^2 = 2(\rho^2 + z^2 - \sigma^2)$, $r_2^2 - r_1^2 = 4i\sigma z$, $r_1^2 r_2^2 = (\rho^2 + z^2 - \sigma^2)^2 + 4\sigma^2 z^2$. Using these identities, simple calculations give $z = \xi\eta\sigma$ and $\rho = \sigma\sqrt{(\xi^2 + 1)(1 - \eta^2)}$. To calculate the differential length element, the simplest procedure is to start from its expression in terms of cylindrical coordinates $dl^2 = d\rho^2 + \rho^2 d\phi^2 + dz^2$. Switching from the cylindrical coordinates to elliptic coordinates was explained previously and, after a long but straightforward calculation, one obtains finally:

$$dl^2 = \sigma^2(\xi^2 + \eta^2)\left[\frac{d\xi^2}{1 + \xi^2} + \frac{d\eta^2}{1 - \eta^2}\right] + \sigma^2(1 + \xi^2)(1 - \eta^2)d\phi^2.$$

3. The kinetic energy is obtained easily from the square of the length element since $T = \frac{1}{2}mv^2 = \frac{1}{2}m(dl/dt)^2 = \frac{1}{2}m(dl^2)/(dt^2)$. The potential is also obtained easily since

$$V = -\frac{1}{2}K\left(\frac{1}{r_1} + \frac{1}{r_2}\right) = -K\frac{r_1 + r_2}{2r_1 r_2}.$$

With $r_1 + r_2 = 2\sigma\xi$ and $r_1 r_2 = \sigma^2(\xi^2 + \eta^2)$, the potential is obtained as: $V = -K\xi/\left[\sigma(\xi^2 + \eta^2)\right]$. Finally, the Lagrangian is obtained from its definition $L = T - V$, that is:

$$L = \frac{1}{2}m\sigma^2\left[\frac{\xi^2 + \eta^2}{1 + \xi^2}\dot\xi^2 + \frac{\xi^2 + \eta^2}{1 - \eta^2}\dot\eta^2 + (1 + \xi^2)(1 - \eta^2)\dot\phi^2\right] + \frac{K\xi}{\sigma(\xi^2 + \eta^2)}.$$

Problem Solutions

The momenta may now be calculated with the usual recipe:
$$p_\xi = \partial_{\dot\xi} L = m\sigma^2(\xi^2+\eta^2)\,\dot\xi/(1+\xi^2);$$
$$p_\eta = \partial_{\dot\eta} L = m\sigma^2(\xi^2+\eta^2)\,\dot\eta/(1-\eta^2);$$
$$p_\phi = \partial_{\dot\phi} L = m\sigma^2(1+\xi^2)(1-\eta^2)\,\dot\phi.$$

Lastly, the Hamiltonian is obtained by the Legendre transform, $H = p_\xi \dot\xi + p_\eta \dot\eta + p_\phi \dot\phi - L$. We arrive at the result:

$$H = \frac{1}{2m\sigma^2}\left[\frac{1+\xi^2}{\xi^2+\eta^2}p_\xi^2 + \frac{1-\eta^2}{\xi^2+\eta^2}p_\eta^2 + \frac{1}{(1+\xi^2)(1-\eta^2)}p_\phi^2\right] - \frac{K\xi}{\sigma(\xi^2+\eta^2)}.$$

We now introduce the reduced action $\tilde S(\xi,\eta,\phi,E)$. It obeys the characteristic Hamilton–Jacobi equation: $H(\xi,\eta,\phi,\partial_\xi \tilde S,\partial_\eta \tilde S,\partial_\phi \tilde S) = E$. With the previous Hamiltonian, the characteristic equation is written in the form:

$$2m\sigma^2 E = \frac{1}{\xi^2+\eta^2}\left[(1+\xi^2)\left(\frac{\partial \tilde S}{\partial \xi}\right)^2 + (1-\eta^2)\left(\frac{\partial \tilde S}{\partial \eta}\right)^2 - 2Km\sigma\xi\right]$$
$$+ \frac{1}{(1+\xi^2)(1-\eta^2)}\left(\frac{\partial \tilde S}{\partial \phi}\right)^2.$$

4. This equation is separable. One remarks first that the ϕ coordinate is cyclic and this allows us to choose the action in the form $\tilde S(\xi,\eta,\phi,E) = \hat S(\xi,\eta,E) + \alpha\phi$. The characteristic equation is transformed into:

$$2m\sigma^2 E = \frac{1}{\xi^2+\eta^2}\left[(1+\xi^2)\left(\frac{\partial \hat S}{\partial \xi}\right)^2 + (1-\eta^2)\left(\frac{\partial \hat S}{\partial \eta}\right)^2 - 2Km\sigma\xi\right]$$
$$+ \frac{\alpha^2}{(1+\xi^2)(1-\eta^2)}.$$

Lastly, one employs the identity

$$\frac{1}{(1+\xi^2)(1-\eta^2)} = \frac{1}{(\xi^2+\eta^2)}\left[\frac{1}{1-\eta^2} - \frac{1}{1+\xi^2}\right]$$

and multiplies the previous characteristic equation by $(\xi^2+\eta^2)$. Grouping the terms suitably, the equation can be recast in the form:

$$\left[(1+\xi^2)\left(\frac{\partial \hat S}{\partial \xi}\right)^2 - 2Km\sigma\xi - \frac{\alpha^2}{1+\xi^2} - 2m\sigma^2 E\xi^2\right]$$
$$+ \left[(1-\eta^2)\left(\frac{\partial \hat S}{\partial \eta}\right)^2 + \frac{\alpha^2}{1-\eta^2} - 2m\sigma^2 E\eta^2\right] = 0.$$

As such, the equation is manifestly separable and one is led to set

$$\hat{S}(\xi, \eta, E, \alpha) = S_\xi(\xi, E, \alpha) + S_\eta(\eta, E, \alpha).$$

This equation then becomes:

$$\left[(1+\xi^2) \left(\frac{dS_\xi}{d\xi} \right)^2 - 2Km\sigma\xi - \frac{\alpha^2}{1+\xi^2} - 2m\sigma^2 E\xi^2 \right]$$
$$+ \left[(1-\eta^2) \left(\frac{dS_\eta}{d\eta} \right)^2 + \frac{\alpha^2}{1-\eta^2} - 2m\sigma^2 E\eta^2 \right] = 0.$$

Partial derivatives have been replaced by total derivatives so that it is clear that the first bracket is a function of ξ alone, whereas the second one is a function of η alone. In order for this equation to be satisfied, one must identify the first bracket with a constant $-\beta$ and the second one with the constant β so that finally, we have to solve two distinct differential equations:

$$(1+\xi^2) \left(\frac{dS_\xi}{d\xi} \right)^2 - 2Km\sigma\xi - \frac{\alpha^2}{1+\xi^2} - 2m\sigma^2 E\xi^2 = -\beta$$

$$(1-\eta^2) \left(\frac{dS_\eta}{d\eta} \right)^2 + \frac{\alpha^2}{1-\eta^2} - 2m\sigma^2 E\eta^2 = \beta.$$

The solutions are obtained through integrals. Assembling all the contributions, one obtains the complete reduced action $\tilde{S}(\xi, \eta, \phi, E, \alpha, \beta)$ and, thereby, the total action $S = \tilde{S} - Et$. Explicitly, one has:

$$S(\xi, \eta, \phi, t, E, \alpha, \beta) = -Et + \alpha\phi$$
$$\pm \int \frac{d\eta}{1-\eta^2} \sqrt{(1-\eta^2)(\beta + 2m\sigma^2 E\eta^2) - \alpha^2}$$
$$\pm \int \frac{d\xi}{1+\xi^2} \sqrt{(1+\xi^2)(-\beta + 2m\sigma^2 E\xi^2 + 2Km\sigma\xi) + \alpha^2}.$$

Using Jacobi's theorem, and differentiating this expression with respect to the constants E, α, β and equating respectively these derivatives to other constants, one completely solves the problem.

Much as in the problem concerned with the Stark effect, there exists, in the plane (ξ, η), a rectangular allowed region ($\xi_m \leq \xi \leq \xi_M$, $-\eta_m \leq \eta \leq \eta_m$) for the motion of the satellite. ξ_m, ξ_M are the turning points of the effective potential $W_\xi(\xi)$ and $\pm\eta_m$ the turning points of the effective potential $W_\eta(\eta)$. In the plane (ρ, z), the corresponding region is comprised between two nested ellipses ($\xi = \xi_m$, $\xi = \xi_M$) and a hyperbola ($\eta = \eta_m$). A possible trajectory is represented in Fig. 5.8.

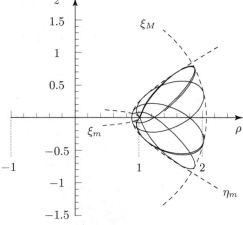

Fig. 5.8 Trajectory of the satellite in the plane (ρ, z). The parameters that have been used are the following: $\sigma = m = 1$, $K = 2$, $\alpha = 0.5$, $E = -1$. The allowed region is comprised between two nested ellipses ($\xi = \xi_m = 0.135$, $\xi = \xi_M = 1.81$) and a hyperbola ($\eta = \eta_m = 0.464$)

5.11. Phase and Group Velocities
[Statement and Figures p. 251]

1. For a free particle, $E = \boldsymbol{p}^2/(2m)$ and $\boldsymbol{p} = m\dot{\boldsymbol{r}}$. The phase velocity is thus $v_\varphi = E/|\boldsymbol{p}| = |\boldsymbol{p}|/(2m)$. On the other hand, the group velocity is given by $v_g = |\dot{\boldsymbol{r}}| = |\boldsymbol{p}|/m$. We have the obvious relationship:

$$v_g = 2v_\varphi.$$

2. With the given form relative to the plane wave, the Schrödinger equation $i\hbar\partial_t\psi = -\hbar^2/(2m)\,\Delta\psi$ leads to the relation $\hbar^2 k^2/(2m) = \hbar\omega$. Thanks to the de Broglie relation, this equation is equivalent to the non-relativistic expression $E = \boldsymbol{p}^2/(2m)$. The phase velocity is given by

$$v_\varphi = \frac{\omega}{k} = \frac{\hbar k}{2m}$$

and the group velocity by

$$v_g = \frac{d\omega}{dk} = \frac{d(\hbar\omega)}{d(\hbar k)} = \frac{\hbar k}{m}.$$

One sees that here also:

$$v_g = 2v_\varphi.$$

We thus have the same relationship. This is not surprising since E is identified with ω and p with k with the same proportionality constant $\hbar$.

3. In the case of a particle in a potential,
$$E = \frac{p^2}{2m} + V(r) \quad \text{and} \quad p = m\dot{r}.$$

Therefore $|p| = \sqrt{2m(E - V(r))}$ and $v_\varphi = E/|p|$, or $v_\varphi |p| = E$; moreover $v_g = |\dot{r}| = |p|/m$. One deduces immediately that

$$v_\varphi v_g = \frac{E}{m}.$$

This quantity is constant along the trajectory.

4. For a free relativistic particle, with velocity $\dot{r}$ and kinematic parameter $\gamma = (1 - \dot{r}^2/c^2)^{-1/2}$, the energy is given by $E = \gamma mc^2$ and the momentum by $p = \gamma m \dot{r}$. The phase velocity is thus

$$v_\varphi = \frac{E}{|p|} = \frac{c^2}{|\dot{r}|} = \frac{c^2}{v_g}.$$

Therefore the relation

$$v_\varphi v_g = c^2$$

always holds. In particular, this is the case for a light wave in a homogeneous medium.

For a relativistic particle in a scalar potential, the Hamiltonian is

$$H(r,p) = \sqrt{(mc^2 + V(r))^2 + p^2 c^2}$$

which is identified with the energy on the trajectory. The phase velocity is given by $v_\varphi = E/|p|$. On the other hand, Hamilton's equation gives

$$v_g = |\dot{r}| = |\nabla_p H| = \frac{|p| c^2}{H} = \frac{|p| c^2}{E} = \frac{c^2}{v_\varphi}.$$

We thus still have the relation:

$$v_\varphi v_g = c^2$$

which is therefore more general than the corresponding relation for the free particle. It remains true in the case of a scalar potential.

Chapter 6
Integrable Systems

Summary

A system is said to be integrable if it has a regular or predictable behaviour. In nature, only a few examples are known (among them is the famous Kepler problem) but they are practically very important. Furthermore, numerous systems, although not integrable, are "close" to integrable systems. A good understanding of integrable systems is then a preliminary condition to the study of this latter category.

6.1. Basic Notions

We consider here only autonomous systems.

As an illustrative example, we consider a particle subject to a central potential in a two-dimensional space.

6.1.1. Some Definitions

The Poisson bracket of two functions $F(q, p)$ and $G(q, p)$, defined in phase space, is given by the expression:

$$\{F, G\} = \sum_i \left(\frac{\partial F(q,p)}{\partial q_i}\right)\left(\frac{\partial G(q,p)}{\partial p_i}\right) - \left(\frac{\partial F(q,p)}{\partial p_i}\right)\left(\frac{\partial G(q,p)}{\partial q_i}\right). \quad (6.1)$$

The Poisson bracket of a function F and the Hamiltonian H is of primary importance. A function $F(q, p)$ defined in phase space is a **first integral** if its Poisson bracket with the Hamiltonian vanishes: $\{F, H\} = 0$. Of course, the Hamiltonian itself is a first integral. For a time independent Hamiltonian, the notion of a first integral is identified with the already familiar notion of a constant of the motion.

A n dimensional system is **integrable** if it has n independent first integrals which are in involution.[1]

This is, in particular, the case of a one-dimensional autonomous system with the energy as a first integral. This is also the case for a two or three dimensional central potential for which, as first integrals, in addition to the energy, we have also the projections of the angular momentum.[2]

Let us denote the n first integrals by $F_m(q, p)$ ($m = 1, ..., n$) (with $F_1(q, p) = H(q, p)$). They are independent; this means that none of them can be expressed in terms of the $n-1$ others. In other words, each function F_m is not a function of the others in the whole phase space. They are in involution; this means that the Poisson bracket of any pair of them vanishes: $\{F_m, F_p\} = 0$, $\forall m, p$.

Since they are independent, one can, at least theoretically, express the momenta as a function of the coordinates and of the first integrals: $p(q, F)$.

Illustration: For a two-dimensional system in a central force field, the two first integrals are the energy $F_1(q, p) = H(q, p) = E$ and the angular momentum $F_2(q, p) = p_\phi = \sigma$ so that the momenta are given by:

$$p_\rho(\rho, \phi, E, \sigma) = \sqrt{2m\left(E - V(\rho) - \frac{\sigma^2}{2m\rho^2}\right)};$$

$$p_\phi(\rho, \phi, E, \sigma) = \sigma.$$

Consequently, in a phase space with dimension $2n$, the system evolves on a manifold[3] with dimension n. If, on this manifold, none of the coordinates can reach infinity and if the first integrals are in involution, the topology of this manifold is that of a torus which means that each point on the manifold M_f is specified[4] by n angles modulo 2π, as illustrated in Fig. 6.1.

---

[1] The n first integrals $F_m(q, p)$ are said to be in involution if they fulfill the $n(n-1)/2$ relations $\{F_m, F_p\} = 0$, $\forall m, p = 1, \ldots, n$. These relations are much more constraining than the n relations $\{F_m, H\} = 0$.

[2] With Cartesian coordinates in a three-dimensional space, we have 4 independent first integrals! But they are not in involution.

[3] It is in a manifold or on a manifold.

[4] This is not the case for a sphere for which, at the poles, the azimuthal angle is meaningless.

Summary

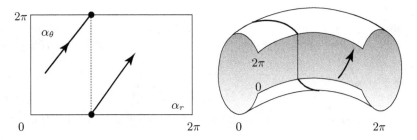

Fig. 6.1 On the left hand side, the state of the system in M_f is specified by two angles α_r and α_θ (modulo 2π). The values 0 and 2π must be considered as describing the same state. The full line corresponds to a motion in M_f. To achieve visual continuity, one must attach the sides of the rectangle which have the values 0 and 2π for both angles. By continuous deformation, this procedure generates the visual representation of the torus on the right hand side. The trajectory is continuous on the torus

For systems with two degrees of freedom, two first integrals are automatically in involution. Indeed, if $F_2(q,p)$ is a first integral different from the Hamiltonian $H(q,p) = F_1(q,p)$, its derivative with respect to time vanishes by definition and this property implies, using Hamilton's equations,

$$\dot{F}_2(q,p) = 0 = (\partial_q F_2(q,p))\,\dot{q} + (\partial_p F_2(q,p))\,\dot{p}$$
$$= (\partial_q F_2(q,p))\,(\partial_p H(q,p)) - (\partial_p F_2(q,p))\,(\partial_q H(q,p)).$$

This last equality is nothing more than $\{F_2, H\} = 0$.

6.1.2. Good Coordinates: The Angle–Action Variables

Instead of using q and p to define the state of a system in phase space, one can imagine other sets of coordinates[5] $Q(q,p), P(q,p)$. If the passage from one set to the other preserves the Liouville invariant (sum of the areas of the projection of a parallelepiped on the planes (q_i, p_i)), the transformation is said to be **canonical**.

We will encounter, in the section devoted to complements, a very simple condition which characterizes a canonical transformation.

---

[5] This transformation, called a contact transformation, can depend on time. We will develop this point in the complements.

A consequence of the canonicity is that the flow of the Hamiltonian $H(q,p)$, which is called, more precisely, the generator of time translations, coincides with the flow of the new Hamiltonian $K(Q,P) = H(q(Q,P), p(Q,P))$ that is to say:

$$\dot{Q} = \partial_P K(Q,P); \quad \dot{P} = -\partial_Q K(Q,P).$$

In other words, a canonical transformation preserves the form of Hamilton's equations.

If one chooses as new momenta P the first integrals F_m, then two very interesting properties result:

- First, from the second Hamilton equation, the generator of the time translation or Hamiltonian does not depend on the new coordinates[6] Q: $K(Q,P) = K(P)$; the fact that it is possible to find a set of n cyclic coordinates is precisely a property specific to integrable systems.

- Second, from the first Hamilton equation, these new coordinates have a constant velocity $\omega(P) = \partial_P K(P)$ and thus vary linearly in time. These new coordinates are automatically angles since no coordinate can reach infinity.

In practice, one does not take as new momenta the first integrals themselves, but rather combinations of them called "**action variables**" which are defined precisely as:

$$I_m(F) = \frac{1}{2\pi} \oint_{\Gamma_m} p(q, F)\, dq. \tag{6.2}$$

The Γ_m are the n closed contours based on the torus, all different and non reducible to a single point. The involution property for the first integrals implies that the integral (6.2) takes the same value whatever the contours provided they can be identified with each other by a continuous deformation. This property is proved in the case of a two-dimensional system in Problem 6.8.

The action variables being functions of first integrals are themselves first integrals. Inverting Equation (6.2), one obtains $F_m(I)$ and, in particular, one can express the Hamiltonian F_1 in terms of the action variables only: $K(I)$.

Illustration: One of the irreducible contours corresponds to a forward and back variation for the radius without a change of angle: the radius varies from the perihelion $\rho_{\min}$ to the aphelion $\rho_{\max}$ and returns to the perihelion.

---

[6] They are all cyclic or ignorable coordinates.

Summary

The action variable is thus:

$$I_\rho(E, \sigma) = \frac{2}{2\pi} \int_{\rho_{\min}}^{\rho_{\max}} \sqrt{2m\left(E - V(\rho) - \sigma^2/2m\rho^2\right)}\, d\rho.$$

The other contour corresponds to a rotation without change in the radius and the associated action variable is:

$$I_\phi = \frac{1}{2\pi} \int_0^{2\pi} \sigma\, d\phi = \sigma.$$

Importance of the action variables

In many situations, one cannot avoid a quantum approach for a mechanical system. This is obviously the case for physical phenomena with atomic scales, but not only. The rules for quantum mechanics are precise but rather involved to realize practically. An economic way to insure quantization in an approximate way, which is often satisfactory (one speaks of a semi-classical approximation), starts from a fully classical description in which the Hamiltonian is expressed in terms of action variables. The quantization recipe, known as the rule EBK (Einstein-Brillouin-Kramers), stipulates that the action variables can take only integer values[7] of $\hbar$, the reduced Planck action:

$$I_i = n_i \hbar \tag{6.3}$$

The expression $K(I)$ leads thus to the quantization of energy states.

In case of an integrable system, the problem is entirely solved!

Why choose as a first set of new momenta precisely the quantities I_m, which have the dimension of an action? The reason is that the conjugate coordinates $\alpha\,(\alpha_1, \alpha_2, \ldots, \alpha_n)$ associated through the canonical transformation are angles. The advantage is that when one of these is changed by a multiple of 2π, we come back at the same point in phase space, namely $q(\alpha, I) = q(\alpha + 2\pi k, I)$, $p(\alpha, I) = p(\alpha + 2\pi k, I)$ where k is an array of n arbitrary integers. The canonical transformation $(p, q) \leftrightarrow (\alpha, I)$ is very interesting since the Hamiltonian depends only on action variables and not on angle variables.

Consequently, after the canonical transformation, the Hamilton equations are expressed as:

$$\dot\alpha_m = \frac{\partial K(I)}{\partial I_m} = \omega_m(I); \quad \dot I_m = 0 \tag{6.4}$$

---

[7] Sometimes half-integer.

which can be integrated to give $I_m = \text{const}$, $\alpha_m(q, I) = \omega_m(I)t + \alpha_{m0}$. Using the inverse canonical transformation, one obtains $q(\alpha(t), I)$ and $p(\alpha(t), I)$. The mechanical problem is entirely solved.

Regularity, periodicity and quasi-periodicity

In the space of angles, two close initial conditions give trajectories that deviate from one another in a regular manner: the angular velocities $\omega_m(I) = \partial_{I_m} H(I)$ are close for close actions. The motion is regular and predictable. This is not the case for chaotic systems.

Each coordinate, and more generally each function in phase space, is a function of the angle variables through periodic functions. Thus, if one analyzes their variation in time using Fourier series, one will find basic angular frequencies and their harmonics, as well as combinations of these. One speaks in this case of quasi-periodicity. If these basic frequencies are commensurable, every function in phase space exhibits a periodic dependence in time. In this case, the torus is said to be **resonant**.

6.2. Complements

6.2.1. Building the Angle Variables

We saw that the canonical transformation to the angle-action variables is especially interesting since the equations of motion are very simple: the actions are constant and the angles vary linearly with time. Building the action variables is summarized in Formula (6.2). But how do we build the angle coordinates?

One must first introduce the reduced action function, as explained in Chapter 5,

$$\tilde{S}(q, I) = \int p(q, F(I)) \cdot dq$$

which, once more, takes the same value for contours that can be transformed one to the other by a continuous deformation.

The angle variables are obtained after differentiation:

$$\alpha_m(q, I) = \frac{\partial \tilde{S}(q, I)}{\partial I_m}. \tag{6.5}$$

Illustration: We already calculated the reduced action for this system which is separable:

$$\tilde{S}(\rho,\phi,I_\rho,I_\phi) = \int \sqrt{2m\left(E(I_\rho,I_\phi) - V(\rho) - I_\phi^2/(2m\rho^2)\right)}\, d\rho + \int I_\phi\, d\phi.$$

Applying formula (6.5), one obtains:

$$\alpha_\rho(\rho,I_\rho,I_\phi) = \omega_\rho \sqrt{2m} \int \frac{d\rho}{2\sqrt{\left(E(I_\rho,I_\phi) - V(\rho) - I_\phi^2/(2m\rho^2)\right)}}$$

with $\omega_\rho = \partial_{I_\rho} E(I_\rho, I_\phi)$,

$$\alpha_\phi(\rho,\phi,I_\rho,I_\phi) = \int \frac{(2m\omega_\phi - I_\phi/\rho^2)\, d\rho}{2\sqrt{2m\left(E(I_\rho,I_\phi) - V(\rho) - I_\phi^2/(2m\rho^2)\right)}} + \phi$$

with $\omega_\phi = \partial_{I_\phi} E(I_\rho, I_\phi)$.

In principle, by inversion of (6.5), one can obtain $q(\alpha(t), I)$. Thus, with the help of integrals and inversion of functions only, the problem can be solved entirely, whence the name of integrable systems.

6.2.2. Flow/Poisson Bracket/Involution

Let us consider an arbitrary function $F(q,p)$ defined in phase space. The integral curves of the differential equations:

$$\frac{dq}{d\tau} = \frac{\partial F(q,p)}{\partial p}; \qquad \frac{dp}{d\tau} = -\frac{\partial F(q,p)}{\partial q}.$$

are referred to as the flow of the generator $F(q,p)$

The τ variable defined by the above equalities is called the **flow parameter**.

We already met the generator for time translations, namely the Hamiltonian $H(q,p)$. In Problem 6.18, we will build the generator of translations with a translation parameter, as well as the generator of rotations with a rotation parameter.

The variation of an arbitrary function $F(q,p)$ in phase space along the flow of $G(q,p)$ is given by a very simple expression:

$$dF(q,p) = \{F,G\}\, d\tau \qquad (6.6)$$

in which appears naturally the Poisson bracket $\{F,G\}$, as defined by (6.1).

In the case for which $G = H$, the flow parameter is the time and the previous definition implies: $\dot{F}(q,p) = \{F, H\}$. One sees immediately that if F is a first integral (it commutes with the Hamiltonian) it is also a constant of the motion, as was emphasized in the preceding section. This property explains why the two notions of "first integral" and "constant of the motion" are often identified.

Nevertheless, some caution is required. The preceding statement is true in the case of a function F which does not depend explicitly on time. In the most general case, we have rather the relation:

$$\frac{dF(q,p,t)}{dt} = \{F, H\} + \frac{\partial F}{\partial t}. \tag{6.7}$$

Strictly speaking, F is no longer a constant of the motion.

Another way to define the involution of two generators consists in requiring commutativity in the order of their flows, as illustrated in Fig. 6.2.

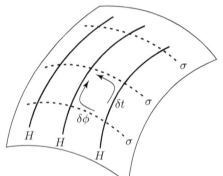

Fig. 6.2 Flows of the Hamiltonian H and of the angular momentum $p_\phi = \sigma$. The flow parameters are the time t for H and the polar angle ϕ for p_ϕ

6.2.3. Criterion to Obtain a Canonical Transformation

The canonical transformations, which preserve the form of Hamilton's equations, are of first importance and it is crucial to check the canonicity of a contact transformation in phase space.

For a system with one degree of freedom, the transformation must insure area conservation in phase space. In other words the Jacobi matrix for this transformation must be equal to unity, so that:

$$(\partial_q Q(q,p,t))\,(\partial_p P(q,p,t)) - (\partial_p Q(q,p,t))\,(\partial_q P(q,p,t)) = 1.$$

This equality expresses the important property concerning the Poisson bracket $\{Q, P\} = 1$.

More generally, in the case of an arbitrary number of degrees of freedom, the condition for canonicity becomes:

$$\{Q_i, Q_j\} = 0 \; ; \; \{Q_i, P_j\} = \delta_{i,j} \; ; \; \{P_i, P_j\} = 0, \quad \forall i, j. \quad (6.8)$$

There exists a very simple way to build a canonical transformation, even when it depends on time; it relies on the **generating functions**. If one starts from the original coordinates (q, p) and original Hamiltonian $H(q, p, t)$, the new coordinates (Q, P) and the new Hamiltonian $K(Q, P, t)$ can be linked by partial differential equations that use four different types of functions $G_1(Q, q, t)$, $G_2(P, q, t)$, $G_3(p, Q, t)$, $G_4(P, p, t)$, called generating functions. The defining relations are the following:

$$p = \frac{\partial G_1}{\partial q}; \quad P = -\frac{\partial G_1}{\partial Q}; \quad K = H + \frac{\partial G_1}{\partial t}$$

$$p = \frac{\partial G_2}{\partial q}; \quad Q = \frac{\partial G_2}{\partial P}; \quad K = H + \frac{\partial G_2}{\partial t}$$

$$q = -\frac{\partial G_3}{\partial p}; \quad P = -\frac{\partial G_3}{\partial Q}; \quad K = H + \frac{\partial G_3}{\partial t}$$

$$q = -\frac{\partial G_4}{\partial p}; \quad Q = \frac{\partial G_4}{\partial P}; \quad K = H + \frac{\partial G_4}{\partial t}. \quad (6.9)$$

If there exists any one of the four generating functions satisfying the corresponding partial differential equations, the canonicity of the transformation is established with certainty. This notion, which is quite difficult to grasp, will be discussed in detail in several problems (in particular, Problems 6.9, 6.10, 6.11).

Problem Statements

6.1. Expression of the Period for a One-Dimensional Motion [Solution p. 305]   ★

Relation between the action function and the action variable

We introduced in Chapter 5 the *reduced action function*.

1. Show that this function increases by $2\pi I(E)$ ($I(E)$ being the *action variable*) when the system returns to the same point in phase space, that is after one period.

2. Relying on the results of Chapter 5, show that the angular frequency ω is given by: $\omega(E) = dE/dI = 1/(\partial_E I(E))$ and check the agreement with the results of this chapter and the Hamilton equations with the set of angle-action coordinates.

6.2. One-Dimensional Particle in a Box
[Solution and Figure p. 306] ★ ★

A very simple integrable system

A particle with mass m is considered as enclosed in a "linear box", if it is free to move in the interval $[-a, a]$ but is not allowed to pass over the borders of this interval. Strictly speaking, the Hamiltonian for such a system does not exist (the potential would be null inside the interval and infinite outside; one sometimes speaks of a square well). However, a power-law potential $V = (q/a)^\lambda$ looks quite similar to that of the box in the limit of a very large power $\lambda \to \infty$.

1. Write down the Hamiltonian of this system with the power-law potential.

2. Let E be the energy of the system. Plot the phase portrait; calculate the action variable $I(E)$ (in the limit $\lambda \to \infty$) and the action variable $\alpha(q, E)$.

3. Express the energy as a function of the action. Perform the quantization following the EBK rule (see 6.3). The solution of the corresponding Schrödinger equation gives a spectrum in the form $E_n = \left(n^2 \hbar^2 \pi^2\right) / \left(8ma^2\right)$. Compare to the semi-classical EBK treatment.

6.3. Ball Bouncing on the Ground
[Solution and Figure p. 308] ★ ★

Quantization in a constant gravitational field

A ball of mass m is placed in a constant gravitational field g. It is released from a given height with a null initial speed and it bounces elastically on the ground; the motion of the ball is along the vertical.

1. Plot the phase portrait; calculate the action variable $I(E)$ and the angle variable $\alpha(q, E)$.

2. Write down the energy as a function of the action $E(I)$.

3. Quantize the problem according to the EBK rule (see 6.3).

   The quantum formula gives approximately:

   $$E_n = m^{1/3} \left(\frac{3\pi}{2\sqrt{2}} (n - 1/4) \hbar g \right)^{2/3}.$$

   What may we conclude?

Problem Statements

6.4. Particle in a Constant Magnetic Field
[Solution p. 310] ★★★

A problem of classical electromagnetism studied from another point of view

We consider once more the problem of a particle with mass m and charge q_e subject to a constant magnetic field B directed along the Oz axis, but now viewed as an integrable system. This system has already been discussed in Problems 2.9 and 5.7; we will choose the same gauge and the same Lagrangian.

1. Write down the Hamiltonian of the system.

2. Find two first integrals in involution.

3. This system is integrable; moreover it is separable. Solve the Hamilton-Jacobi equation (do not attempt to evaluate the integral on the y variable) to find the action function.

4. Using Jacobi's theorem, give the solution of the problem in the form $t(y)$ and $x(y)$. Show that, in configuration space, the trajectory $x(y)$ is a circle.

5. Plot the phase portrait (y, p_y) in phase space. Notice the analogy with the harmonic oscillator problem. Show that this phase portrait is an ellipse the parameters of which are to be determined. Find the action variable I_y (you can avoid involved calculations, using the simple expression πab for the area of the ellipse).

6. Quantize the system according to the EBK rule (see 6.3).

7. There exists another very useful expression of the action which requires only the magnetic flux Φ across the trajectory. Give this expression.

8. We superpose on the magnetic field B a constant electric field E (not to be confused with the energy E) directed along Oy.

   Give the Hamiltonian of this new system.

9. Is this still an integrable system?

10. Give the corresponding action function.

11. Solve the problem using the Hamilton-Jacobi method. To obtain the trajectory, you will fix the unknown speed v in order that the integrant giving $x - vt$ is easily calculable. Interpret your result.

12. Show that the phase portrait is still an ellipse. Calculate the action variable and recover the results of Question 5. Is this surprising?

6.5. Actions for the Kepler Problem
[Solution p. 314] ★ ★ ★

A classical but nevertheless important integrable system

One considers the Kepler motion. This is a particle of mass m subject to a central force; one knows that the trajectory is planar. In this problem one works from the very beginning in the plane of the trajectory and employs the polar coordinates (ρ, ϕ). The potential is denoted $V(\rho) = -K/\rho$.

1. Recall the expressions of the Lagrangian and of the Hamiltonian.

2. Denote by (E, σ) the first integrals and give the action variable I_ϕ corresponding to the angle ϕ.

3. We consider the case of an attractive potential ($K > 0$) and a negative energy ($E < 0$). Give the expression of the radial momentum p_ρ as a function of the two first integrals and the integral expression for the radial action variable I_ρ.

4. The integral providing I_ρ is not easy to calculate. It is convenient to proceed indirectly. Thus, you will calculate the simpler integral corresponding to $\partial_E I_\rho(E, \sigma)$, which is the inverse of the radial angular frequency ω_ρ. To do this you will need to change the variable and switch to the θ variable defined as $\rho = \frac{1}{2}[\rho_M + \rho_m + (\rho_M - \rho_m)\cos\theta]$, where we introduce the perihelion ρ_m and the aphelion ρ_M of the trajectory.

5. Give, up to a constant, the expression of the action variable I_ρ. Using the Hamilton equation $\dot{p}_\rho = -\partial_\rho H$, show that the energy for a circular orbit is $E_c = -mK^2/(2\sigma^2)$. Determine the constant in the expression for the radial action and give the energy in terms of both actions $E(I_\rho, I_\phi)$.

6. Deduce the energies of the levels for the hydrogen atom (here $K = e^2/(4\pi\epsilon_0)$), applying the EBK rule (see 6.3). This is the Bohr model. Set $I_\rho = n_\rho \hbar$, $I_\phi = l\hbar$ and introduce the "fine structure constant" $\alpha = K/(\hbar c)$, a dimensionless quantity whose value is approximately $1/137$. The quantized values for the energy will be expressed as a function of the principal quantum number $n = n_\rho + l$.

Remark: In this problem, we considered a particle in a central force field. For the application to the hydrogen atom, this corresponds to an infinite mass for the proton. Nevertheless, we saw in Chapter 2, that a finite mass M_p for the proton can be dealt with exactly provided that the electron mass m in the previous expressions is replaced by the reduced mass of the system $\mu = M_p m/(M_p + m)$.

6.6. The Sommerfeld Atom
[Solution p. 316] ★ ★ ★

Relativistic correction to the Bohr atom

The atomic model for the hydrogen atom, due to N. Bohr, was very successful at the beginning of the elaboration of quantum mechanics. It relies on the quantization of type EBK for the classical Kepler orbits concerning an electron rotating around a proton under the action of the Coulomb force (see problem 6.5). The spectrum of the resulting electromagnetic emission agrees well with experimental data, but it suffers nevertheless from some drawbacks; in particular, it does not explain the so-called "fine structure" which is observed experimentally.

To cure these drawbacks, Sommerfeld introduced relativity in the formalism, the motion of the electron around the proton occurring with a speed that is not completely negligible with respect to the speed of light. In this problem, we study the effect of a relativistic treatment on the quantization of the states. We assume that the proton has an infinite mass whereas the electron, with mass m, moves in a central Coulomb force field. We use the convenient notation: $K = e^2/(4\pi\epsilon_0)$.

The force being central, we know that the motion occurs in a plane; the system has two degrees of freedom and the use of polar coordinates (ρ, ϕ) is suitable.

1. Recall the expression of the relativistic Lagrangian and the corresponding Hamiltonian.

2. Denote by E and σ the first integrals and give the action variable I_ϕ corresponding to the angle ϕ.

3. Give the expression of the radial momentum p_ρ as a function of the two first integrals, and the integral expression of the radial action variable I_ρ.

4. Calculate, as in the non-relativistic case of Problem 6.5, the simpler integral $\partial_E I_\rho(E, \sigma)$, which is the inverse of the radial angular frequency ω_ρ. To do this, you will make the same change of variable $\rho = \frac{1}{2}[\rho_M + \rho_m + (\rho_M - \rho_m)\cos\theta]$, using the perihelion ρ_m and the aphelion ρ_M of the trajectory. In the non-relativistic limit, find again the results for the Bohr atom.

5. Using the relation $\int (1-x^2)^{-3/2} dx = x(1-x^2)^{-1/2}$, obtain, up to a constant, the expression for the action variable I_ρ. Relying on Hamilton's equation $\dot{p}_\rho = -\partial_\rho H$, show that the energy of the circular orbit is

$$E_c = mc^2\sqrt{1 - (K/(\sigma c))^2}.$$

Determine the value of the constant appearing in the previous expression of I_ρ and, lastly, give the expression of the energy as a function of the two actions $E(I_\rho, I_\phi)$.

6. Let us introduce the "fine structure constant" $\alpha = K/(\hbar c)$, a dimensionless quantity whose value is approximately $1/137$. Quantize the expression of the energy obtained in the previous question using the EBK prescription (see 6.3); the quantization condition imposes $I_\rho = n_\rho \hbar$ (n_ρ integer ≥ 0) and $I_\phi = l\hbar$ (l integer ≥ 0).

7. Performing a truncated expansion to second order in α^2, calculate the quantized expression of the binding energy $B = E - mc^2$. The result will be expressed in terms of the principal quantum number $n = n_\rho + l$ and the angular quantum number l. It is very useful to introduce the Rydberg constant $R_\infty = mc\alpha^2/(2h)$ and $hcR_\infty = 13.6$ eV (for historical reasons it was the Planck constant h that appears in the Rydberg constant and not the normalized constant $\hbar$).

The term in α^2 leads to the non-relativistic Bohr formula, which depends on n only and implies level degeneracy. The term in α^4 is a relativistic correction depending on both numbers n and l and which, in consequence, splits the level degeneracy; this effect is known as the fine structure of the atom. Sommerfeld's formula looks very similar to Dirac's formula obtained in a relativistic quantum mechanical treatment. The tiny difference comes from spin effects.

Important remark: In the case of a finite mass M_p for the proton, the correct separation of the center of mass motion, which can be considered as fixed, and the relative motion of a fictive particle is valid only in a non-relativistic treatment and cannot be invoked in the present relativistic framework. Nevertheless, replacing the electron mass m by the reduced mass of the system $\mu = M_p m/(M_p + m)$ in the definition of the Rydberg constant provides a very good approximation.

6.7. Energy as a Function of Actions
[Solution p. 318] ★ ★ ★

The two most classical problems in mechanics and their semi-classical quantization

This problem investigates the expression of the energy as a function of actions for two traditional systems: the hydrogen atom and the harmonic oscillator; the purpose is to obtain a quantum expression for the energy.

Let us consider, in a two-dimensional space, a particle with mass m subject to a central potential. This problem is integrable and separable if one chooses as generalized coordinates the polar coordinates (ρ, ϕ).

You can check that the action variables are respectively the angular momentum $I_\phi = p_\phi = \sigma$ and the surface area (to within 2π) given by the phase portrait in the plane (ρ, p_ρ), namely

$$2\pi I_\rho(E, \sigma) = \oint p_\rho \, d\rho.$$

The aim of the problem is to determine the expression of the Hamiltonian as a function of the actions. Of course, the result depends on the choice of the potential $V(\rho)$. The answer is easy for the most classical periodic motions (Kepler's problem $V(\rho) = -K/\rho$, where, for the hydrogen atom, $K = e^2/(4\pi\epsilon_0)$, and the harmonic oscillator $V(\rho) = \frac{1}{2}m\omega^2\rho^2$).

A – Kepler's problem

1. Plot schematically the trajectories in spaces (ρ, p_ρ) and $(\phi, p_\phi = \sigma)$. Compare the angular frequencies for the motion concerning the angle ω_ϕ and the radius ω_ρ (how many revolutions of the trajectory in space (ρ, p_ρ) take place when one revolution is performed in space $(\phi, p_\phi = \sigma)$). One can take advantage of the knowledge we have concerning the orbit of the motion.

2. Using Hamilton's equations, deduce that the Hamiltonian is a function of both actions I_ρ, I_ϕ only in an expression of the form

$$H(I_\rho, I_\phi) = H(I_\rho + I_\phi).$$

B – Harmonic oscillator

1. Examine the same questions as the preceding case and deduce that the Hamiltonian for a harmonic oscillator takes the form

$$H(I_\rho, I_\phi) = H(2I_\rho + I_\phi).$$

C – Semi-classical quantization

1. For these two cases, consider the circular orbit. Using Newtonian mechanics (centrifugal force = central force), calculate the corresponding energy $E_c(\sigma)$.

2. Deduce the expression for Hamilton's function as a function of the action for both problems, and give the energy levels of the corresponding quantum problems, using the EBK rule (see 6.3).

6.8. Invariance of the Circulation Under a Continuous Deformation
[Solution p. 322] ★ ★ ★

Demonstration of a fundamental geometric property in the two-dimensional case; importance of involution

The aim of this problem is to demonstrate, in the case of an integrable system in two dimensions, the invariance of the circulation of the momentum on a path belonging to phase space which is defined by two functions F_1 and F_2 that are in involution: $\{F_1, F_2\} = 0$. We consider in a phase space with four dimensions, a point of which is specified by the Cartesian coordinates (x, y, p_x, p_y), a two-dimensional surface M_f which is the intersection of two hypervolumes in 3 dimensions: $F_1(x, y, p_x, p_y) = f_1$ and $F_2(x, y, p_x, p_y) = f_2$. After inversion of these relations, one can express, **on the surface**, the momenta as a function of the coordinates, namely $p_x(x, y; f_1, f_2)$ and $p_y(x, y; f_1, f_2)$.

Let Γ be a closed path entirely belonging to M_f and let us consider the reduced action calculated on Γ, that is the circulation of $p(q; f)$ (see Fig. 6.3):

$$\oint_\Gamma p(q; f) \cdot dq = \oint_\Gamma \left(p_x(x, y; f_1, f_2) \, dx + p_y(x, y; f_1, f_2) \, dy \right).$$

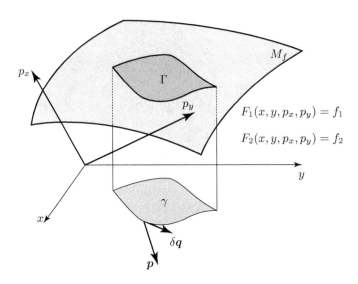

Fig. 6.3 Circulation, in configuration space, of the quantity $p \cdot dq$ for two paths, with the same extremities, chosen on the surface M_f

The property to be proved is the following: whatever the considered contour Γ, reducible to a single point, the circulation vanishes. This is equivalent to the property that, for two arbitrary paths Γ_1 and Γ_2 connecting the same two points, and in the same sense, the circulation is the same.

1. Differentiating the two constants of the motion F_1 and F_2 and choosing, for each of these equations, paths on M_f with constant x and y coordinates, deduce four partial differential equations connecting p to the partial derivatives of F.

2. Eliminating correctly the quantities $\partial_x p_x$ and $\partial_y p_y$ from these four equations, find the resulting systems of equations, and calculate the quantities $\partial_x p_y$ and $\partial_y p_x$ as a function of the partial derivatives of F.

3. Use the involution property to prove the relation, valid on M_f: $\partial_x p_y - \partial_y p_x = 0$.

4. With the help of Stokes' theorem, deduce that the integral $\int (p_x dx + p_y dy)$, on the M_f surface, does not depend on the path chosen to calculate it.

6.9. Ball Bouncing on a Moving Tray
[Solution p. 324]   ★

Change of the Hamiltonian in a time dependent canonical transformation

We wish to study the vertical motion of a ball subject to gravity in a non-Galilean frame (in an elevator for instance) by adding an inertial force. In this problem, we employ a more complicated, but very instructive, method using a time dependent canonical transformation.

Subject to a constant gravitational field g, a ball of mass m bounces on a tray (the floor of the elevator) whose altitude varies following a law $h(t)$. Rather than the altitude q above the ground, it is better to take as generalized coordinate the height Q above the tray.

One thus considers the time dependent contact transformation:

$$Q(q,t) = q - h(t); \quad P(p,t) = p - m\dot{h}(t).$$

1. Show that this transformation is a canonical transformation.

2. Find the generating function of the second type $G_2(P, q, t)$ for this transformation.

3. Deduce the new Hamiltonian $K(Q, P, t)$. Notice that a pure time dependence in the Hamilton function does not modify the equations of motion.

6.10. Harmonic Oscillator with a Variable Frequency [Solution p. 324] ★ ★

A further variant on the theme of the harmonic oscillator

In this problem, we are concerned with a time dependent canonical transformation. The system under consideration is a one-dimensional harmonic oscillator with a variable angular frequency, the Hamiltonian of which is given by:

$$H(q,p,t) = \frac{p^2}{2m} + \frac{1}{2}m\omega(t)^2 q^2.$$

This type of Hamiltonian mimics for example a pendulum whose length varies in time (for small amplitudes) or a heated spring with a constant that varies in time.

Perform the canonical transformation with the generating function of the first type:

$$G_1(Q,q,t) = \frac{1}{2}m\omega(t)q^2 \cot(Q).$$

Find the new Hamiltonian and write down Hamilton's equations for the new set of coordinates.

The advantage of this canonical transformation is that the equation giving $Q(t)$ is a differential equation of first order (albeit non-linear), whereas the Lagrange equation giving $q(t)$ is of second order.

6.11. Choice of the Momentum [Solution p. 325] ★ ★

For a one-dimensional system, it is always possible to choose a momentum proportional to the Hamiltonian

For an autonomous system with one degree of freedom (phase space with two dimensions), it is always possible to perform a canonical transformation such that the new momentum $P(q,p)$ is a function of the Hamiltonian.

To illustrate this property we consider the harmonic oscillator and choose the new momentum P proportional to the Hamiltonian:

$$P(q,p) = \lambda H = \frac{1}{2}\lambda\omega(p^2 + q^2).$$

1. Calculate the old coordinate q in terms of the old momentum p and the new momentum P. Which type of generating function occurs naturally? Calculate this generating function (do not try to evaluate the primitive).

2. Deduce the new coordinate Q, first as a function of the (P, p) variables, and then as a function of the old variables (q, p).

3. Adjust the constant λ in order that one period for the motion corresponds to one revolution along the trajectory. Calculate the corresponding action variable.

You can adopt this approach for any Hamiltonian, for instance the Hamiltonian corresponding to free fall.

6.12. Invariance of the Poisson Bracket Under a Canonical Transformation
[Solution p. 326] ★

A calculation that you should perform once in your life

Let (q, p) be conjugate variables for a one-dimensional system and $F(q, p)$ and $G(q, p)$ two arbitrary functions defined in phase space. Let us perform a contact transformation $Q(q, p)$, $P(q, p)$ for which the inverse transformation is denoted by $q(Q, P)$ and $p(Q, P)$.

Express, in terms of the (q, p) variables, the quantity $(\partial_Q F)(\partial_P G)$ using the functions $F(q(Q, P), p(Q, P))$ and $G(q(Q, P), p(Q, P))$, as well as the analogous expression obtained by permuting F and G. Demonstrate the relation:

$$\{F, G\}_{Q,P} = [(\partial_Q q(Q, P))(\partial_P p(Q, P)) - (\partial_Q p(Q, P))(\partial_P q(Q, P))]\{F, G\}_{q,p}.$$

Deduce the invariance of the Poisson bracket under a canonical transformation.

6.13. Canonicity for a Contact Transformation [Solution p. 327] ★

Generalization of the previous problem to a system with n degrees of freedom

For a system with one degree of freedom, checking for canonicity is simple: the area in phase space is conserved.

1. In the case of a system with n degrees of freedom, generalize the method of Problem 6.12 to find the following necessary and sufficient conditions:

$$\sum_k \left(\frac{\partial Q_i}{\partial q_k} \frac{\partial Q_j}{\partial p_k} - \frac{\partial Q_i}{\partial p_k} \frac{\partial Q_j}{\partial q_k} \right) = 0;$$

$$\sum_k \left(\frac{\partial P_i}{\partial q_k} \frac{\partial P_j}{\partial p_k} - \frac{\partial P_i}{\partial p_k} \frac{\partial P_j}{\partial q_k} \right) = 0;$$

$$\sum_k \left(\frac{\partial Q_i}{\partial q_k} \frac{\partial P_j}{\partial p_k} - \frac{\partial Q_i}{\partial p_k} \frac{\partial P_j}{\partial q_k} \right) = \delta_{i,j}.$$

valid for any pair (i, j).

Using the Poisson bracket for the (q, p) variables, these equations read more simply

$$\{Q_i, Q_j\} = 0; \qquad \{P_i, P_j\} = 0; \qquad \{Q_i, P_j\} = \delta_{i,j}.$$

In other words, the answer implies that it is necessary and sufficient that all the elementary Poisson brackets are invariant under a canonical transformation.

In fact, this result also holds for time dependent transformations. The difference between the two cases lies in the expression of the new Hamiltonian.

2. When the transformation is canonical, demonstrate the invariance of the Poisson brackets for any two functions of phase space:

$$\{F, G\}_{Q,P} = \{F, G\}_{q,p}.$$

6.14. One-Dimensional Free Fall
[Solution p. 329] ★ ★

An involved, but instructive, way to obtain the equation of motion for a simple problem

A particle of mass m is submitted to a vertical constant gravitational field g. It is specified by its altitude q above the ground. The Hamiltonian for the system is $H(q, p) = p^2/(2m) + mgq$. We wish to find a canonical transformation which provides a cyclic coordinate. To do this, it is useful to define a new momentum as $P = \lambda H$.

1. Adopting any method which seems to you well suited, show that a conjugate variable can be chosen under the form $Q = \mu p$.

2. Determine the unknown constants λ, μ in order to obtain a canonical transformation.

3. Solve the corresponding equations and recover the traditional equations expressed in terms of the original variables.

6.15. One-Dimensional Free Fall Again
[Solution p. 330] ★ ★

Angle-action variables for the free fall problem, using a generating function

Consider again the problem of the ball of mass m, subject to a constant gravitational field g, bouncing on the ground (see Problem 6.3). In this example, we will use a generating function.

Let us investigate a canonical transformation which gives a cyclic coordinate. The original Hamiltonian $H(q, p) = p^2/(2m) + mgq$ must be expressed in term of the new momentum P only. The new Hamiltonian is denoted by $H(P)$; it is an arbitrary function that can be simply identified with P.

1. Give $Q(t)$. Find again the equation of motion in terms of the original coordinates. It will be useful to introduce a generating function of the fourth type $G_4(P, p)$ to obtain $q(t)$ and $p(t)$.

2. Calculate the action as a function of the energy.

3. Determine the angular frequency and deduce the corresponding angle variable. Check your results against those of Problem 6.3.

6.16. Scale Dilation as a Function of Time
[Solution p. 332] ★ ★ ★

Motion in a time-dependent potential

Consider a particle of mass m moving on a straight line Ox, trapped in a potential whose range varies in time according to the law $V(x/l(t))$. We set $l'(t) = dl/dt$, $l''(t) = d^2l/dt^2$.

1. Write down the Lagrangian using as generalized coordinate the variable $q(x, t) = x/l(t)$ (even for a particle at rest and subject to no force, the coordinate varies in time).

2. Deduce the Hamilton function.

3. Perform a time dependent canonical transformation based on the generating function

$$G_2(P,q,t) = Pq + \frac{1}{2}mq^2 l(t) l'(t).$$

Give the new Hamiltonian.

This result is very important. Indeed, it shows that one can deal quite easily with a linear change of scale ($l''(t)=0$) provided we work with new variables. In the general case, this means that one must add to the original potential a harmonic potential (repulsive or attractive).

4. Notice that the kinetic term depends on time through the function l^2. We proved (see Problem 4.4) that a new system, whose Hamiltonian is a function of the Hamiltonian of an old system, has the same families of trajectories as the old system, but described with a different temporal law. What is the new temporal law $\tau(t)$ and the new Hamiltonian $K(Q,P,\tau)$?

6.17. From the Harmonic Oscillator to Coulomb's Problem [Solution p. 333]   ★ ★

The two most classical problems are related by a canonical transformation

In a plane, a particle of mass m is specified by its polar coordinates (ρ, ϕ). We recall that the Hamiltonian of an isotropic oscillator is given by

$$H_{HO}(\rho, \phi, p_\rho, p_\phi) = \frac{p_\rho^2}{2m} + \frac{p_\phi^2}{2m\rho^2} + \frac{1}{2}m\omega^2 \rho^2.$$

Remember also that, the ϕ coordinate being cyclic, the angular momentum p_ϕ is a constant of the motion.

We perform the contact transformation $Q = \rho^2/l$; the length l is arbitrary but constant and its role is to make Q dimensionally equal to a length, as is ρ.

1. What should be the expression of the new momentum P in order to make the transformation canonical? A contact transformation is automatically canonical, provided that we take as momentum the expression $P = \partial_{\dot{Q}} L$. Check this property in this particular case.

2. Give the new Hamiltonian $K(Q, \phi, P, p_\phi)$.

3. The system is conservative so that the value E_{ho} of the Hamiltonian is a constant.

The expression obtained for the Hamiltonian looks similar to that of the Hamiltonian $H_c(Q, \alpha, P, p_\alpha)$ encountered in Kepler's problem with an attractive potential $V(\rho) = -K/\rho$. What are the identities which must be imposed on the constant K, the energy E_c, and the angular momentum p_α to pass from H_{HO} to H_c?

4. The analogy can be pursued by seeking a relationship between the polar angles ϕ and α. Comparing the values $d\rho/d\phi$ in the case of the harmonic oscillator and $dQ/d\alpha$ in the Coulomb case, show that one must have $\alpha = 2\phi$. This relation is not surprising since, between two perihelia, the polar angle ϕ varies by π in the case of the harmonic oscillator, and the polar angle α varies by 2π in the case of Kepler's problem (this property was underlined in Problem 6.7).

We will profit from these analogies to obtain the features of the trajectory for the isotropic harmonic oscillator based on the properties of the Kepler motion. We remind you that the trajectory of the Kepler problem is an ellipse with a focus at the center of force, whose polar equation is $Q = p/[1 + e\cos\alpha]$, where the parameter p is given by $p = p_\alpha^2/(mK)$ and the eccentricity by

$$e = \sqrt{1 + (2E_c p_\alpha^2)/(mK^2)}.$$

5. Show that the trajectory for the harmonic oscillator is an ellipse with its center at the center of force. Give the values of the semi-major axis a and the semi-minor axis b.

6. Check that the value of the Hamiltonian $H_{HO}(\rho, \phi, p_\rho, p_\phi)$, which may be calculated for instance at the perihelion, is precisely E_{oh}.

7. An interesting feature of this approach is the calculation of the radial action: the transformation being canonical, the result is the same if one makes the correct substitutions.

The radial action was calculated in Problem 6.5:

$$I_\rho(E_c, p_\alpha) = \frac{1}{2}\left(K\sqrt{2m/|E_c|} - 2p_\alpha\right).$$

With the substitutions proposed in Question 3, deduce the radial action for the harmonic oscillator. Apply the EBK quantization (see 6.3) in the case of a two-dimensional harmonic oscillator.

6.18. Generators for Fundamental Transformations [Solution p. 336] ★ ★

Study of the generators for the most classical transformations and their flow

A – Translations

Let us consider a constant vector $\boldsymbol{a} = (a_1, a_2, a_3)$ in a three-dimensional space.

1. Calculate the variation of a function $F(\boldsymbol{r}, \boldsymbol{p})$ defined in phase space under the effect of this translation. Be careful!: this transformation affects the coordinates but not the momenta which are proportional to the velocity.

2. Consider a translation by a along the Ox_i axis. Using formula (6.6), give the flow parameter and the generator T_{a_i} for this translation.

3. Show that generators for translations along different axes commute, that is $\{T_{a_i}, T_{a_j}\} = 0$.

B – Rotations in a three-dimensional space

We examine the motion of a particle moving in a three-dimensional space (phase space with six dimensions). In this phase space, we consider the function $L_z = xp_y - yp_x$.

1. Calculate the variation of a function $F(\boldsymbol{r}, \boldsymbol{p})$ for a rotation by an infinitesimal angle $d\phi$ around the Oz axis.

2. Plot the flow of L_z in ordinary space and determine the flow parameter.

3. Defining L_x and L_y from L_z by a circular permutation, demonstrate the following property: $\{L_x, L_y\} = L_z$. One deduces that rotations around non parallel axes do not commute.

C – Galilean transformation in a one-dimensional space

The Galilean transformation connects the same point, seen from two frames R and R' moving relatively one to the other at a constant speed v. In other words $q' = q - vt$. One can consider this operation as an increase by v of the velocity, the so-called boost.[8]

One deals with a particle of mass m moving on a straight line.

1. Calculate the variation of a function $F(q, p)$ defined in phase space for a small relative velocity v between the two frames.

2. Calculate, using Formula (6.6), the generator for the Galilean transformation $G(q, p, t)$ noticing that the flow parameter is the speed v.

D – Lorentz transformation in a one-dimensionial space

In relativity, the time loses its universal status and is no longer privileged as compared to space; thus it is convenient to consider it (multiplied par the

---

[8] To consider the same system from two distinct frames is known as the "passive point of view". To consider, in the same frame, the action on the system (as a displacement or an increase of the velocity) is called the "active point of view".

Problem Solutions

speed of light c) as a coordinate. The conjugate variable is the negative of the energy divided by c. The dimension of phase space is increased by two and a point in phase space is defined by $(q_1, q_2, p_1, p_2) = (q, ct, p, -E/c)$.

The Lorentz transformation can be seen as the transformation of the coordinates of the same point defined in two frames which move with a relative speed v. If you have forgotten the form of a Lorentz transformation, look it up in a textbook on relativity.

1. Show that, for an infinitesimal speed $\beta = dv/c$, the change of coordinates is given by:

$$q' = q - \beta ct; \qquad ct' = ct - \beta q$$
$$p' = p - \beta \frac{E}{c}; \qquad \frac{E'}{c} = \frac{E}{c} - \beta p.$$

2. Calculate the variation of a function defined in phase space for the Lorentz transformation.

3. Calculate, using formula (6.6), the generator of the Lorentz transformation $U(q, t, p, E)$ noticing that the flow parameter is the speed v.

Problem Solutions

6.1. Expression of the Period for a One-Dimensional Motion [Statement p. 289]

Let us recall the definition of the reduced action

$$\tilde{S}(q_1, q, E) = \int_{q_1}^{q} p(E, \tilde{q}) \, d\tilde{q},$$

where $\tilde{q}$ denotes the generalized coordinate along the trajectory. Since the one-dimensional motion is periodic, after one period T, the system resets its coordinate. Let us substitute $q = q_1$ in the previous expression:

$$\tilde{S}(q_1, q_1, E) = \oint p(E, \tilde{q}) \, d\tilde{q}.$$

But this integral is precisely 2π times the action variable $I(E)$. Thus $\tilde{S}(q_1, q_1, E) = 2\pi I(E)$. On the other hand, the travel time along the trajectory is given by (see (5.8)) $t - t_1 = \partial_E \tilde{S}(q_1, q, E)$. If one chooses the end point to coincide with the starting point after one period, this time is precisely the period $T = \partial_E \tilde{S}(q_1, q_1, E) = 2\pi(\partial_E I(E))$ so that the angular frequency ω which is related to the period by $\omega = 2\pi/T$ is $\omega = 1/(\partial_E I(E)) = \partial_I E$:

$$\omega(E) = \frac{dE}{dI} = \frac{1}{(dI/dE)}$$

Using angle-action variables, the angular frequency is given by $\omega = dE(I)/dI$ and the period of the motion by $T = 2\pi/\omega = 2\pi/(dE(I)/dI)$, which is identical with $2\pi dI(E)/dE$.

6.2. One-Dimensional Particle in a Box
[Statement p. 290]

1. The kinetic energy of the particle is $T = \frac{1}{2}m\dot{q}^2$, and the Lagrangian $L = T - V$. One deduces the momentum $p = m\dot{q}$ and the Hamiltonian $H = p\dot{q} - L = p^2/(2m) + V$. With the proposed potential:

$$H(q,p) = \frac{p^2}{2m} + \left(\frac{q}{a}\right)^\lambda.$$

2. The phase portrait is given by the curve $p^2/(2m) + (q/a)^\lambda = E$. It is depicted in Fig. 6.4 for several values of the energy and for the value $\lambda = 10$.

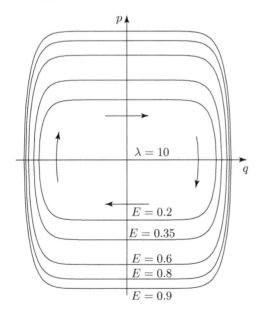

Fig. 6.4 Phase portrait for a square well that is "softened" to a power-law potential with a parameter $\lambda = 10$ and for five different values of the energy

The action variable is defined by

$$I(E) = \frac{1}{2\pi} \oint p(q, E)\, dq,$$

or, taking into account the symmetries and with the proposed potential:

$$I(E) = \frac{4}{2\pi} \int_0^{aE^{1/\lambda}} \sqrt{2m\left(E - (q/a)^\lambda\right)}\, dq.$$

In the integral, let us make the change of variable $x = q/(aE^{1/\lambda})$ to rewrite

$$I(E) = \frac{2}{\pi}\sqrt{2mE}\, aE^{1/\lambda} J(\lambda),$$

with the expression $J(\lambda) = \int_0^1 \sqrt{1 - x^\lambda}\, dx$. Let the potential tend to the square well, that is let us take the limit $\lambda \to \infty$. We obtain $E^{1/\lambda} \to 1$ and $J(\lambda) \to 1$. Consequently, we are left with the expression:

$$I(E) = \frac{2a\sqrt{2mE}}{\pi}.$$

For a given energy E, since the particle inside the well is free, its speed is given by $v = p/m = \sqrt{2E/m}$. During half a period $T/2$, the particle travels over a distance $2a$. One deduces that $T = 4a/v$ and that the angular frequency $\omega = 2\pi/T = \pi v/(2a) = (\pi/a)\sqrt{E/(2m)}$. The same expression could have been obtained from the general relation $\omega = 1/(dI/dE)$.

The angle variable evolves linearly in time $\alpha = \omega t$. But it has to be expressed in terms of the coordinate q and energy E, a procedure that does not pose any particular problem since $t = q/v$ and thus $\alpha = (\pi v/2a)(q/v) = \pi q/(2a)$.

$$\alpha(q, E) = \frac{\pi}{2a} q$$

3. The EBK quantization, which gives the energy as a function of the quantum number n, is obtained very simply by substituting for the action, in the expression $E(I)$, the value $I = n\hbar$. From the result given in the first question, one deduces $E = \pi^2 I^2/(8ma^2)$ or, after performing the above substitution, the value for the energy:

$$E_n = \frac{\pi^2 \hbar^2}{8ma^2} n^2$$

This semi-classical expression is identical with the corresponding quantum expression.

6.3. Ball Bouncing on the Ground
[Statement p. 290]

1. The Hamiltonian of the system is $H(q,p) = p^2/(2m) + mgq$; it corresponds to a constant of the motion whose value is E. The ball is released at rest from an altitude h. For $q = h$, we have $p = mv = 0$. Therefore we obtain the relation $h = E/mg$. The phase portrait results from the curve given by $p^2/(2m) + mgq = E$, or:

$$p(q, E) = \pm\sqrt{2m(E - mgq)}.$$

From $q = h$, $p = 0$, the particle begins to fall ($p < 0$ with our conventions concerning the axes); the phase portrait is the portion of the preceding parabola corresponding to the $-$ sign (phase 1 for the motion). At ground level, one has $q = 0$, $p = -\sqrt{2mE}$; the particle bounces so that the sign of the velocity changes instantaneously $q = 0$, $p = \sqrt{2mE}$ (phase 2). Then the particle rises following the phase portrait corresponding to the $+$ sign, which is still a portion of parabola symmetric to the previous one (phase 3). At the point $q = h$, $p = 0$, it has achieved a cycle and it then repeats this periodic motion. The corresponding phase portrait is depicted in Fig. 6.5.

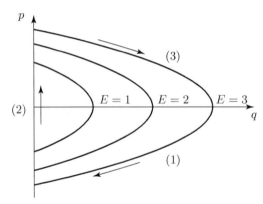

Fig. 6.5 Phase portrait of the ball bouncing on the ground, for three values of the energy. The rebound corresponds to the vertical portion of each curve with an instantaneous inversion of the velocity. The three phases of the motion (1), (2), (3) are described in the text

The action variable is defined as

$$I(E) = \frac{1}{2\pi} \oint p(q, E)\, dq$$

which, owing to the symmetry of the up and down journey, is written in the form

$$I(E) = \frac{1}{\pi} \int_0^h \sqrt{2m(E - mgq)}\, dq.$$

This integral is elementary and is calculated more easily by performing the change of variable $x = (mg/E)\,q$. The final result is

$$I(E) = \frac{2}{3\pi g}\sqrt{\frac{2}{m}}\, E^{3/2}.$$

The fall time $\sqrt{2h/g}$ corresponds to a half-period so that $T = 2\sqrt{2h/g}$; the angular frequency is given by

$$\omega = \frac{2\pi}{T} = \pi\sqrt{\frac{g}{2h}}, \quad \text{or} \quad \omega = \pi g\sqrt{\frac{m}{2E}}.$$

This quantity can also be obtained from the general formula $\omega = dE/dI$. The angle variable varies linearly in time $\alpha = \omega t$, but one must replace the time by the energy; this relation can be obtained from the temporal law $q = -\frac{1}{2}gt^2 + h$ which gives

$$t = \sqrt{\frac{2(h-q)}{g}}.$$

Finally, replacing ω and h by their values in terms of the energy, the angle variable is:

$$\alpha(q, E) = \pi\sqrt{1 - \frac{mg}{E}q}$$

2. Writing the energy as a function of the action is simply achieved by inverting the relation giving the action as a function of the energy, which was obtained above. This is easy to do and leads to the desired equation:

$$E(I) = \left[\frac{3\pi g I}{2}\right]^{2/3}\left(\frac{m}{2}\right)^{1/3}.$$

3. To quantize this value using the semi-classical EBK rule, it is sufficient to replace the action in the previous equation by an integer multiple of the elementary action $\hbar$: $I = n\hbar$. We thus obtain the following semi-classical formula,

$$E_n = (m)^{1/3}\left[\frac{3\pi g\hbar}{2\sqrt{2}}n\right]^{2/3}$$

which is identical to the quantum formula if one makes the substitution $n \to n - 1/4$.

6.4. Particle in a Constant Magnetic Field
[Statement p. 291]

1. The magnetic field is constant and directed along the Oz axis. On several occasions we saw that the vector potential can be chosen in the form $\mathbf{A}(-yB, 0, 0)$. In our case, the Lagrangian of the system is $L = \frac{1}{2}m(\dot{x}^2 + \dot{y}^2 + \dot{z}^2) - q_e B \dot{x} y$. One deduces the momenta

$$p_x = \partial_{\dot{x}} L = m\dot{x} - q_e B y$$
$$p_y = \partial_{\dot{y}} L = m\dot{y}$$
$$p_z = \partial_{\dot{z}} L = m\dot{z}.$$

Since $\partial_z L = 0$, one has $p_z = const$ which can be chosen as a null quantity (it is enough to change the inertial frame); one then deduces $z = const$ which can be taken null as well (translation of the axes). This means that the motion occurs in the plane $z = 0$ and the problem is two-dimensional. The Hamiltonian is the Legendre transform of the Lagrangian $H = \dot{x} p_x + \dot{y} p_y - L$, which, expressed in terms of coordinates and momenta, is written as

$$H(x, y, p_x, p_y) = \frac{(p_x + q_e B y)^2}{2m} + \frac{p_y^2}{2m}.$$

2. The Hamiltonian is time-independent, so that it is a first integral. Furthermore, it does not depend on the x variable, so that p_x is also a first integral. These two first integrals are independent and in involution (case $n = 2$).

3. The system possesses two degrees of freedom and two first integrals are in involution; it is integrable. The action function is given by $S(x, y, t) = -Et + \tilde{S}(x, y, E)$, where the reduced action obeys the characteristic equation $E = H(x, y, \partial_x \tilde{S}, \partial_y \tilde{S})$, or

$$E = \frac{1}{2m} \left(\partial_x \tilde{S} + q_e B y \right)^2 + \frac{1}{2m} \left(\partial_y \tilde{S} \right)^2 / (2m).$$

The x variable is cyclic and one can set $\tilde{S}(x, y, E) = Px + S_2(y, E)$, where P is the value taken by the first integral p_x. The characteristic equation now gives

$$\frac{dS_2}{dy} = \pm\sqrt{2mE - (P + q_e B y)^2}.$$

In summary, we obtain the total action in the complete form:

$$S(x, y, t; E, P) = -Et + Px \pm \int \sqrt{2mE - (P + q_e B y)^2}\, dy.$$

4. Jacobi's theorem (5.10), applied to the energy constant, gives

$$\partial_E S(x,y,t;E,P) = \text{const.}$$

Suitably choosing the time origin so as to cancel this constant, we arrive at the temporal equation:

$$t(y) = \pm m \int \frac{dy}{\sqrt{2mE - (P + q_e By)^2}}.$$

Applying the same theorem to the momentum constant, one has:

$$\partial_P S(x,y,t;E,P) = \text{const} = x \mp \int \frac{(P + q_e By)}{\sqrt{2mE - (P + q_e By)^2}} dy.$$

The calculation of the integral is elementary and provides, denoting by x_c the integration constant, the equation of the trajectory

$$x = \pm \sqrt{2mE - (P + q_e By)^2}/(q_e B) + x_c.$$

Squaring, it can be recast in the form:

$$(x - x_c)^2 + \left(y + \frac{P}{q_e B}\right)^2 = \frac{2mE}{q_e^2 B^2}.$$

One recognizes the equation of a circle with center $(x_c, -P/(q_e B))$ and radius $\sqrt{2mE}/|q_e B|$.

5. Since, on the trajectory, one has $H = E$ and $p_x = P$, the expression obtained in question 1 allows us to write: $2mE = (P+q_e By)^2 + p_y^2$. This is precisely the expression of the phase portrait which can be translated into the more explicit form:

$$\frac{(y + P/(q_e B))^2}{(2mE)/(q_e^2 B^2)} + \frac{p_y^2}{2mE} = 1.$$

The phase portrait is an ellipse with its center displaced by $-P/(q_e B)$ on the Oy axis, with the semi-major (or semi-minor) axis $a = \sqrt{2mE}/|q_e B|$ and semi-minor (or semi-major) axis $b = \sqrt{2mE}$.

The action I_y is given by $(1/2\pi)\times$ (Area of the trajectory in phase space). The area of the ellipse is πab, and, with the data concerning the axes given above, we arrive at the value for the action variable:

$$I_y = \frac{mE}{|q_e B|}.$$

6. The energy depends only on the action I_y through the formula $E = |q_e B| I_y / m$. The EBK quantization rule stipulates that one must replace the action by an integer multiple of the elementary action $I_y = n\hbar$. We thus obtain the quantized Landau levels:

$$E_n = \frac{n\hbar |q_e B|}{m}.$$

7. Let us remind you that the trajectory is a circle of radius

$$R = \sqrt{2mE}/|q_e B|.$$

The magnetic flux intercepted by the trajectory is thus $\Phi = \pi R^2 B = 2\pi m E/(q_e^2 B)$ so that $E = q_e^2 B \Phi/(2\pi m)$. Substituting this value in the expression for the action leads to:

$$I_y = \frac{|q_e \Phi|}{2\pi}.$$

8. One must add the kinetic energy, already obtained in Question 1, to the electric potential energy to obtain the total Hamiltonian:

$$H(x, y, p_x, p_y) = \frac{(p_x + q_e By)^2}{2m} + \frac{p_y^2}{2m} + q_e Ey.$$

9. As before, the system is conservative and the energy is a first integral $H = E$. Moreover, x is a cyclic coordinate so that p_x is a second first integral. They are independent and in involution (case $n = 2$). The system is integrable.

10. With a reasoning analogous to that developed in the preceding case (x being still cyclic, the problem remains separable), the expression of the action function is found to be:

$$S(x, y, t; E, P) = -Et + Px \pm \int \sqrt{2m(E - q_e Ey) - (P + q_e By)^2} \, dy.$$

11. The equations of motion are obtained from Jacobi's theorem:

$$\partial_E S(x, y, t; E, P) = \text{const} = -t \pm \int \frac{m \, dy}{\sqrt{2m(E - q_e Ey) - (P + q_e By)^2}};$$

$$\partial_P S(x, y, t; E, P) = \text{const} = x \mp \int \frac{(P + q_e By)}{\sqrt{2m(E - q_e Ey) - (P + q_e By)^2}} \, dy.$$

Problem Solutions

The constants can be chosen as null quantities without loss of generality. Let us combine these two equations to form:

$$x - vt = \pm \int \frac{(P + q_e By) - mv}{\sqrt{2m(E - q_e Ey) - (P + q_e By)^2}} \, dy.$$

Let us now take $v = -E/B$; the integration is then immediate and gives the equation of the trajectory:

$$(x - vt - x_c)^2 + (y - y_c)^2 = \frac{2mE^*}{q_e^2 B^2}$$

where x_c is an integration constant,

$$y_c = -\frac{P - mv}{q_e B} \quad \text{and} \quad E^* = E + \frac{1}{2}mv^2 - vP.$$

It is easy to check that $E = E^* + \frac{1}{2}mv^2 + q_e E y_c$ and interpret E^* as the kinetic energy of the particle in a frame that drifts in a direction perpendicular to the electric and magnetic fields with a speed $v = -E/B$. In this frame, the trajectory is the cyclotron circle with radius $(2mE^*)/(q_e^2 B^2)$ and in the original frame, it is represented by the equation of the curve given above.

12. The phase portrait is given by the equation

$$\frac{p_y^2}{2m} + \frac{(P + q_e By)^2}{2m} + q_e Ey = E$$

which can be recast in the more friendly form:

$$\frac{(y - y_c)^2}{(2mE^*)/(q_e^2 B^2)} + \frac{p_y^2}{2mE^*} = 1.$$

The phase portrait is still an ellipse, with its center displaced by y_c on the Oy axis, with semi-major (semi-minor) axis $a = \sqrt{2mE^*}/|q_e B|$ and semi-minor (semi-major) axis $b = \sqrt{2mE^*}$. The action variable has the same expression as that obtained in the absence of an electric field $I_y = \pi ab/(2\pi)$, which is the magnetic flux through the drifting circle multiplied by $q_e/(2\pi)$.

This conclusion is not surprising. One knows from electromagnetism that an observer moving with respect to the magnetic field sees an electric field perpendicular to its velocity and to the magnetic field itself. It is thus possible to cancel the electric field if one works in a frame that drifts perpendicularly to this field. In this frame, we recover the well known situation corresponding to the cyclotron motion alone.

6.5. Actions for the Kepler Problem
[Statement p. 292]

1. Using polar coordinates (ρ, ϕ), the Lagrangian L and the Hamiltonian H are given respectively by:

$$L(\rho, \phi, \dot\rho, \dot\phi) = \frac{1}{2}m\left(\dot\rho^2 + \rho^2\dot\phi^2\right) + \frac{K}{\rho};$$

$$H(\rho, \phi, p_\rho, p_\phi) = \frac{p_\rho^2}{2m} + \frac{p_\phi^2}{2m\rho^2} - \frac{K}{\rho}.$$

2. The Hamiltonian is conservative; there exists a first integral which is the energy $H = E$. Moreover, ϕ is a cyclic coordinate so that there exists a second first integral: the angular momentum $p_\phi = \sigma$. The action variables are defined by

$$2\pi I_i = \oint_{\Gamma_i} p \cdot dq = \oint_{\Gamma_i} (p_\rho\, d\rho + p_\phi\, d\phi).$$

On the torus, it is possible to choose a contour with $\rho = \text{const}$. The corresponding action is $I_\phi = 1/(2\pi) \oint p_\phi\, d\phi = \sigma$. Thus

$$I_\phi = \sigma.$$

3. On the other hand, the radial momentum is easily obtained from the value of the energy

$$p_\rho = \pm\sqrt{2m\left(E + \frac{K}{\rho} - \frac{\sigma^2}{2m\rho^2}\right)}.$$

Let us consider an elliptic orbit; the turning points are the perihelion ρ_m and the aphelion ρ_M; at these points the radial momentum vanishes. Moreover, the energy is negative $E = -|E|$. The radial action I_ρ is defined, owing to the symmetry properties, by:

$$2\pi I_\rho = \oint p_\rho\, d\rho = 2\int_{\rho_m}^{\rho_M} \sqrt{2m\left(K/\rho - |E| - \sigma^2/(2m\rho^2)\right)}\, d\rho.$$

4. This integral would be more easily computable if the square root were present in the denominator. It can be put under this form if we differentiate it with respect to energy. This procedure provides the inverse of the angular frequency. Thus:

$$\frac{1}{\omega_\rho} = \frac{dI_\rho}{dE} = \frac{\sqrt{2m}}{2\pi\sqrt{|E|}}\int_{\rho_m}^{\rho_M} \frac{\rho\, d\rho}{\sqrt{(\rho - \rho_m)(\rho_M - \rho)}}.$$

Let us introduce the mean radius

$$\bar{\rho} = \frac{\rho_m + \rho_M}{2} = \frac{K}{2|E|}$$

and half the difference $\delta = (\rho_M - \rho_m)/2$. Now perform the change of variable $\rho = \bar{\rho} + \delta \cos\theta$. The integral is transformed into

$$\frac{\sqrt{2m}}{2\pi\sqrt{|E|}} \int_0^\pi \frac{(\bar{\rho} + \delta\cos\theta)\,\delta\sin\theta\,d\theta}{\delta\sin\theta} = \frac{\sqrt{2m}K}{4|E|^{3/2}}.$$

To summarize:
$$\frac{dI_\rho}{dE} = \frac{\sqrt{m}K}{|2E|^{3/2}}.$$

5. The previous expression is the derivative of

$$\frac{\sqrt{2m}K}{2|E|^{1/2}} + \text{const} = I_\rho.$$

The constant is determined if one considers the circular orbit. Hamilton's equation gives the radius of the circular orbit as $\rho_c = \sigma^2/(mK)$, which, inserted in the Hamiltonian, gives the energy $E_c = -mK^2/(2\sigma^2)$. Moreover, for the circular orbit, $p_\rho = 0$ and thus $I_\rho = 0$. One deduces the value of the constant $-\sqrt{2m}K/(2|E_c|^{1/2}) = -\sigma$. The value of the radial action follows:

$$I_\rho = \frac{1}{2}\left[K\sqrt{2m/|E|} - 2\sigma\right].$$

Finally the energy is written in terms of the actions as:

$$E = -\frac{mK^2}{2(I_\rho + I_\phi)^2}.$$

6. The EBK quantization prescription consists in replacing, in the energy, each action by an integer multiple of the elementary action $\hbar$. Therefore, one sets $I_\rho = n_\rho \hbar$, $I_\phi = l\hbar$ and one obtains the semi-classical expression for the energy, which corresponds to the Rydberg expression:

$$E_n = -\frac{mK^2}{2\hbar^2(n_\rho + l)^2} = -\frac{\alpha^2 mc^2}{2n^2}.$$

This expression is identical to the quantum formula.

It is worthwhile to point out that the quantum numbers n_ρ and l appear through their sum only, the so-called principal quantum number $n = n_\rho + l$. This remarkable property is specific to the Coulomb potential and leads to degeneracies among the energy levels.

6.6. The Sommerfeld Atom [Statement p. 293]

1. Using polar coordinates (ρ, ϕ), the expression for the velocity is $v^2 = \dot{\rho}^2 + \rho^2\dot{\phi}^2$. As we saw in Chapter 2, the free relativistic Lagrangian is simply $L_0(\mathbf{r}, \dot{\mathbf{r}}) = -mc^2\sqrt{1 - \dot{r}^2/c^2}$. Subtracting the Coulomb potential $V(\rho) = -K/\rho$, we obtain the total Lagrangian of the system:

$$L(\rho, \phi, \dot{\rho}, \dot{\phi}) = -mc^2\sqrt{1 - \frac{1}{c^2}\left(\dot{\rho}^2 + \rho^2\dot{\phi}^2\right)} + \frac{K}{\rho}.$$

The Hamiltonian is deduced from the Legendre transform and is expressed as:

$$H(\rho, \phi, p_\rho, p_\phi) = c\sqrt{p_\rho^2 + \frac{p_\phi^2}{\rho^2} + m^2c^2} - \frac{K}{\rho}.$$

2. The system is autonomous; the Hamiltonian is a first integral whose value is the energy E. The ϕ coordinate is cyclic and the momentum

$$p_\phi = \partial_{\dot{\phi}} L = \frac{m\rho^2\dot{\phi}}{\sqrt{1 - \left(\dot{\rho}^2 + \rho^2\dot{\phi}^2\right)/c^2}}$$

is also a first integral. As is often the case, it is identified with the angular momentum, denoted usually as σ. Furthermore, it also corresponds to the angular action variable $I_\phi = \oint p_\phi \, d\phi/(2\pi)$. In summary:

$$p_\phi = I_\phi = \sigma.$$

The system, with two degrees of freedom, possesses two first integrals in involution: it is integrable.

3. From the expression of the Hamiltonian, we easily deduce the radial momentum:

$$p_\rho(E, \sigma) = \pm\sqrt{(E + K/\rho)^2/c^2 - m^2c^2 - \sigma^2/\rho^2}.$$

The radial action is given by its usual expression $I_\rho = \oint p_\rho \, d\rho/(2\pi)$. Making use of the symmetry of the trajectory and introducing the perihelion ρ_m and the aphelion ρ_M, the explicit value of the action is obtained easily:

$$I_\rho(E, \sigma) = \frac{1}{\pi}\int_{\rho_m}^{\rho_M}\sqrt{(E + K/\rho)^2/c^2 - m^2c^2 - \sigma^2/\rho^2}\, d\rho.$$

The perihelion and the aphelion are the roots of the equation

$$(m^2c^4 - E^2)\rho^2 - 2EK\rho + \sigma^2c^2 - K^2 = 0.$$

Remember that, for a bound state, $0 < E < mc^2$.

Problem Solutions

4. Exactly as in the Problem 6.5, in order to calculate this integral, it is expedient to first compute its derivative with respect to the energy, which is precisely:

$$\partial_E I_\rho(E,\sigma) = \frac{1}{\pi c^2} \int_{\rho_m}^{\rho_M} \frac{E + K/\rho}{\sqrt{(E+K/\rho)^2/c^2 - m^2c^2 - \sigma^2/\rho^2}} \, d\rho.$$

which can be rewritten in the more amenable form:

$$\partial_E I_\rho(E,\sigma) = \frac{1}{\pi c \sqrt{m^2 c^4 - E^2}} \int_{\rho_m}^{\rho_M} \frac{E\rho + K}{\sqrt{(\rho_M - \rho)(\rho - \rho_m)}} \, d\rho.$$

The proposed change of variable is $\rho = \bar{\rho} + \delta \cos\theta$ with the definitions $\bar{\rho} = (\rho_m + \rho_M)/2 = EK/(m^2c^4 - E^2)$ and $\delta = (\rho_M - \rho_m)/2$, with which the integral is transformed into

$$\frac{1}{\pi c \sqrt{m^2 c^4 - E^2}} \int_{\rho_m}^{\rho_M} (E(\bar{\rho} + \delta \cos\theta) + K) \, d\theta =$$

$$\frac{E\bar{\rho} + K}{c\sqrt{m^2c^4 - E^2}} = \frac{m^2 c^3 K}{(m^2c^4 - E^2)^{3/2}}.$$

Finally: $$\partial_E I_\rho(E,\sigma) = \frac{m^2 c^3 K}{(m^2c^4 - E^2)^{3/2}}.$$

5. To find the action, it is sufficient to calculate the primitive of the previous quantity with respect to energy:

$$I_\rho(E,\sigma) = \frac{KE}{c\sqrt{m^2c^4 - E^2}} + \text{const.}$$

Concerning the circular orbit, the Hamilton equation $\dot{p}_\rho = -\partial_\rho H = 0$ provides the radius $\rho_c = (\sigma^2)\sqrt{1 - (K/(\sigma c))^2}/(mK)$ which is smaller than the non-relativistic equivalent. Inserting this value in the Hamiltonian allows us to obtain the energy for this particular orbit $E_c = mc^2\sqrt{1 - (K/(\sigma c))^2}$. Owing to the fact that, for the circular orbit, $I_\rho = 0$, it is easy to determine the integration constant which is $-\sqrt{\sigma^2 - (K/c)^2}$. We deduce the exact expression of the action as a function of the two first integrals and, then, after inversion, the energy in terms of the actions;

$$E(I_\rho, I_\phi) = \frac{mc^2}{\sqrt{1 + \left(\frac{(K/c)}{I_\rho + \sqrt{I_\phi^2 - (K/c)^2}}\right)^2}}.$$

6. The EBK quantization procedure allow us to write $I_\phi = \sigma = l\hbar$ (l integer ≥ 1) and $I_\rho = n_\rho \hbar$ (n_ρ integer ≥ 0). Inserting these values and introducing the fine structure constant $\alpha = K/(\hbar c)$ in the preceding expression for the energy, one arrives at the semi-classical expression of the energy for the Sommerfeld atom:

$$E(n_\rho, l) = \frac{mc^2}{\sqrt{1 + \left(\alpha/(n_\rho + \sqrt{l^2 - \alpha^2})\right)^2}}.$$

7. Restricting to second order in α^2, one has

$$\sqrt{l^2 - \alpha^2} \approx l - \frac{\alpha^2}{2l} - \frac{\alpha^4}{8l^3},$$

then

$$n_\rho + \sqrt{l^2 - \alpha^2} \approx n - \frac{\alpha^2}{2l} - \frac{\alpha^4}{8l^3},$$

then

$$\alpha^2 \left[n_\rho + \sqrt{l^2 - \alpha^2}\right]^{-2} \approx \frac{\alpha^2}{n^2}\left[1 + \frac{\alpha^2}{nl}\right]$$

and lastly

$$\left[1 + \frac{\alpha^2}{\left(n_\rho + \sqrt{l^2 - \alpha^2}\right)^2}\right]^{-1/2} \approx 1 - \frac{\alpha^2}{2n^2} - \frac{\alpha^4(n/l - 3/4)}{2n^4}.$$

This equality allows us to find the truncated expansion of E, and thence the binding energy of the atom $B = E - mc^2$. In the final expression, it is convenient to introduce the Rydberg constant $R_\infty = mc\alpha^2/(2h)$. An elementary calculation leads to the final result:

$$B = -\frac{hcR_\infty}{n^2}\left[1 + \frac{\alpha^2}{n^2}\left(\frac{n}{l} - \frac{3}{4}\right)\right].$$

The first term of the truncated expansion corresponds to the classical value found by N. Bohr. The correction in α^2 has a relativistic origin and contributes to the so-called fine structure of the atomic levels.

6.7. Energy as a Function of Actions
[Statement p. 294]

For a two-dimensional system subject to a central force, the motion occurs in a plane, and it is natural to choose the polar coordinates (ρ, ϕ) as generalized coordinates. The Hamiltonian reads

$$H(\rho, \phi, p_\rho, p_\phi) = \frac{p_\rho^2}{2m} + \frac{p_\phi^2}{2m\rho^2} + V(\rho).$$

Problem Solutions 319

The system is conservative; there exists a first integral which is the energy $H = E$. Moreover, the ϕ coordinate is cyclic and there exists another first integral: the angular momentum $p_\phi = \sigma$. The action variables are defined by

$$2\pi I_i = \oint_{\Gamma_i} p \cdot dq = \oint_{\Gamma_i} (p_\rho \, d\rho + p_\phi \, d\phi).$$

On the torus, one can choose a contour with $\rho = \text{const}$. The corresponding action is $I_\phi = 1/(2\pi) \oint p_\phi \, d\phi = \sigma$. Therefore:

$$I_\phi = \sigma.$$

On the other hand, one has $p_\rho^2/(2m) + \sigma^2/(2m\rho^2) + V(\rho) = E$, whence the radial momentum:

$$p_\rho = \sqrt{2m(E - V(\rho)) - \sigma^2/\rho^2}.$$

The radial action variable is obtained taking $\phi = \text{const}$ on the torus:

$$I_\rho = \frac{1}{2\pi} \oint p_\rho \, d\rho = \frac{1}{2\pi} \oint \sqrt{2m(E - V(\rho)) - \sigma^2/\rho^2} \, d\rho$$

or, more explicitly, $I_\rho(E, \sigma)$ or $I_\rho(E, I_\phi)$. Inverting this expression allows us to obtain $E(I_\rho, I_\phi)$ which is the desired expression. One cannot proceed further if the form of $V(\rho)$ remains unknown.

A – Kepler's problem

1. In the part (ϕ, p_ϕ) of phase space, the phase portrait is very simple; it is just the straight line $p_\phi = \sigma$. In the part (ρ, p_ρ) the phase portrait is given by $p_\rho(\rho) = \pm\sqrt{2m(E - K/\rho) - \sigma^2/\rho^2}$. This momentum vanishes at the two turning points, the perihelion $\rho_{\min}$, and the aphelion $\rho_{\max}$. These remarks are illustrated in the Fig. 6.6.

The trajectory for the Kepler problem is an ellipse. Assume that for time $t = 0$, one has $\phi = 0$ and $\rho = \rho_{\min}$; after half a period $t = T/2$, one has $\phi = \pi$ and $\rho = \rho_{\max}$. Lastly after one period $t = T$, the system is to be found in its original position in phase space $\phi = 2\pi \equiv 0$ and $\rho = \rho_{\min}$.

Denoting by α_i the angle variable associated with the action I_i, we can assert that $\alpha_i = \omega_i t$ (be careful! $\alpha_\phi \neq \phi$, because the polar angle does not rotate with constant angular velocity). After one period, we are back to the same point of phase space so that $\alpha_\rho(T) = 2\pi$ and $\alpha_\phi(T) = 2\pi$. In consequence, we find the equality between the angular and radial frequencies:

$$\omega_\rho = \omega_\phi.$$

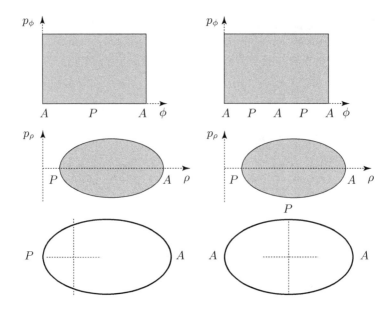

Fig. 6.6 Phase portraits for angular coordinates (top) and radial coordinates (middle). We also plot the trajectory (bottom) and indicate the center of force. The notation A and P indicates "aphelion" and "perihelion". The left part corresponds to Kepler's problem whereas the right part corresponds to the harmonic oscillator problem

2. The Hamilton equations written in terms of angle-action variables provide the relations $\dot{\alpha}_\rho = \omega_\rho = \partial_{I_\rho} H$ and $\dot{\alpha}_\phi = \omega_\phi = \partial_\sigma H$. The equality between the two frequencies just proved above leads to $\partial_{I_\rho} H = \partial_\sigma H$. Since the system is integrable, the Hamilton function is expressed in terms of actions only: $H(I_\rho, \sigma)$. Introducing the new independent variables $u = I_\rho + \sigma$, $v = I_\rho - \sigma$, the preceding equality concerning the partial derivatives implies that $\partial_v H = 0$, and thus H is a function of u only. Thus:

$$H(I_\rho, I_\phi) = H(I_\rho + I_\phi).$$

B – Harmonic problem

1. We are dealing now with the potential $V(\rho) = \frac{1}{2}k\rho^2 = \frac{1}{2}m\omega^2\rho^2$. We know that the trajectory is an ellipse with its center placed precisely at the center of force. The phase portrait is also represented in the figure 6.6. Assume again that for time $t = 0$, one has $\phi = 0$ and $\rho = \rho_{\min}$; after half a period $t = T/2$, one has $\phi = \pi$, and again $\rho = \rho_{\min}$ ($\rho = \rho_{\max}$ is reached after a quarter of period only).

Lastly, after one period $t = T$, the system returns to its original position in phase space $\phi = 2\pi \equiv 0$ and $\rho = \rho_{\min}$. In terms of angle variables, one has $\alpha_\phi(T) = 2\pi$ (after one period the angular coordinate recovers its original value) but $\alpha_\rho(T/2) = 2\pi$ (after a half-period the radial coordinate recovers its original value), that is $\alpha_\rho(T) = 4\pi$. In consequence we have the following relations concerning the angular frequencies:

$$\omega_\rho = 2\omega_\phi.$$

In this case, the Hamilton equations provide the relations $\partial_{I_\rho} H = 2\partial_\sigma H$. Introducing the new independent variables $u = 2I_\rho + \sigma$, $v = 2I_\rho - \sigma$, the equality concerning the preceding partial derivatives implies $\partial_v H = 0$, so that H is a function of u only:

$$H(I_\rho, I_\phi) = H(2I_\rho + I_\phi).$$

C – Semi-classical quantization

1. It is time now to find the u dependence of $H(u)$ in both cases. In order to do this, it is interesting to work with the circular orbit $\rho = \text{const} = R$, $p_\rho = 0 = I_\rho$, $u = \sigma$. One has $\sigma = mR^2\omega$.

   For Kepler's problem, Newton's equation provides $mR\omega^2 = K/R^2$, whence $1/R = mK/\sigma^2$. Therefore $E = \sigma^2/(2mR^2) - K/R = -mK^2/(2\sigma^2)$ (this value is also easily obtained from the virial theorem). For a circular orbit $p_\rho = 0$, so that $I_\rho = 0$ and $\sigma = u$. The Hamiltonian takes the required form: $H(u) = -mK^2/(2u^2)$. Following the result of Question 2, one can write more generally:

$$H_{(K)}(I_\rho, I_\phi) = -\frac{mK^2}{2(I_\rho + I_\phi)^2}.$$

Concerning the harmonic oscillator, $E = \sigma^2/(2mR^2) + \frac{1}{2}m\omega^2 R^2$. With the expression of the angular momentum $\sigma = mR^2\omega$, each contribution takes the same value $\omega\sigma/2$ (this is again the consequence of the virial theorem) and thus $E = \omega\sigma$. For the same reason as before, we have $H(u) = \omega u$. In the general case and following Question 3, we have:

$$H_{(HO)}(I_\rho, I_\phi) = \omega(2I_\rho + I_\phi).$$

2. The EBK quantization procedure consists in replacing each action by an integer multiple of the elementary action $\hbar$. Consequently, we set $I_\rho = n\hbar$, $I_\phi = \sigma = l\hbar$; the semi-classical formulae for quantized energies follow immediately:

$$E_{nl}(\text{Kepler}) = -\frac{mK^2}{2\hbar^2(n+l)^2}; \qquad E_{nl}(\text{HO}) = \hbar\omega(2n + l).$$

6.8. Invariance of the Circulation Under a Continuous Deformation
[Statement and Figure p. 296]

1. In four-dimensional phase space (x, y, p_x, p_y), one considers two first integrals $F_1(x, y, p_x, p_y) = f_1$ and $F_2(x, y, p_x, p_y) = f_2$ in involution

$$\{F_1, F_2\} = 0.$$

For given values of the constants f_1 and f_2, they define a two-dimensional manifold $M_f = (f_1, f_2)$ which is a subset of the phase space. In this manifold, one considers a path Γ beginning at $q_0 = (x_0, y_0, p_{x_0}, p_{y_0})$ and ending at $q_1 = (x_1, y_1, p_{x_1}, p_{y_1})$. We wish to demonstrate the path independence of the quantity

$$S(q_0, q_1, f_1, f_2) = \int_\Gamma p_x(x, y, f_1, f_2)\, dx + p_y(x, y, f_1, f_2)\, dy.$$

For simplicity, we set $u = p_x$, $v = p_y$. Then differentiating the equation for the surface labelled $i = 1$ or 2 we write:

$$(\partial_x F_i)\, dx + (\partial_y F_i)\, dy + (\partial_u F_i)\, du + (\partial_v F_i)\, dv = 0.$$

This equality is valid whatever the elementary displacements dx, dy, du, $dv \in M_f$. In particular, one can set $dy = 0$ and then divide by dx. In this case, it should be clear that du/dx can be identified with $\partial_x u$, since we are working at $y = \text{const}$. Performing the same treatment for $dx = 0$, we arrive at:

$$(\partial_u F_i)(\partial_x u) + (\partial_v F_i)(\partial_x v) + \partial_x F_i = 0$$
$$(\partial_u F_i)(\partial_y u) + (\partial_v F_i)(\partial_y v) + \partial_y F_i = 0$$

Explicitly, we find the four partial differential equations:

(1) $(\partial_u F_1)(\partial_x u) + (\partial_v F_1)(\partial_x v) + \partial_x F_1 = 0$
(2) $(\partial_u F_1)(\partial_y u) + (\partial_v F_1)(\partial_y v) + \partial_y F_1 = 0$
(3) $(\partial_u F_2)(\partial_x u) + (\partial_v F_2)(\partial_x v) + \partial_x F_2 = 0$
(4) $(\partial_u F_2)(\partial_y u) + (\partial_v F_2)(\partial_y v) + \partial_y F_2 = 0.$

2. We now calculate $(1) \times (\partial_u F_2) - (3) \times (\partial_u F_1) = 0$. Simple algebra gives:

$$\frac{\partial v}{\partial x} = \frac{(\partial_x F_1)(\partial_u F_2) - (\partial_u F_1)(\partial_x F_2)}{(\partial_u F_1)(\partial_v F_2) - (\partial_v F_1)(\partial_u F_2)}.$$

Problem Solutions

A similar treatment for $(2) \times (\partial_v F_2) - (4) \times (\partial_v F_1)$ leads to:
$$\frac{\partial u}{\partial y} = \frac{(\partial_v F_1)(\partial_y F_2) - (\partial_y F_1)(\partial_v F_2)}{(\partial_u F_1)(\partial_v F_2) - (\partial_v F_1)(\partial_u F_2)}.$$

3. We now calculate the quantity $I = \partial_x v - \partial_y u$ which, with the values obtained in the preceding question, can be written as:

$$\begin{aligned} I &= \frac{(\partial_x F_1)(\partial_u F_2) + (\partial_y F_1)(\partial_v F_2) - (\partial_u F_1)(\partial_x F_2) - (\partial_v F_1)(\partial_y F_2)}{(\partial_u F_1)(\partial_v F_2) - (\partial_v F_1)(\partial_u F_2)} \\ &= \frac{\{F_1, F_2\}}{(\partial_u F_1)(\partial_v F_2) - (\partial_v F_1)(\partial_u F_2)}. \end{aligned}$$

The two first integrals are in involution and this property implies that the numerator of the previous expression vanishes. The result is explicitly:

$$\frac{\partial p_y}{\partial x} - \frac{\partial p_x}{\partial y} = 0.$$

4. Let us choose for Γ a closed path; Stokes theorem then implies

$$\oint \boldsymbol{p} \cdot d\boldsymbol{l} = \oint p_x \, dx + p_y \, dy = \int\!\!\!\int_\Sigma (\boldsymbol{\nabla} \times \boldsymbol{p}) \cdot d\boldsymbol{n} = \int\!\!\!\int_\Sigma (\partial_x p_y - \partial_y p_x) \, dn_z.$$

The integrant of the last integral is null as we proved just above; this is also the case for its flux and consequently the circulation of the momentum vanishes :

$$\oint \boldsymbol{p} \cdot d\boldsymbol{l} = 0.$$

Let Γ_1 and Γ_2 be two paths beginning at the same point and ending at the same point on the manifold. We now form a closed path using the outward path Γ_1 and the return path Γ_2. Since this latter contribution is the negative of the value for going directly from the starting point to the end point following Γ_2, one has the trivial equality

$$\oint \boldsymbol{p} \cdot d\boldsymbol{l} = 0 = \int_{\Gamma_1} \boldsymbol{p} \cdot d\boldsymbol{l} - \int_{\Gamma_2} \boldsymbol{p} \cdot d\boldsymbol{l}.$$

Thus we arrived at the desired relation:

$$\int_{\Gamma_1} p_x \, dx + p_y \, dy = \int_{\Gamma_2} p_x \, dx + p_y \, dy.$$

6.9. Ball Bouncing on a Moving Tray
[Statement p. 297]

1. The Hamiltonian for the ball is $H(q,p) = p^2/(2m) + mgq$, where q is the coordinate of the ball above the ground. The most suitable coordinate for our problem is the altitude above the tray $Q = q - h(t)$. Let us take as the new momentum $P = p - m\dot{h}(t)$ (this is dimensionally consistent). Now calculate the Jacobian of the transformation in phase space $J = (\partial_q Q)(\partial_p P) - (\partial_q P)(\partial_p Q) = (1) \times (1) - (0) \times (0) = 1$. Although the transformation depends on time, it conserves areas; it is canonical.

2. The generating function $G_2(P, q, t)$ must be such that $p = \partial_q G_2$ and $Q = \partial_P G_2$. The first equation implies $\partial_q G_2 = P + m\dot{h}$ which can be integrated to give $G_2 = q(P + m\dot{h}) + f(P, t)$. The second equation implies $Q = q - h = q + \partial_P f$ which leads to $\partial_P f = -h$ whence $f(P, t) = -hP$. Finally, the generating function reads:

$$G_2(P, q, t) = q\left(P + m\dot{h}(t)\right) - Ph(t).$$

The explicit existence of a generating function once again proves the canonicity of the transformation.

3. The new Hamiltonian is given by

$$K(Q, P, t) = H(q(Q, P), p(Q, P)) + \partial_t G_2(P, q, t).$$

A simple calculation relying on the previous remarks allows us to write

$$K(Q, P, t) = \frac{P^2}{2m} + m(g + \ddot{h})Q + mgh + mh\ddot{h} + \frac{1}{2}m\dot{h}^2.$$

Adding to the Hamiltonian a function that depends on time only does not change the Hamilton equations, and thus the description of the system. In consequence one can add to the Hamiltonian K the function $-(mgh + mh\ddot{h} + \frac{1}{2}m\dot{h}^2)$, to obtain a simpler equivalent Hamiltonian:

$$K(Q, P, t) = \frac{P^2}{2m} + m\left(g + \ddot{h}(t)\right)Q.$$

6.10. Harmonic Oscillator with a Variable Frequency [Statement p. 298]

One considers a harmonic oscillator whose angular frequency is time-dependent and whose Hamiltonian is given by: $H(q, p, t) = p^2/(2m) + \frac{1}{2}m\omega(t)^2 q^2$.

Let us introduce the type 1 generating function:

$$G_1(Q,q,t) = \frac{1}{2}m\omega(t)\,q^2\,\cot(Q).$$

We deduce the expressions for the momenta

$$p = \partial_q G_1 = m\omega(t)\,q\,\cot(Q), \qquad P = -\partial_Q G_1 = m\omega(t)\,q^2/(2\sin^2 Q)$$

and the new Hamiltonian

$$K = H + \partial_t G_1 = H + \frac{1}{2}m\dot\omega(t)\,q^2\,\cot(Q).$$

From previous relations, one obtains $m\omega\,q^2 = 2P\sin^2 Q$, $p^2 = 2m\omega P\cos^2 Q$ and these equations allow us to express first the original Hamiltonian and then the new one in terms of the new coordinates. Finally, one has:

$$K(Q,P,t) = P\left(\omega(t) + \frac{\dot\omega(t)\,\sin(2Q)}{2\omega(t)}\right).$$

Since the transformation is canonical, one easily deduces the new Hamilton equations $\dot Q = \partial_P K$ and $\dot P = -\partial_Q K$, or explicitly:

$$\dot Q = \omega(t) + \frac{\dot\omega(t)\,\sin(2Q)}{2\omega(t)}; \qquad \dot P = -\frac{\dot\omega(t)P\cos(2Q)}{\omega(t)}.$$

6.11. Choice of the Momentum
[Statement p. 298]

1. The original Hamiltonian for the harmonic oscillator is $H = \frac{1}{2}\omega(p^2 + q^2)$ (with a suitable choice of coordinates). One decides to perform a canonical transformation taking as new momentum the quantity

$$P(q,p) = \lambda H = \frac{1}{2}\lambda\omega(p^2 + q^2).$$

One deduces $q = \pm\sqrt{(2P/(\lambda\omega)) - p^2}$. As for the choice of the generating function, one can take either G_3 with $q = -\partial_p G_3(Q,p)$ or G_4 with $q = -\partial_p G_4(P,p)$. Since $q = q(P,p)$, it is the G_4 function that appears naturally. It is obtained by taking the primitive of $q = q(P,p)$. In summary:

$$q = \pm\sqrt{\frac{2P}{\lambda\omega} - p^2}; \qquad G_4(P,p) = \mp\int dp\sqrt{\frac{2P}{\lambda\omega} - p^2} + F(P).$$

2. The relationship between the new coordinate and the generating function is $Q = \partial_P G_4(P, p)$, or

$$Q = \mp \frac{1}{\lambda \omega} \int dp / \sqrt{\frac{2P}{\lambda \omega} - p^2} + F'(P).$$

One chooses the particular solution such that $F'(P) = 0$ and one performs the change of variable $p = \sqrt{2P/(\lambda \omega)} \cos u$. This leads to

$$Q = \pm u/(\lambda \omega) = \pm \frac{1}{\lambda \omega} \arccos\left(p\sqrt{\frac{\lambda \omega}{2P}}\right).$$

Finally, substituting for P in terms of the original variables, one obtains the desired expression:

$$Q(q, p) = \pm \frac{1}{\lambda \omega} \arccos\left(\frac{p}{\sqrt{q^2 + p^2}}\right).$$

3. From the preceding equation, one deduces $p/\sqrt{q^2 + p^2} = \cos(\lambda \omega Q)$. Q must be an angle such that if it is increased by 2π, we return to the same point in phase space for which $p/\sqrt{q^2 + p^2}$ retains its value. Consequently, one must have $2\pi \lambda \omega = 2\pi$, that is:

$$\lambda = 1/\omega.$$

Finally, the canonical transformation reads

$$P = \frac{1}{2}(p^2 + q^2) = \frac{E}{\omega} \quad \text{and} \quad Q = \arccos\left(p/\sqrt{q^2 + p^2}\right).$$

Since Q is the angle variable, which increases by 2π to return to the same point in phase space, its conjugate variable P is the action variable so that:

$$I(E) = \frac{E}{\omega}$$

6.12. Invariance of the Poisson Bracket Under a Canonical Transformation
[Statement p. 299]

We work in phase space described by the variables (q, p) in which we define two functions $F(q, p)$ and $G(q, p)$. We now perform a contact transformation $Q(q, p), P(q, p)$, which can be inverted to give $q(Q, P), p(Q, P)$. The functions F and G become functions of the new coordinates $F(Q, P) = F(q(Q, P), p(Q, P))$ with a similar relation for G.

Problem Solutions   327

Caution: The functional form of F in terms of the variables Q, P is different from its form in terms of the variables q, p; in spite of this, its value is obviously identical if one calculates it at the same point in phase space. This is the reason for maintaining the same notation for this function.

The rules concerning partial differentiation for functions provide the relations

$$\partial_Q F = (\partial_q F)(\partial_Q q) + (\partial_p F)(\partial_Q p);$$
$$\partial_P G = (\partial_q G)(\partial_P q) + (\partial_p G)(\partial_P p).$$

They allow the calculation of $(\partial_Q F)(\partial_P G)$ and then, inverting the roles of F and G, the quantity $(\partial_Q G)(\partial_P F)$.

It is easy to obtain the Poisson bracket with respect to variables Q, P from its definition $\{F, G\}_{(Q,P)} = (\partial_Q F)(\partial_P G) - (\partial_Q G)(\partial_P F)$ in the form given in the statement:

$$\{F, G\}_{(Q,P)} = \left(\frac{\partial q(Q,P)}{\partial Q} \frac{\partial p(Q,P)}{\partial P} - \frac{\partial p(Q,P)}{\partial Q} \frac{\partial q(Q,P)}{\partial P} \right) \{F, G\}_{(q,p)}.$$

The Jacobian of the contact transformation occurs naturally. If the transformation is canonical, the Jacobian is unity and both expressions for the Poisson bracket are identical. In practice, we always consider canonical transformations and, in this case, the Poisson bracket acquires a universal status which allows us to avoid specifying the variables with respect to which it is calculated.

6.13. Canonicity for a Contact Transformation [Statement p. 299]

1. We begin with a contact transformation $Q(q,p), P(q,p)$. Let us differentiate the Q_i coordinate with respect to time and apply Hamilton's equations to the original Hamiltonian $H(q,p)$:

$$\dot{Q}_i = \sum_l \left[(\partial_{q_l} Q_i)(\partial_{p_l} H) - (\partial_{p_l} Q_i)(\partial_{q_l} H) \right].$$

We now express the original Hamiltonian in terms of the new form of the Hamiltonian $H(q,p) = K(Q(q,p), P(q,p))$. The calculation of the partial derivatives of H with respect to the original coordinates is translated now in terms of the partial derivatives of H with respect to the new coordinates. Making this substitution in the previous expression, we obtain:

$$\dot{Q}_i = \sum_j \left[(\partial_{Q_j} K) \{Q_i, Q_j\} + (\partial_{P_j} K) \{Q_i, P_j\} \right],$$

where we introduced the Poisson bracket
$$\{Q_i, Q_j\} = \sum_l \left[(\partial_{q_l} Q_i)(\partial_{p_l} Q_j) - (\partial_{q_l} Q_j)(\partial_{p_l} Q_i)\right],$$
and an analogous expression for $\{Q_i, P_j\}$. The transformation is canonical if the traditional Hamilton equations are fulfilled in terms of the new variables. In particular, $\dot{Q}_i = \partial_{P_i} K$ implies
$$\{Q_i, Q_j\} = 0 \quad \text{and} \quad \{Q_i, P_j\} = \delta_{i,j}.$$
A similar calculation based on $\dot{P}_i$ leads to
$$\dot{P}_i = \sum_j \left[(\partial_{Q_j} K)\{P_i, Q_j\} + (\partial_{P_j} K)\{P_i, P_j\}\right].$$

The canonicity for the transformation requires $\dot{P}_i = -\partial_{Q_i} K$, which leads to $\{P_i, P_j\} = 0$ and $\{P_i, Q_j\} = -\delta_{i,j}$. This latter Poisson bracket can also be deduced easily from $\{P_i, Q_j\} = -\{Q_j, P_i\} = -\delta_{i,j}$. In summary, for the transformation to be canonical it is necessary and sufficient that we have the following conditions:
$$\{Q_i, Q_j\} = 0; \qquad \{P_i, P_j\} = 0; \qquad \{Q_i, P_j\} = \delta_{i,j}.$$

2. Consider two functions
$$F(q, p) = F(Q(q, p), P(q, p)) \quad \text{and} \quad G(q, p) = G(Q(q, p), P(q, p)).$$
Let calculate the Poisson bracket
$$\{F, G\}_{(q,p)} = \sum_i \left[(\partial_{q_i} F)(\partial_{p_i} G) - (\partial_{p_i} F)(\partial_{q_i} G)\right].$$
Expressing the partial derivatives with respect to (q, p) in terms of the partial derivatives with respect to (Q, P) and substituting in the preceding expression, we arrive at
$$\{F, G\}_{(q,p)} = \sum_{k,l}(\partial_{Q_k} F)(\partial_{Q_l} G)\{Q_k, Q_l\} + \sum_{k,l}(\partial_{Q_k} F)(\partial_{P_l} G)\{Q_k, P_l\}$$
$$+ \sum_{k,l}(\partial_{P_k} F)(\partial_{Q_l} G)\{P_k, Q_l\} + \sum_{k,l}(\partial_{P_k} F)(\partial_{P_l} G)\{P_k, P_l\}.$$
Using the expressions for the Poisson brackets obtained in the first question, we are left with:
$$\{F, G\}_{(q,p)} = \sum_{k,l}(\partial_{Q_k} F)(\partial_{P_l} G)\delta_{k,l} - \sum_{k,l}(\partial_{P_k} F)(\partial_{Q_l} G)\delta_{k,l}$$
$$= \sum_k \left[(\partial_{Q_k} F)(\partial_{P_k} G) - (\partial_{P_k} F)(\partial_{Q_{kl}} G)\right].$$

Problem Solutions

In the latter expression, one recognizes the definition of the Poisson bracket $\{F,G\}_{(Q,P)}$. Finally:

$$\{F,G\}_{(q,p)} = \{F,G\}_{(Q,P)}.$$

This equation expresses the invariance of the Poisson brackets under a canonical transformation. It is therefore licit to consider the Poisson brackets as an intrinsic property of phase space, which are independent of the manner we choose the coordinates, provided that the contact transformations are canonical. In consequence, under these conditions, it is not necessary to specify in the indices the variables with respect to which the Poisson brackets are evaluated.

6.14. One-dimensional Free Fall
[Statement p. 300]

1. The original Hamiltonian is $H(q,p) = p^2/(2m) + mgq$. It is time-independent and its value on the trajectory is constant and equal to the energy E. We perform a canonical transformation and choose as new momentum a variable which is proportional to the Hamiltonian

$$P(q,p) = \lambda H(q,p) = \lambda \left[\frac{p^2}{2m} + mgq\right].$$

One can choose as new coordinate a variable proportional to the old momentum $Q(q,p) = \mu p$.

2. We require this contact transformation to be canonical. This is achieved by imposing the condition $\{Q,P\} = 1$. Thus,

$$\{Q,P\} = \lambda\mu \left[\frac{1}{2m}\{p^2,p\} + mg\{p,q\}\right] = -\lambda\mu mg = 1$$

so that we have the relation $\lambda\mu = -1/(mg)$. This condition can also be found by considering the conservation of area which imposes a unit Jacobian.

We still have some arbitrariness and we make the final choice:

$$\lambda = 1; \qquad \mu = -\frac{1}{mg}.$$

3. As the canonical transformation does not depend on time, the new Hamiltonian coincides with the old one $K = H$ and thus, with the constant values obtained in question 2, we have $K(Q,P) = P$. A first Hamilton equation gives $\dot{P} = -\partial_Q K = 0$, or $P = $ const. Since $P = H$, this constant is simply the energy: $P = E$.

The second Hamilton equation gives $\dot{Q} = \partial_P K = 1$, which can be integrated at once to provide $Q = t + Q_0 = t - p_0/(mg)$ or $p = -mgQ = -mgt + p_0$. Then

$$\frac{p_0^2}{2m} + mgq_0 = E = \frac{p^2}{2m} + mgq = \frac{(-mgt + p_0)^2}{2m} + mgq.$$

Rearranging the terms, we are led to the equation of the trajectory:

$$q(t) = -\frac{1}{2}gt^2 + \frac{p_0}{m}t + q_0$$

which is precisely as expected.

6.15. One-dimensional Free Fall Again
[Statement p. 301]

1. Expressed with the original coordinates, the Hamiltonian reads $H(q,p) = p^2/(2m) + mgq$. We decide to take as the new momentum the Hamiltonian itself $P = H$. The Hamiltonian and the new coordinate do not depend on time so that the new Hamiltonian $K(Q,P)$ is identical to the original Hamiltonian $K(P) = P = p^2/(2m) + mgq$. The first Hamilton equation leads to $\partial_Q K = -\dot{P} = 0$, thus $P = const$, a result that we already know since this quantity is identified with the energy. The second Hamilton equation gives $\dot{Q} = \partial_P K = 1$, which is integrated easily:

$$Q(t) = t + Q(0).$$

Let us choose a generating function of the fourth type $G_4(P,p)$. From $P = p^2/(2m) + mgq$, one deduces

$$q = -\frac{1}{mg}\left[\frac{p^2}{2m} - P\right].$$

On the other hand, one has also $q = -\partial_p G_4$. The preceding expression allows us to obtain, after integration, the generating function (up to a function of P that can be chosen as null):

$$G_4(P,p) = \frac{1}{mg}\left(\frac{p^3}{6m} - Pp\right).$$

The rest follows naturally; from $Q = \partial_P G_4 = -p/(mg)$, one deduces $p = -mgQ$, or:

$$p(t) = -mgt + p_0$$

Problem Solutions

where we set $p_0 = -mgQ(0)$. Since P is constant, it can be identified with $P = p_0^2/(2m) + mgq_0$. Lastly, from

$$q = -\frac{1}{mg}\left[\frac{p^2}{2m} - P\right],$$

and from the expressions of $p(t)$ and P given previously, we obtain the temporal law:

$$q(t) = -\frac{1}{2}gt^2 + \frac{p_0}{m}t + q_0.$$

We have obtained the usual equations of motion, in a more involved way than in the formalism "a la Newton". The purpose was to illustrate the fact that the method based on canonical transformations can be applied without any restriction.

2. Since the ball bounces on the ground, it follows a periodic motion from $q = 0$ to $q_{max} = E/(mg)$ on the upward path and a symmetric motion going down. The action variable is defined by

$$I(E) = \frac{1}{2\pi}\oint p(q, E)\, dq,$$

or, in our particular case,

$$I = \frac{2}{2\pi}\int_0^{q_{max}} \sqrt{2m(E - mgq)}\, dq = \frac{\sqrt{2gm}}{\pi}\int_0^{q_{max}} \sqrt{q_{max} - q}\, dq.$$

The integral is elementary and gives $I = 2\sqrt{2m}\sqrt{gq_{max}^3}/(3\pi)$ which can be expressed as a function of the energy (see also the Problem 6.3):

$$I(E) = \frac{2}{3\pi g}\sqrt{\frac{2}{m}}E^{3/2}.$$

3. The angular frequency is obtained by the general formula

$$\omega(E) = \frac{1}{(dI/dE)},$$

which, using the expression of the action given above, provides the relation:

$$\omega(E) = \pi g\sqrt{\frac{m}{2E}}.$$

Lastly, the angle variable is obtained from

$$\alpha = \omega t = \omega \int dq/\dot{q} = \omega m \int dq/p = \omega m \int dq/\sqrt{2m(E - mgq)}.$$

The primitive is elementary and we arrive at:

$$\alpha(q, E) = -\pi\sqrt{1 - \frac{mg}{E}q}.$$

6.16. Scale Dilation as a Function of Time
[Statement p. 301]

1. Let us denote $l'(t) = dl/dt$, $l''(t) = d^2l/dt^2$ and the new coordinate $q = x/l(t)$. One verifies rapidly that $\dot{x} = \dot{q}l + ql'$. Substituting this expression in the kinetic energy $T = \frac{1}{2}m\dot{x}^2$, this last quantity becomes $T = \frac{1}{2}m(\dot{q}l + ql')^2$ and the Lagrangian $L = T - V$ reads:

$$L(q, \dot{q}, t) = \frac{1}{2}m\left(\dot{q}l(t) + ql'(t)\right)^2 - V(q).$$

This Lagrangian depends on time.

2. First one calculates the momentum $p = \partial_{\dot{q}}L = ml(\dot{q}l + ql')$, then Hamilton's function $H = p\dot{q} - L$ which must be expressed in terms of coordinate and momentum. Explicitly one obtains:

$$H(q, p, t) = \frac{1}{l(t)^2}\left(\frac{p^2}{2m} - pql(t)l'(t)\right) + V(q).$$

The Hamiltonian also depends on time.

3. We employ a canonical transformation relying on the generating function $G_2(P, q, t) = Pq + \frac{1}{2}mq^2l(t)l'(t)$. One deduces $Q = \partial_P G_2 = q$ and $p = \partial_q G_2 = P + mqll'$.

We thus easily obtain the canonical transformation. The new Hamiltonian $K(Q, P, t)$ is derived from the general formula $K = H + \partial_t G_2$ which must be expressed in terms of the new variables. A simple calculation leads to the result:

$$K(Q, P, t) = \frac{P^2}{2ml(t)^2} + V(Q) + \frac{1}{2}mQ^2l(t)l''(t).$$

Although the expression of the Hamiltonian is rather simple, it exhibits a rather cumbersome time dependence of the kinetic energy.

4. Let $\tau(t)$ be the primitive of $1/l^2$, so that $dt/d\tau = l^2$. Hamilton's equations are $\dot{Q} = \partial_P K = P/(2ml^2)$ and $\dot{P} = -\partial_Q K = -V'(Q) - mQll''$. Instead of considering the variables (Q, P) as functions of t, they are considered as functions of τ. The equations of motion now become (the trajectory is unchanged but is covered with a different temporal law): $dQ/d\tau = P/(2m)$, $dP/d\tau = -l^2 V'(Q) - mQl^3 l''$. They can be considered as Hamilton's equations arising from the new Hamiltonian:

$$\tilde{K}(Q, P, \tau) = \frac{P^2}{2m} + l(t(\tau))^2 V(Q) + \frac{1}{2} m Q^2 l(t(\tau))^3 l''(t(\tau)).$$

The kinetic energy no longer exhibits the awkward scale dependence. In particular, for a linear dependence $l'' = 0$, the new Hamiltonian presents a very simple form.

6.17. From the Harmonic Oscillator to Coulomb's Problem [Statement p. 301]

1. We begin with the Hamiltonian for the harmonic oscillator

$$H_{HO}(\rho, \phi, p_\rho, p_\phi) = \frac{p_\rho^2}{2m} + \frac{p_\phi^2}{2m\rho^2} + \frac{1}{2} m\omega^2 \rho^2.$$

The ϕ coordinate is cyclic and the corresponding momentum, the angular momentum σ, is a first integral $p_\phi = \sigma$. In (ρ, p_ρ) phase space, one takes as new coordinate the quantity $Q = \rho^2/l$. Let P be the new momentum. In order for the transformation to be canonical, one must satisfy the conservation of area, a property that implies a unit Jacobian

$$J = 1 = (\partial_\rho Q)(\partial_{p_\rho} P) - (\partial_\rho P)(\partial_{p_\rho} Q).$$

With $(\partial_\rho Q) = 2\rho/l$, $(\partial_{p_\rho} Q) = 0$, this relation is equivalent to $(\partial_{p_\rho} P) = l/(2\rho)$ which is integrated easily to give:

$$P(\rho, p_\rho) = \frac{l p_\rho}{2\rho}.$$

One can check that, if the Lagrangian is written in terms of $(Q, \dot{Q})$ variables,

$$L(Q, \dot{Q}) = \frac{1}{2} m \left(\frac{l \dot{Q}^2}{4Q} + \frac{\sigma^2}{m^2 l Q} - \omega^2 l Q \right),$$

one has $P = m l \dot{\rho}/(2\rho) = m l \dot{Q}/(4Q)$ which is identified with $P = \partial_{\dot{Q}} L$.

2. This canonical transformation does not depend on time and the new Hamiltonian $K(Q, \phi, P, p_\phi)$ is simply identical with the old one, but expressed in terms of the new coordinates, $H_{HO}(\rho(Q), \phi, p_\rho(Q, P), p_\phi)$. A simple calculation leads to the following expression:

$$K(Q, \phi, P, p_\phi) = \frac{2QP^2}{ml} + \frac{p_\phi^2}{2mlQ} + \frac{1}{2}m\omega^2 lQ.$$

3. The system being conservative, the value of this Hamiltonian on the trajectory is a constant equal to the energy

$$\frac{2QP^2}{ml} + \frac{p_\phi^2}{2mlQ} + \frac{1}{2}m\omega^2 lQ = E_{ho}.$$

Let us divide this equation by $4Q/l$ to obtain:

$$\frac{P^2}{2m} + \frac{p_\phi^2}{8mQ^2} - \frac{lE_{ho}}{4Q} = -\frac{1}{8}m\omega^2 l^2.$$

For a Coulomb potential, one has $V(Q) = -K/Q$, and the Hamiltonian for the system is

$$H_c(Q, \alpha, P, p_\alpha) = \frac{P^2}{2m} + \frac{p_\alpha^2}{2mQ^2} - \frac{K}{Q} = E_c.$$

One notices that it can be written in the same form as the previous Hamiltonian provided one makes the following substitutions:

$$K = \frac{lE_{ho}}{4}; \qquad E_c = -\frac{m\omega^2 l^2}{8}; \qquad p_\alpha = \frac{p_\phi}{2}.$$

4. The Kepler trajectory is given by $dQ/d\alpha = \dot{Q}/\dot{\alpha}$. Moreover, Hamilton's equations provide $\dot{Q} = \partial_P H_c = P/m$ and $\dot{\alpha} = \partial_{p_\alpha} H_c = p_\alpha/(mQ^2)$ so that $\dot{Q}/\dot{\alpha} = PQ^2/p_\alpha = \rho^3 p_\rho/(lp_\phi)$. The trajectory for the harmonic oscillator is given by $d\rho/d\phi = \dot{\rho}/\dot{\phi}$. In this case, Hamilton's equations are $\dot{\rho} = \partial_{p_\rho} H_{HO} = p_\rho/m$, and $\dot{\phi} = \partial_{p_\phi} H_{HO} = p_\phi/(m\rho^2)$ which lead to $d\rho/d\phi = \rho^2 p_\rho/p_\phi$. Comparing these two expressions

$$\frac{dQ}{d\alpha} = \frac{\rho}{l}\frac{\rho^2 p_\rho}{p_\phi} = \frac{\rho}{l}\frac{d\rho}{d\phi} = \frac{1}{2}\frac{d(\rho^2/l)}{d\phi} = \frac{1}{2}\frac{dQ}{d\phi},$$

whence $d\alpha = 2\,d\phi$ and thus, with a judicious choice of the origin:

$$\alpha = 2\phi.$$

5. Let us start from the Kepler trajectory given by
$$\frac{1}{Q} = \frac{1}{p}[1 + e\cos\alpha]$$
with $p = p_\alpha^2/(mK)$, $e = \sqrt{1 + 2E_c p_\alpha^2/(mK^2)}$. We make the substitutions of question 3 to find the equation of the trajectory for the harmonic oscillator:
$$\frac{l}{\rho^2} = \frac{1}{p} + \frac{1}{p}e\cos(2\phi).$$
Using the property $\cos(2\phi) = \cos^2\phi - \sin^2\phi$ and the Cartesian coordinates $x = \rho\cos\phi$, $y = \rho\sin\phi$, we arrive at the following equation:
$$\frac{x^2}{[pl/(1+e)]} + \frac{y^2}{[pl/(1-e)]} = 1.$$
This is the equation of an ellipse with its center at the origin, i.e., at the center of force, with a semi-minor axis $b = \sqrt{pl/(1+e)}$, and semi-major axis $a = \sqrt{pl/(1-e)}$. We present in Fig. 6.7, the correspondence between the two types of trajectories.

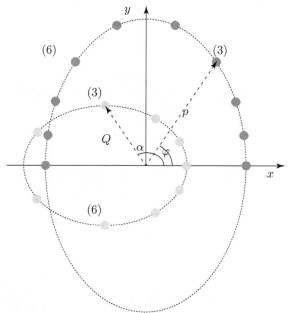

Fig. 6.7 Passage from the Kepler trajectory (in light grey) to the harmonic oscillator trajectory (dark grey) using the proposed contact transformation.
The corresponding positions are displayed as full circles. A complete revolution around the Kepler ellipse corresponds to half a revolution on the harmonic oscillator ellipse

6. It is easy to check that

$$p = \frac{p_\phi^2}{mlE_{ho}}, \qquad 1 - e^2 = \frac{\omega^2 p_\phi^2}{E_{ho}^2}.$$

The energy being constant along the trajectory, it is convenient to calculate it at the most suitable point, the perihelion. At that point $(p_\rho = 0)$, the kinetic energy is

$$T = \frac{p_\phi^2}{2mb^2} = \frac{1}{2}E_{ho}(1+e)$$

and the potential energy $V = \frac{1}{2}m\omega^2 b^2 = \frac{1}{2}E_{ho}(1-e)$. Consequently $H_{HO} = T + V = E_{ho}$, as expected.

$$H_{HO} = T + V = E_{ho}.$$

7. Let us start from the radial action obtained in the Kepler problem $2I_\rho = K\sqrt{2m/|E|} - 2p_\alpha$. Substituting the quantities suggested in Question 3, we arrive at:

$$2I_\rho = \frac{E_{ho}}{\omega} - p_\phi.$$

Performing the EBK quantization procedure, one makes the substitutions $I_\rho = n\hbar$ and $I_\phi = \lambda\hbar$ to obtain the semi-classical quantized value for the harmonic oscillator:

$$E_{ho} = \hbar\omega(2n + \lambda).$$

6.18. Generators for Fundamental Transformations [Statement p. 303]

A – Translations

1. By definition, the variation of the function in phase space is:

$$dF = F(x_1+a_1, x_2+a_2, x_3+a_3, p_{x_1}, p_{x_2}, p_{x_3}) - F(x_1, x_2, x_3, p_{x_1}, p_{x_2}, p_{x_3}),$$

since the translation does not affect momenta which are proportional to velocities. For infinitesimal transformations, this variation depends on the first derivatives of F. Explicitly:

$$dF = \sum_{i=1}^{3} a_i \frac{\partial F}{\partial x_i}.$$

Problem Solutions

2. Let consider the translation along the first axis Ox_1 by an infinitesimal quantity a. Let T_a be the corresponding generator. From the first question, one has $dF = a(\partial_{x_1} F)$. On the other hand, relying on (6.6), we also have $dF = \{F, T_a\} a$, since a is the flow parameter. Expliciting the Poisson bracket and identifying with the previous expression, one obtains $(\partial_{p_{x_1}} T_a) = 1$, $(\partial_{p_{x_2}} T_a) = 0$, $(\partial_{p_{x_3}} T_a) = 0$, $(\partial_{x_i} T_a) = 0$. These equations allow us to obtain the generator for translations along the Ox_1 axis: $T_{a_1} = p_{x_1}$. An analogous reasoning for the other axes would give similar results:

$$T_{a_i} = p_{x_i}.$$

3. From the results of the preceding question $\{T_{a_i}, T_{a_j}\} = \{p_{x_i}, p_{x_j}\}$. Developing the latter Poisson bracket, we easily find that it vanishes:

$$\{T_{a_i}, T_{a_j}\} = 0.$$

This property implies that translation operations along two different axes commute.

B – Rotations

1. Let $F(x, y, z, p_x, p_y, p_z)$ be a function defined at a point in a six-dimensional phase space. One calculates the variation of this function, considering a neighbouring point $(x + dx, y + dy, z + dz, p_x + dp_x, p_y + dp_y, p_z + dp_z)$ obtained by a rotation by an angle $d\phi$ around the Oz axis. Since r and p are two vectors of ordinary three-dimensional space, they transform under rotation with the same law. Their Cartesian components fulfill the relations $dx = -y\, d\phi$, $dy = x\, d\phi$, $dz = 0$ and similar relations for the components of the momentum. Therefore one can write:

$$dF = \left[x(\partial_y F) - y(\partial_x F) + p_x(\partial_{p_y} F) - p_y(\partial_{p_x} F)\right] d\phi.$$

Furthermore,

$$\{F, L_z\} = \sum_{i=1}^{3} (\partial_{r_i} F)(\partial_{p_i} L_z) - (\partial_{p_i} F)(\partial_{r_i} L_z).$$

The only non vanishing derivatives are $(\partial_x L_z) = p_y$, $(\partial_y L_z) = -p_x$, $\partial_{p_x} L_z = -y$ and $\partial_{p_y} L_z = x$. Consequently

$$\{F, L_z\} = x(\partial_y F) - y(\partial_x F) + p_x(\partial_{p_y} F) - p_y(\partial_{p_x} F),$$

so that:

$$dF = \{F, L_z\}\, d\phi.$$

2. With Cartesian coordinates, the flow of L_z is given by $(\partial_{p_x} L_z, \partial_{p_y} L_z, \partial_{p_z} L_z, -\partial_x L_z, -\partial_y L_z, -\partial_z L_z)$, or $(-y, x, 0, -p_y, p_x, 0)$. In ordinary space, the flow is given by $(-y, x, 0)$, that is the vector field $-\rho \sin\phi \, \hat{e}_x + \rho \cos\phi \, \hat{e}_y = \rho \hat{e}_\phi$. The integral lines of this flow are circles with radius ρ, centered on the Oz axis with z constant.

3. One defines the generators of rotations around the Ox axis and Oy axis respectively by $L_x = yp_z - zp_y$ and $L_y = zp_x - xp_z$. Then

$$\{L_x, L_y\} = \sum_{i=1}^{3} (\partial_{r_i} L_x)(\partial_{p_i} L_y) - (\partial_{p_i} L_x)(\partial_{r_i} L_y).$$

Owing to the definition of the generators, one is left only with the difference $(\partial_z L_x)(\partial_{p_z} L_y) - (\partial_{p_z} L_x)(\partial_z L_y)$, or

$$(-p_y)(-x) - (y)(p_x) = xp_y - yp_x = L_z.$$

The same calculation can be performed alternatively relying on the elementary Poisson brackets. Thus

$$\{L_x, L_y\} = L_z.$$

Rotations around non parallel axes do not commute.

C – Galilean transformation

1. For a free particle moving on a straight line, one has $p = m\dot{q}$. In a Galilean transformation, with infinitesimal speed v, the relation between the coordinates of the particle seen in both frames is $q' = q - vt$. The link between corresponding momenta is obtained by differentiation and leads to $p' = p - mv$. The variation of a function $F(q, p, t)$ under a Galilean transformation is expressed as $dF = F(q', p', t') - F(q, p, t)$, or, thanks to the previous relations and to the universality of time $t' = t$: $dF = F(q - vt, p - mv, t) - F(q, p, t)$. Expanding this expression to first order in the velocity, which is the flow parameter, one obtains:

$$dF = -[t(\partial_q F) + m(\partial_p F)] \, v.$$

2. Let us denote by G the generator of the free particle for a Galilean transformation. From the Definition (6.6), we have $dF = \{F, G\} v$. On the other hand, the Poisson bracket is translated as $\{F, G\} = (\partial_q F)(\partial_p G) - (\partial_p F)(\partial_q G)$. Identifying this equation with the previous one, we arrive at the relations $(\partial_p G) = -t$, $(\partial_q G) = m$. These equations can be integrated to provide the generator of the Galilean transformation (in this generator the time is a parameter, but not a component in phase space):

$$G(q, p, t) = mq - pt.$$

D – Lorentz Transformation

1. The one-dimensional Lorentz transformation is written generally as $q' = \gamma(q - \beta ct)$, $ct' = \gamma(ct - \beta q)$ for the event quadrivector and $p' = \gamma(p - \beta E/c)$, $E'/c = \gamma(E/c - \beta p)$ for the momentum quadrivector. We set, as usual, $\beta = v/c$ and $\gamma = (1 - \beta^2)^{-1/2}$.

   If the relative velocity between the two frames is infinitesimal $\beta \ll 1$, one can restrict oneself to first order in β in the previous transformation; in this case $\gamma \approx 1$ and the Lorentz transformation reduces to:

$$q' = q - \beta ct; \qquad ct' = ct - \beta q;$$
$$p' = p - \beta E/c; \qquad E'/c = E/c - \beta p.$$

2. The phase space is determined by the coordinates ($q_1 = q, q_2 = ct$) and by the momenta ($p_1 = p, p_2 = p_{ct}$). We saw several times (in particular in Problem 4.9) that the momentum conjugate to time is the negative of the energy: $p_t = -E$; thus the conjugate momentum for the coordinate ct is, naturally: $p_{ct} = -E/c$. Consequently, a function in phase space is chosen in the form $F(q_1, q_2, p_1, p_2) = F(q, ct, p, -E/c)$. The variation of this function $dF = F(q', ct', p', -E'/c) - F(q, ct, p, -E/c)$ for the infinitesimal Lorentz transformation given previously leads to the following relation:

$$dF = [-ct(\partial_{q_1} F) - q(\partial_{q_2} F) - (E/c)(\partial_{p_1} F) + p(\partial_{p_2} F)]\, \beta.$$

3. Relying on (6.6), the preceding expression is the variation of the function along the flow of the generator of the Lorentz "boost" U. One recognizes that the velocity $v = c\beta$ is the flow parameter and that the term in brackets should be identified with $c\{F, U\}$. This identification implies the relations $(\partial_{p_1} U) = -t = -q_2/c$, $(\partial_{p_2} U) = -q/c = -q_1/c$, $(\partial_{q_1} U) = E/c^2 = -p_2/c$, $(\partial_{q_2} U) = -p/c = -p_1/c$. After integration, one deduces the desired generator $U = -(p_1 q_2 + q_1 p_2)/c$.

   One can remark the analogy with the expression of the component of the angular momentum L_z (to within a sign). Substituting for the conjugate variables by their more traditional expressions, we obtain the final form of the generator for the Lorentz transformation (in this case time is no longer a parameter, but a full coordinate):

$$U(q, t, p, E) = \left(q\frac{E}{c^2} - pt\right).$$

   Comparing the relativistic expression, U, with the non-relativistic one, G, for the generator concerning a change of frame, one notices the famous relation: $E = mc^2$.

Chapter 7
Quasi-Integrable Systems

Summary

7.1. Introduction

A quasi-integrable system differs only slightly from an integrable system; one uses the properties of the latter system to have an idea of the properties of the original one. Schematically, one distinguishes two types of quasi-integrable systems.

1. First we consider systems whose Hamiltonian differs from that of an integrable system by the addition of a function in phase space, known as a perturbation, which we can imagine as weak as we wish. This is the essence of *perturbation theory*. This method has been largely employed in celestial mechanics to study the motion of planets. The integrable system is the couple Sun-Planet and the perturbation is the influence of all other planets.

2. Second we consider integrable systems whose Hamiltonian depends on parameters. If one or several of these parameters vary in time these systems are generally no longer integrable. However, if this variation is sufficiently slow, one can obtain important results. This is the essence of the *theory of adiabatic invariants*. This approach is very useful in the study of the motion of particles embedded in slowly varying electromagnetic fields.

7.2. Perturbation Theory

The full Hamiltonian is $H(q,p) = H_0(q,p) + V(q,p)$ and we know the solution $q_0(t)$, $p_0(t)$ of Hamilton's equations for H_0. Perturbation theory consists in seeking the solution of the more general problem governed by the Hamiltonian $H_\varepsilon = H_0 + \varepsilon V$ under the form of an expansion in powers of ε: $q_\varepsilon(t) = q_0(t) + \varepsilon q_1(t) + \varepsilon^2 q_2(t) + \ldots$ with an analogous expansion for $p_\varepsilon(t)$.

These expansions are inserted in Hamilton's equations $\dot{q}_\varepsilon(t) - \partial_p H_\varepsilon = 0$ and $\dot{p}_\varepsilon(t) + \partial_q H_\varepsilon = 0$, which become polynomials in ε with infinite degree. The unknown functions $q_i(t)$, $p_i(t)$ are then chosen in order to cancel the coefficients of this polynomial with order less or equal to m. They are obtained generally by a chain of cascades order by order starting from the known solution $q_0(t)$, $p_0(t)$. Finally, it is sufficient to set $\varepsilon = 1$ in the expressions of $q_\varepsilon(t)$ and $p_\varepsilon(t)$ to obtain the approximate solution of the perturbation up to order m.

7.3. Canonical Perturbation Theory

For quasi-integrable systems, the correct choice for coordinates in phase space is the set of angle-action variables (α, I) of the integrable system the Hamiltonian of which is denoted $H_0(I)$ (angle independent). Thus the Hamiltonian of the quasi-integrable system is written as:

$$H(\alpha, I) = H_0(I) + V(\alpha, I). \tag{7.1}$$

As explained previously, one uses the intermediate Hamiltonian $H_\varepsilon = H_0 + \varepsilon V$.

The canonical perturbation theory is an elegant manner of using the general perturbation theory. The idea is the following: one seeks a canonical transformation to new angle-action variables (ϕ, J) in the form of an expansion to arbitrary order in ε, but in a rather peculiar way. One expands the **new angle**[1] ϕ and the **old action** I as functions of **old angle** α and the **new action** J.

To illustrate this principle, let restrict ourselves to the study of a one-dimensional system. For the **first order** perturbation theory, we define:

$$\phi(\alpha, J) = \alpha + \varepsilon \phi^{(1)}(\alpha, J); \tag{7.2}$$

$$I(\alpha, J) = J + \varepsilon I^{(1)}(\alpha, J). \tag{7.3}$$

---

[1] For a simpler typographical notation, we omit the index ε for the variables (ϕ, J) which should be written more rigorously as $(\phi_\varepsilon, J_\varepsilon)$.

Summary

This unexpected way to perform the expansions allows us to impose the canonical condition for the transformation[2] more easily.

The original Hamiltonian is then written in terms of the old angle and the new action in the following form:

$$H_0(I) + \varepsilon V(\alpha, I) = H_0(J) + \varepsilon \left[I^{(1)}(\alpha, J)(\partial_J H_0(J)) + V(\alpha, J) \right] + O(\varepsilon^2).$$

The unknown function $I^{(1)}(\alpha, J)$ can be chosen in order that the first order term in ε is independent of the angle and is written as $\varepsilon H^{(1)}(J)$.

As soon as $I^{(1)}$ is determined, the canonicity of the transformation allows us to specify $\phi^{(1)}(\alpha, J)$ more precisely. The problem of first order perturbation theory is solved by setting $\varepsilon = 1$. The solution of Hamilton's equations for the Hamiltonian $H(J) = H_0(J) + H^{(1)}(J)$ differs from the exact solution in second order.

This solution is simple: the new Hamiltonian $H(J)$ is integrable; the new actions are first integrals $J = $ const and the new angles ϕ vary with a constant angular frequency $w(J)$:

$$\dot{\phi}(J) = \partial_J \left(H_0(J) + H^{(1)}(J) \right) = w_0(J) + w^{(1)}(J) = w(J).$$

To retrieve the initial variables, if necessary...

The old angle α is given indirectly to first order by

$$\alpha = w(J)t - \phi^{(1)}(\alpha, J) = w(J)t - \phi^{(1)}(w(J)t, J) + O(\varepsilon^2)$$

and the old action I directly by

$$I(\alpha, J) = J + I^{(1)}(\alpha, J).$$

It can be proved generally that
1. The first correction to the Hamiltonian is the perturbation averaged over each periodic motion $\overline{V}(J)$

$$H^{(1)}(J) = \frac{1}{(2\pi)^n} \int V(\alpha, J) \, d\alpha_1 \ldots d\alpha_n = \overline{V}(J). \qquad (7.4)$$

2. For a problem with one degree of freedom

$$I^{(1)}(\alpha, J) = \frac{H^{(1)}(J) - V(\alpha, J)}{w_0(J)}; \qquad (7.5)$$

$$\phi^{(1)}(\alpha, J) = \int \left(\partial_J I^{(1)}(\alpha, J) \right) d\alpha. \qquad (7.6)$$

---

[2] The readers who were courageous enough to study the last elements of Chapter 5.11 will realize that it is sufficient to find a generating function of type 2 in the form $S_\varepsilon(\alpha, J) = \alpha J + \varepsilon S^{(1)}(\alpha, J)$.

3. For a problem with more than one degree of freedom, the preceding formulae can be generalized in the following form[3]:

$$I_i^{(1)}(\alpha, J) = -\sum_m m_i e^{im\cdot\alpha} \frac{V_m(J)}{m\cdot\omega_0(J)}; \qquad (7.7)$$

$$\phi_i^{(1)}(\alpha, J) = -\sum_m e^{im\cdot\alpha} \partial_{J_i}\left(\frac{V_m(J)}{im\cdot\omega_0(J)}\right) \qquad (7.8)$$

in which use is made of the following quantities:
- An array of integer numbers (positive and negative) m to specify the set of n numbers $(m_1, m_2, \ldots, m_n)$;
- the "scalar products" $m\cdot\omega_0(J)$ to symbolize $m_1\omega_{01}(J) + m_2\omega_{02}(J) + \ldots + m_n\omega_{0n}(J)$ and $m\cdot\alpha$ for $m_1\alpha_1 + m_2\alpha_2 + \ldots + m_n\alpha_n$;
- the Fourier coefficients of the potential $V_m(J)$ defined as usual by

$$V(\alpha, J) = \sum_m e^{im\cdot\alpha} V_m(J).$$

Second order perturbations
- One can adopt the same approach and postulate a new canonical transformation with angle-action variables which are close to their initial values up to second order and, then, require the Hamiltonian to be angle independent for the terms in ε and ε^2.
- One can also remark that $H - H_0(J) - \varepsilon H^{(1)}(J)$, expressed as a function of the new variables, differs from H, up to terms in ε^2, and can be considered as an even smaller perturbation which can be treated at first order as was done previously.

In addition to the averaged value $\overline{V}(J)$ already present in first order, the second order theory requires the calculation of the average value for the square of the potential $\overline{V^2}(J)$ defined as:

$$\overline{V^2}(J) = \frac{1}{(2\pi)^n}\int (V(\alpha, J))^2 \, d\alpha_1 \ldots d\alpha_n.$$

In this case, the second order correction to the Hamiltonian is written:

$$H^{(2)}(J) = \partial_J\left(\frac{\overline{V}(J)^2 - \overline{V^2}(J)}{2\omega_0(J)}\right) \qquad (7.9)$$

---

[3] With $\omega_0(J) = \partial_J H_0(J) = (\omega_{01}(J), \ldots, \omega_{0n}(J)) = ((\partial_{J_1} H_0(J)), \ldots, (\partial_{J_n} H_0(J)))$, as we saw in Chapter 5.11.

and the second order correction to the angular frequency is $\omega^{(2)}(J) = \partial_J H^{(2)}(J)$.

The perturbative treatment up to second order will be presented in more detail in Problem 7.4. We will give, in this problem, the second order corrections to the angles and to the actions whose expressions are much more involved.

The catastrophe of small denominators

The treatment with the help of the canonical theory of perturbation is very useful, but it is especially important because it emphasizes what is called the catastrophe of small denominators or resonance phenomenon. This is apparent in the Expression (7.7) in the case where $m \cdot \omega_0(J) = 0$.

Let us give an example in the case of a two-dimensional system. There may exist initial conditions for J values such that the angular frequencies are commensurable, for instance $2\omega_{01}(J) \cong 3\omega_{02}(J)$. In other words, after each three periods for the variable 1 and each two periods for the variable 2, the system resets its state; the perturbation acts in an identical and cumulative way: we speak of a 3:2 resonance. There exists a divergence in the Expansion (7.7) for the term $m_1 = 2, m_2 = -3$. There exist many examples in celestial mechanics: for instance, the daily period of Mercury is commensurable with its period of revolution.

This property is expressed mathematically by the smallness of the denominator in the corrective terms which no longer correspond to an approximation, and there is no reason to think that this problem can be cured by increasing the order of perturbation. In special cases, one can get rid of these undesirable effects with the help of a further canonical transformation. This failure of the perturbation theory close to the resonance lies at the origin of the fundamental works of Poincaré, which will be discussed in the next chapter.

7.4. Adiabatic Invariants

One considers now an integrable Hamiltonian $H(q, p, \lambda)$ which depends on one (or several) fixed parameter(s) denoted λ. There exists a canonical transformation $q(\alpha, I, \lambda)$, $p(\alpha, I, \lambda)$ or, inversely, $\alpha(q, p, \lambda)$, $I(q, p, \lambda)$ such that the Hamiltonian can be written as $K(I, \lambda)$.

Let us assume now that the parameter depends on time : $\lambda(t)$. The preceding statements remain valid provided that we add a term to the Hamiltonian (see the part of Chapter 5.11 dealing with the time dependent canonical transformations); moreover $I(q, p, \lambda(t))$ is no longer constant.

It can be shown that, for a variation of the parameter $\Delta\lambda$ (not necessarily small), the action variable differs from its initial value by terms of order inversely proportional to the time T required for this variation of the parameter.

The action variable is said to be an **adiabatic invariant**.

This very important result can be understood quite easily.

The variation rate of the action is given, as is the case for any function of phase space and time (see Chapter 5.11), by

$$\dot{I} = \{I, K\} + \partial_t I(q, p, \lambda(t))|_{q,p}.$$

However the Poisson bracket[4] $\{I, K\}$ vanishes. Using the angle-action variables, the previous relation can be written as:

$$\dot{I} = \dot{\lambda}(\partial_\lambda I(q, p, \lambda))|_{q,p} = \dot{\lambda}(t) g(\alpha, I, \lambda(t)), \qquad (7.10)$$

a relation which defines a periodic (and this is an important point) function $g(\alpha, I, \lambda(t))$ of the angle. The angle α has an angular velocity which differs slightly from $\partial_I H(I, \lambda(t))$. If, over a period, the variation of the parameter λ is small the average in time for $g(\alpha, I, \lambda(t))$ will be close to the average over the angle. Yet this average is null for a periodic function[5]. The same thing is true for the average rate of variation of the action variable.

The theory of adiabatic invariants is of prime importance for the motion of charged particles in magnetic fields, as, for example, in the study of the magnetosphere or of particle accelerators.

Let us give the most important result: we saw that the action variable for the cyclotron motion of a particle is just the magnetic flux enclosing the trajectory. This flux is an adiabatic invariant. When the particle moves towards regions with increasing magnetic field, the radius must decrease in order to maintain the flux and so the velocity[6] and the kinetic energy increase. The conservation of kinetic energy leads to a decrease of the component of the velocity parallel to the field, a decrease that can possibly lead to an inversion of the velocity. This property will be studied in more detail in Problem 7.8 (page 354).

---

[4] Let us remind you that the Poisson bracket can be calculated with any set of variables, provided that they are obtained by a canonical transformation.

[5] The average value of a periodic function is not automatically null, but, in the present case, the function $g(\alpha)$ is the derivative with respect to α of another function and this property indeed holds.

[6] See the Problem 6.4 (page 291).

# Problem Statements

7.1. Limits of the Perturbative Expansion
[Solution and Figure p. 358] ★

A differential equation which is solved by a perturbative treatment at all orders

Using the perturbation theory, we wish to solve the differential equation $\dot{q} = q - \varepsilon q^2$ ($\varepsilon > 0$), where $q(t)$ takes the value 1 when $t = 0$. We choose as a starting point the function $q_0(t) = e^t$, and a perturbative expansion of the form

$$q(t) = \sum_0^\infty \varepsilon^n q_n(t).$$

This equation could mimic the time evolution of a population of bacteria which multiply proportionally to their number, with a limitation mechanism proportional to the number of pairs of bacteria.

1. Show that the first correction is $q_1(t) = e^t(1 - e^t)$.

2. Using the technique proposed in the summary, prove that the perturbed term of order n is $q_n(t) = e^t(1 - e^t)^n$.

3. The sum of all corrections is a geometrical series. Give the complete solution and verify that it makes sense only before a critical time which is to be determined.

4. Solve directly the proposed differential equation and compare the result with the perturbed solution.

5. Show that the exact solution tends to a finite limit when $t \to \infty$.

7.2. Non-canonical Versus Canonical Perturbative Expansion
[Solution p. 361] ★★

It may be dangerous to apply perturbation theory with incorrect variables

Let us take, with a suitable set of coordinates, the Hamilton function for a anharmonic oscillator: $H(q,p) = \frac{1}{2}(q^2 + p^2) + \frac{1}{3}\varepsilon q^3$.

The aim of this problem is to show that we must be very cautious if we do not employ the canonical perturbation theory (based on angle-action variables).

1. Derive Newton's equation and write it in the form: $\ddot{q} = -q - \varepsilon q^2$.
2. Find a solution of this equation, corresponding to the initial conditions $q(0) = 1$, $\dot{q}(0) = 0$ in the form of a perturbative expansion in ε with the lowest order $q_0(t) = \cos t$. Give and solve the differential equation for $q_1(t)$. Is there a modification of the period of the motion?
3. Same question for the second order correction. One could notice that the second member of the equation exhibits the natural period; consequently, the solution increases linearly with time. Such a term is known as a secular term.

   This latter solution is no longer periodic (in principle it should be for a given range of energy) and it is therefore valid only for small times. We will see that this drawback disappears if we choose a correct set of coordinates.
4. Use the canonical perturbation theory and prove that, to first order, there is no modification of the period. Calculate the angular frequency to second order as a function of the action (for instance using Formula (7.9) or the results of Problem 7.4).

7.3. First Canonical Correction for the Pendulum [Solution and Figure p. 363] ★ ★ ★

The classical example of a simple pendulum treated by perturbation.

We consider a simple pendulum of mass m and length l. We use as generalized coordinate q, the angle between the direction of the pendulum and the vertical; we remind you that Hamilton's function for this system is: $H(q,p) = p^2/(2ml^2) + mgl(1 - \cos q)$.

1. Find the change of coordinate $Q(q)$ which allows us to transform the Hamiltonian into the form of a harmonic oscillator (corresponding to small amplitudes):

$$H_0(Q, P) = \frac{1}{2}\omega(P^2 + Q^2)$$

   where $\omega = \sqrt{g/l}$ is the unperturbed angular frequency.

2. We wish to take into account the quartic approximation (in q^4), which arises from the expansion of $\cos q$ in the original Hamiltonian. Using the expressions of the angle-action variables adapted for the unperturbed Hamiltonian H_0, write down the Hamilton function $K(\alpha, I)$ in the quartic approximation.

Problem Statements

3. Perform a perturbative treatment to first order to find the expression of the energy $K(J)$ and the new angular frequency $\omega(J)$ in terms of the new set of variables (ϕ, J).

4. Determine, also to first order, the canonical transformation $\alpha(\phi, J)$ and $I(\phi, J)$ between the new and the old set of angle-action variables.

5. Show that the exact period of the pendulum at the quartic approximation (Hamiltonian including the q^2 and q^4 terms) is given by the following expression, where q_m is the maximal elongation:

$$T = \frac{4}{\omega\sqrt{1-q_m^2/12}} \int_0^1 \frac{dq}{\sqrt{(1-q^2)(1-\mu q^2)}} \quad \text{with} \quad \mu = \frac{1}{(12/q_m^2)-1}$$

The integral appearing in this formula is known as the complete elliptic integral of the first kind $K(\mu)$. Its value is given by known numerical algorithms. Particular values are $K(\mu = 0) = \pi/2$ and $K(\mu = 1) = \infty$. For which value of q_m does the period tend to infinity?

6. Show that, for the true pendulum (Hamiltonian depending on $\cos q$), the period is given by the analogous expression:

$$T = \frac{4}{\omega} \int_0^1 \frac{dq}{\sqrt{(1-q^2)(1-\mu q^2)}} \quad \text{with} \quad \mu = \sin^2 \frac{q_m}{2}.$$

For which value of q_m does the period tend to infinity?

7.4. Beyond the First Order Correction
[Solution p. 367]   ★ ★ ★

Canonical theory of perturbation extended up to second order

We wish to apply the method developed in the summary up to second order. We still consider an autonomous one-dimensional system. The formulae dealing with first order are supposed to be known (see Formulae (7.4) and (7.5)).

1. Calculate, as a function of the corrections to the action $I^{(1)}$ and $I^{(2)}$, the correction to second order for the Hamiltonian $K^{(2)}(J)$.

2. Deduce the angular dependence of the second correction to the unperturbed action $I^{(2)}$.

3. Show that the conservation of area for a canonical transformation to second order gives:

$$\int_0^{2\pi} I^{(2)}(\alpha, J)\, d\alpha = 0.$$

4. From the preceding questions and the results concerning first order corrections, calculate the second order correction to the Hamiltonian, as a function of the perturbing potential. You will introduce the average value of the square of the interaction (not to be confused with the square of the average value $\overline{V}(J)^2$):

$$\overline{V^2}(J) = \frac{1}{2\pi} \int_0^{2\pi} V^2(\alpha, J)\, d\alpha.$$

Again using the conservation of area, prove that

$$\partial_\alpha \phi^{(2)}(\alpha, J) = \partial_J I^{(2)}(\alpha, J)$$

and deduce the first part of the canonical transformation $\alpha(\phi, J)$ extended up to second order.

5. Complete the description of the canonical transformation by calculating $I(\phi, J)$ to second order.

7.5. Adiabatic Invariant in an Elevator
[Solution p. 370] ★ ★

A peculiar experiment in an elevator

A ball of mass m bounces elastically on the floor of an elevator, in a constant gravitational field g. We are concerned only with the one-dimensional motion along the vertical. The elevator starts slowly with a null initial acceleration and the height of the floor above the ground is specified by the function $h(t)$. At the end of the ascent, the elevator decelerates slowly and then stops. We assume that the motion can be characterized as adiabatic (the acceleration varies only slightly between two rebounds).

1. Choosing as a generalized coordinate Q, the altitude of the ball above the floor, give the expression of the Lagrangian $L(Q, \dot{Q}, t)$. Find a simpler expression for this function by adding the total time derivative of a well chosen function $F(Q, t)$. Deduce the corresponding Hamiltonian $H(Q, P, t)$.

2. Calculate the action $I(E,a)$ at time t, when the acceleration of the elevator takes the value $\ddot{h}(t) = a$ (one can also refer to Problem 6.3 for this particular point). Give the relation between the energy E and the maximal height Q_m reached by the ball under these conditions.

3. Using the adiabaticity condition, show that the height of the rebound obeys the relation:

$$\frac{Q_{\max}(a)}{Q_{\max}(0)} = \left(\frac{g}{g+a}\right)^{1/3}.$$

4. What is the final height of the rebound after the elevator stops?

7.6. Adiabatic Invariant and Adiabatic Relaxation [Solution and Figure p. 372] ★ ★ ★

Walls do not have "ears" but "wings"

In this problem, we reexamine the system described in Problem 6.2. A particle, of mass m, moves freely on the Ox axis between a "wall" placed at point O ($x = 0$) and a "wall" placed at point A ($x = L$). Striking either of the walls, the particle bounces elastically (change of the sign of the velocity) and moves freely in the opposite direction. The corresponding Hamiltonian is a "square well" which can be simulated by a less singular potential, as was suggested in Problem 6.2 (in the latter problem the abscissae of the walls were $-a$ and a instead of 0 and L; you will make the trivial identification $L = 2a$). In the following, we need to use the fact that the particle velocity is constant between the walls and that it changes sign at the impact on a fixed wall (see Fig. 7.1).

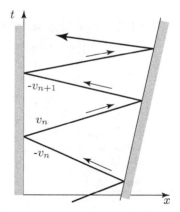

Fig. 7.1 Schematic motion of the particle between a fixed wall on the left side and a wall moving at constant speed on the right side

1. Both walls are assumed to be fixed. Calculate the action of the particle in terms of the velocity v or the energy E.

2. Now assume that the right wall A moves away from the fixed wall O at constant speed u and denote by v_n the modulus of the velocity after the nth impact on the fixed wall. Give the relation between v_{n+1} and v_n. Deduce the expression of v_n as a function of the initial velocity v_0 (at the start of the motion for the moving wall).

3. In this question, one calculates the action I_n, which depends on time, at the moment when **the particle touches the moving wall** after the nth impact on the fixed wall (action calculated between two impacts on the moving wall). This definition is the usual one in the literature. The dimensionless quantity $z_n = u/(v_0 - 2nu)$ is introduced. Give the relation between I_{n+1} and I_n. Show that the action remains constant when the adiabaticity condition ($u \ll v_n$) is fulfilled.

4. In this question, the action I_n is calculated **immediately after the nth impact on the fixed wall** (action calculated between two impacts on the fixed wall). Give the relationship between I_{n+1} and I_n. What can we conclude?

5. One considers now the same particle enclosed in a parallelepipedic box, which expands in time in a homothetic way, that is its length L_i ($i = (x, y, z)$) along the Ox_i axis increases linearly with time: $L_i = u_i t$. One speaks of an adiabatic relaxation, in the sense that the particle still continues bouncing elastically on the walls, without exchanging heat with the surroundings. We remind you that the kinetic energy is proportional to the temperature T. Using the result of Question 3 or 4, what is the relationship between the volume V of the box and the absolute temperature T during an adiabatic process.

7.7. Charge in a Slowly Varying Magnetic Field [Solution and Figure p. 375] ★ ★ ★

More on electromagnetic fields.

We study once more the problem of a particle of mass m and electric charge q_e embedded in a magnetic field $\boldsymbol{B}$ directed along the Oz axis. We assume in a first case that the magnetic field $\boldsymbol{B}$ is uniform and constant. We remind you that, along the direction of $\boldsymbol{B}$, the motion is uniform and that, in the perpendicular plane, the trajectory is a circle covered at constant angular velocity $\omega = q_e B/m$, known as the cyclotron frequency. We are concerned with this two-dimensional motion in the plane Oxy (see also the Problem 2.9 on the cyclotron motion, p. 66).

Using adiabatic invariants, one can extend these results to a less academic situation for which the field may be neither uniform nor constant.

Let us begin with the simplest case of a constant magnetic field. We work from now on in the gauge where the vector potential is given by $\mathbf{A} = (-yB, 0, 0)$. The generalized coordinates are the Cartesian coordinates (x, y).

1. Recall the relation between the generalized momentum and the velocity, and then the expression of the Hamiltonian of the particle with this set of coordinates.

It is useful to perform a canonical transformation switching from the old coordinates (x, y) to new coordinates (ϕ, Y). This canonical transformation is governed by a generating function of type 1 (if necessary, refer to Section 6.2.3):

$$G_1(\phi, Y, x, y) = C\left[\frac{1}{2}(y - Y)^2 \cot(\phi) + xY\right],$$

where C is a constant undetermined for the moment.

2. Express the new variables (coordinates and momenta) in terms of the old ones.

3. Determine the value of the constant C in order to obtain the simplest form for the Hamiltonian expressed in terms of the new set of coordinates. Give its expression. Is the system integrable? What are the angle-action variables?

4. Express the new variables in term of the old ones, as well as the components and the modulus of the velocity. Interpret these new coordinates. Show that p_ϕ is proportional to the magnetic flux Φ through the cyclotron circle, as well as to the magnetic moment μ generated by the cyclotron motion.

5. Show that the energy can be interpreted as the coupling of the magnetic moment with the magnetic field: $H = -\boldsymbol{\mu} \cdot \mathbf{B}$. Prove that $\boldsymbol{\mu}$ and $\mathbf{B}$ are collinear but with opposite sense.

From now on, assume that an operator imposes a time dependence, $\mathbf{B}(t)$, on the field while maintaining its direction along the Oz axis (or that the particle during its motion along the field lines experiences a variation of the field). This dependence is assumed to be adiabatic; this means that an appreciable change of $B(t)$ requires a time which is much larger than the cyclotron period of the same system.

6. What is the quantity which remains constant in time?

This transformation allows us to generalize the notion of drift for the cyclotron orbit (we already met with such a drift by the addition of an electric field in Problem 2.9). Let us imagine a supplementary force $\boldsymbol{F}$ acting on the particle. For simplicity we assume that this force arises from a potential $V(y)$, which depends on the y coordinate only and which varies slowly over a range of order of the cyclotron radius.

7. Write down the complete Hamiltonian, as well as the Hamilton equations relative to the (Y, P_Y) variables. Deduce that, on average over one revolution, the center of the cyclotron circle drifts along the Ox axis with a velocity to be determined.

This result is quite general: there is a drift perpendicular to the force, that is along the equipotential lines, with a velocity $\boldsymbol{V_d} = (\boldsymbol{B} \times \boldsymbol{F})/(q_e B^2)$.

7.8. Illuminations Concerning the Aurora Borealis [Solution and Figure p. 379]   ★ ★

A fascinating natural phenomenon studied in detail. It is strongly advised to solve Problems 2.9 and 7.7 before continuing

To a good approximation, the Earth's magnetism is that of a magnetic dipole. In the magnetic equatorial plane, the magnetic field, which is perpendicular to it, takes the values $B_e = 0.31\,10^{-4}(R_T/R)^3$ Tesla (B_e is the value of B at the equator). In this formula R_T designates the Earth's radius and R the distance to the center of the Earth, for which the field is measured.

An electron of mass m, charge q_e and energy E crosses the equatorial plane, at a distance R from the center of the Earth, its velocity making an angle α with the direction of the magnetic field.

1. Recall the expression of the cyclotron frequency and show that the cyclotron radius (projection of the trajectory onto the equatorial plane) in the non-relativistic limit is:

$$R_c = \frac{\sqrt{2mE \sin^2 \alpha}}{q_e B_e}.$$

In units of the Earth's radius, what is the value of this radius for $\alpha = \pi/4$ and for an energy $E = 60$ keV, at a distance $1.5 R_T$ from the Earth's center. Give also the period of this cyclotron rotation.

Data: $q_e = 1.6\,10^{-19}$ C, $m = 9.11\,10^{-31}$ kg, $R_T = 6367$ km.

2. Show that the action variable (see also Problem 7.7) is:
$$p_\phi = -\frac{mE \sin^2 \alpha}{q_e B_e}.$$
The electron rolls itself up in an helical trajectory around the field line and thus sees a magnetic field which increases as it approaches the pole. The action variable is an adiabatic invariant.

3. Denoting by s the distance covered along the field line and by $B(s)$ the intensity of the magnetic field ($B_e = B(s=0)$), show that the energy associated with the rotational motion is $EB(s)\sin^2\alpha/B_e$. The total energy of the particle is a constant. Why? Deduce that the electron, whatever its energy, cannot explore a region close to the pole for which the magnetic field exceeds the value $B(s_m) = B_e/\sin^2\alpha$.

4. The particle bounces in the regions of strong field close to the poles and performs successive rebounds from one hemisphere to the other.

Prove that the period for this cyclic motion is given by the integral:
$$T = 2\sqrt{\frac{m}{2E}} \int_{s_m}^{s_m} \frac{ds}{\sqrt{1 - B(s)/B(s_m)}}$$
(its value is approximatively 2 s for electrons of 30 keV).

Being close to the poles, thus to the atmosphere, the electric charges trapped by this mechanism (essentially protons and electrons) finally strike and excite the atoms of the higher atmosphere. The decay of these atoms produces a light which generates the aurora borealis.

This bouncing motion from one pole to the other is not limited to the vicinity of the field lines. Globally, the motion drifts easterly for electrons, westerly for protons.

We can understand this phenomenon using the results of Problem 7.7.

One knows that the cyclotron motion generates a current, which itself generates a dipole magnetic moment μ. The rotational energy can be considered as the coupling of the magnetic moment with the magnetic field $E = -\mu \cdot B$; the magnetic field exerts a force and a torque on this magnetic moment.

5. Show that, in our case, this force at the equator, which tends to push the dipole away from the Earth, takes the value $F = 3E\sin^2\alpha/R$. Deduce that the easterly drift velocity for the electron close to the equator is
$$V_d = \frac{3E \sin^2 \alpha}{q_e B_e R}.$$

Additional remarks: This problem is the source of several interesting extensions. The curious reader will easily find complements in an encyclopaedia, or on the Internet. Let us emphasize some important points.

- In fact, the terrestrial magnetic field is not exactly that of a dipole; this induces modifications to the previous behaviour; moreover the solar wind plays a non negligible role.
- There exist two adiabatic invariants associated with two periodic motions: the northerly-southerly motion and the rotation around the Earth due to the drift.
- A further cause of drift is the centrifugal force resulting from the curvature of field lines.

7.9. Bead on a Rigid Wire: Hannay's Phase
[Solution and Figure p. 382] ★ ★ ★

An amusing but simple experiment

In this problem, we will find a result due to Hannay which provides more information on the angle variable.

We consider a bead, of mass m, constrained to slide without friction on a closed, plane and rigid wire, with an arbitrary shape of perimeter L, enclosing a surface S. The position M of the bead on the wire is specified by its curvilinear abscissa $s(t)$, the origin being located at an arbitrary point of the wire.

In a first experiment, the wire is held fixed in the plane and the bead moves at constant speed $\dot{s}_0$ (the only force exerted on the bead is the reaction force normal to the wire which does not produce work and consequently does not modify the kinetic energy, nor the linear velocity). The distance covered during the time t is $D_0(t) = \dot{s}_0 t$. We denote by $\tau = L/\dot{s}_0$ the time needed for the bead to perform a complete revolution. This experiment is the reference experiment.

In a second experiment, we are concerned with a system which is completely identical with the previous one, but in which the wire is forced to rotate in its plane, around a given point O inside the wire (an exterior point could also be considered). The wire performs a complete revolution (rotation by 2π) during the time T. The angular velocity of the wire is denoted by $\dot{\phi}(t)$. We assume that at the beginning $\dot{\phi}(0) = 0$. This value then increases, reaches a maximum, and decreases to vanish again at time T: $\dot{\phi}(T) = 0$. The particular form of this function does not matter, provided that the behaviour is adiabatic, a condition that can be expressed by $\tau \ll T$.

For both experiments, the beads start with the same initial conditions $s_0, \dot{s}_0$. At the same time we start the rotation of the stem for the second experiment. The purpose of the experiments is to measure the relative positions of the two beads after time T, corresponding to a complete revolution of the stem.

As you will see, the second bead will recover its initial velocity $\dot{s}_0$. It will have covered a distance D (which increases as T and $\dot{s}_0$ increase), while the reference bead covered a distance D_0. The striking property that was discovered by Hannay (provided that the adiabaticity condition is fulfilled) is that the difference $D - D_0$ remains constant, giving an unmistakable signature of the rotation of the stem. This property is independent of the position of the rotation axis and of the rotational velocity $\dot{\phi}(t)$; it depends on the shape of the stem only. We say that it is of geometrical nature.

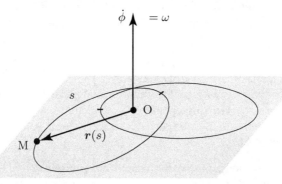

Fig. 7.2 Experiment of a bead sliding on a closed stem in rotation and its reference experiment (see text). The bead is specified by the curvilinear abscissa $s(t)$ with respect to an arbitrary origin on the stem

1. Let $r(s) = \boldsymbol{OM}$, the distance of the bead to the axis of rotation, when it is placed at abscissa s (see Fig. 7.2). Taking into account the relative velocity and the driving velocity, calculate the kinetic energy T_{kin}. For simplicity, it is useful to introduce $\boldsymbol{t}(s)$, the unit vector tangent to the stem at the position $M(s)$ for the bead.

2. Show that, if the square of the driving velocity is neglected, the kinetic energy is given by the formula:
$$T_{kin} = \frac{1}{2} m \left(\dot{s}^2 + 4A'(s)\, \dot{s}\, \dot{\phi} \right),$$
where $A'(s)$ is the derivative with respect to the curvilinear abscissa of the area swept out by the radius vector $r(s)$.

3. With the help of Lagrange's equations, prove that if $\dot{\phi}$ is constant, so is $\dot{s}$. Rederive this result by directly considering the Coriolis force.

Give the expression of the Hamiltonian and show that it does not depend on time. Deduce that the velocity $\dot{s}$ is a constant of the motion.

4. Calculate the action
$$I = \frac{1}{2\pi} \oint p(s)\,ds,$$
performing the integral over a complete revolution of the bead. Give the expression for the conjugate angle variable?

Hint: The angle varies linearly with time, exactly as does the curvilinear abscissa.

5. One assumes now an adiabatic variation for $\dot{\phi}$. Give the expression of the velocity $\dot{s}(t)$.

6. Deduce the difference between the distances covered by the two beads, one without and the other with the rotation of the wire, during a complete revolution time T.

7. On can define an average phase β, corresponding to the position of the bead such that this phase increases by 2π for a revolution on the wire; in other words: $\beta = 2\pi D/L$. The difference of phase for the two experiments is known as the **Hannay phase** β_H. Calculate its value and show that it depends only on the shape of the wire. Check your result in the case of a circular wire.

Important remark: One could consider a complicated shape for the wire exhibiting entangled loops; one must consider in this case S as the algebraic sum of the areas (counted positively if the bead turns on the loop in the same sense as the rotation of the wire, and negatively in the opposite situation). In particular, a wire with an "8" shape leads to a vanishing Hannay phase, the delay accumulated on one loop being exactly compensated by the advance accumulated on the other!

Problem Solutions

7.1. Limits of the Perturbative Expansion
[Statement p. 347]

1. We wish to solve the differential equation $\dot{q} = q - \varepsilon q^2$ with the initial condition $q(0) = 1$. We proceed by perturbation and set
$$q = \sum_{n=0}^{\infty} \varepsilon^n q_n,$$

from which we derive

$$\dot{q} = \sum_{n=0}^{\infty} \varepsilon^n \dot{q}_n, \quad \text{and} \quad \varepsilon q^2 = \sum_{n=1}^{\infty} \varepsilon^n \sum_{i=0}^{n-1} q_i q_{n-i-1}.$$

The differential equation then implies:

$$\dot{q}_0 + \sum_{n=1}^{\infty} \varepsilon^n \dot{q}_n = q_0 + \sum_{n=1}^{\infty} \varepsilon^n \left[q_n - \sum_{i=0}^{n-1} q_i q_{n-i-1} \right].$$

The identification is made term by term for the coefficients of the polynomial to give

$$\dot{q}_0 = q_0 \quad ; \quad \dot{q}_n = q_n - \sum_{i=0}^{n-1} q_i q_{n-i-1}.$$

Taking into account the initial condition $q(0) = 1$, we impose the following constraints $q_0(0) = 1$ and $q_n(0) = 0$, $\forall n$. The first equation at zero order, $\dot{q}_0 = q_0$ can be integrated to give $q_0(t) = e^t$. The first order correction is obtained by setting $n = 1$ in the coupled equations, that is $\dot{q}_1 = q_1 - q_0^2 = q_1 - e^{2t}$. This equation can be integrated easily to give $q_1 = e^t - e^{2t}$. It can be rewritten as:

$$q_1(t) = e^t(1 - e^t).$$

2. For higher order terms, it is useful to adopt a recursion argument. Let us set $q_i(t) = e^t(1 - e^t)^i$, valid for any $i < n$. The first order is satisfied. Inserting this relation in the nth order equation, one obtains: $\dot{q}_n = q_n - ne^{2t}(1 - e^t)^{n-1}$. One is convinced (either integrating directly or just checking a posteriori) that $q_n(t) = e^t(1 - e^t)^n$ is indeed the solution with correct initial condition. The recursion formula is thus proved, and the correction to order n is:

$$q_n(t) = e^t(1 - e^t)^n.$$

3. In this particular case, the entire series

$$q(t) = e^t \sum_{n=0}^{\infty} \left[\varepsilon(1 - e^t) \right]^n$$

can be summed up exactly because it is a geometrical series with a common ratio $\varepsilon(1 - e^t)$.

However, this can be done only if this ratio is less than 1; this condition implies the existence of a critical time t_c obtained when the common ratio is 1. Explicitly:

$$t_c = \ln\left(1 + \frac{1}{\varepsilon}\right).$$

If $t < t_c$, it is legitimate to sum up the series which provides the solution, obtained by perturbation to all orders,

$$q_{\text{pert}}(t) = \frac{e^t}{1 - \varepsilon(1 - e^t)},$$

which can be recast in the simpler form

$$q_{\text{pert}}(t) = \frac{1}{\varepsilon + (1 - \varepsilon)e^{-t}}.$$

Under this form, it can be checked first that the initial condition is fulfilled $q_{\text{pert}}(0) = 1$ and secondly that by setting $\varepsilon = 0$, one recovers the unperturbed solution $q_0(t)$.

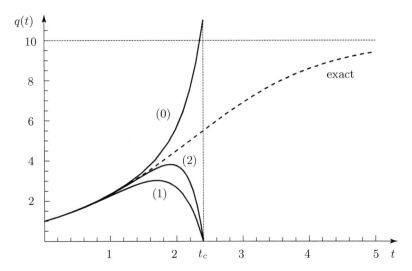

Fig. 7.3 The exact solution of the proposed differential equation, for $\varepsilon = 0.1$, as well as its asymptote, is plotted as a dashed line. The approximations of order (0): q_0, of order (1): $q_0 + q_1$, of order (2): $q_0 + q_1 + q_2$, plotted as full lines, reproduce the exact solution more and more accurately. The critical time t_c is also shown as a dotted line

Problem Solutions

4. The original differential equation is separable and can thus be integrated easily: with the initial condition, the solution is found to be

$$t = \ln\frac{q(1-\varepsilon)}{(1-\varepsilon q)},$$

which, after some rearrangement, can be written as

$$q(t) = \frac{1}{\varepsilon + (1-\varepsilon)e^{-t}}.$$

Consequently $q(t) = q_{\text{pert}}(t)$ so that the two solutions coincide. However the perturbative solution makes sense only if the time is less than the critical time, while the exact solution suffers no restriction. One can consider the exact solution as the analytical extension of the perturbed solution for times larger than the critical time.

In Fig. 7.3, we show how successive approximations compare to the exact solution.

5. With the preceding expression of $q(t)$, one immediately sees that

$$q(t) \to \frac{1}{\varepsilon} \quad \text{when } t \to \infty.$$

7.2. Non-canonical Versus Canonical Perturbative Expansion [Statement p. 347]

1. With the Hamiltonian $H(q,p) = \frac{1}{2}(p^2 + q^2) + \varepsilon\frac{1}{3}q^3$, the Hamilton equations read $\dot{q} = \partial_p H = p$ and $\dot{p} = -\partial_q H = -q - \varepsilon q^2$. Differentiating the first equation with respect to time and substituting in the second one, we obtain Newton's equation:

$$\ddot{q} = -q - \varepsilon q^2.$$

2. Let us perform an expansion to first order for the solution: $q(t) = q_0(t) + \varepsilon q_1(t)$; substituting in Newton's equation and identifying terms of the same order, we obtain the equations: $\ddot{q}_0 = -q_0$ and $\ddot{q}_1 = -q_1 - q_0^2$. The solution of the first equation, which obeys the initial conditions $q_0(0) = 1$, $\dot{q}_0(0) = 0$, is simply $q_0(t) = \cos t$.

Substituting this value in the second differential equation, one obtains the new equation $\ddot{q}_1 + q_1 = -\cos^2 t = -\frac{1}{2}(1 + \cos(2t))$. To fulfill the initial conditions for the complete solution, the initial conditions concerning this first order correction are now $q_1(0) = 0$, $\dot{q}_1(0) = 0$. The solution of the corresponding differential equation is obtained with standard techniques; the general solution of the equation without the second member is $A\sin t + B\cos t$, while a particular solution of the equation is $-\frac{1}{2} + \frac{1}{6}\cos(2t)$. The constants A and B are obtained using the initial conditions. Finally the first order correction reads:

$$q_1(t) = -\frac{1}{2} + \frac{1}{3}\cos t + \frac{1}{6}\cos(2t).$$

One sees that the period of $q_1(t)$ is still 2π, as it is for $q_0(t)$. Thus, to first order the period of the system is not affected.

3. Let us pursue the expansion to second order $q(t) = q_0(t) + \varepsilon q_1(t) + \varepsilon^2 q_2(t)$. Proceeding with the usual technique, we obtain, in addition to the two previous differential equations, the supplementary second order equation $\ddot{q}_2 = -q_2 - 2q_0 q_1$. With the explicit expressions of q_0 and q_1 obtained previously, the differential equation that must be integrated is the following:

$$\ddot{q}_2 + q_2 = -\frac{1}{3} + \frac{5}{6}\cos t - \frac{1}{3}\cos(2t) - \frac{1}{6}\cos(3t).$$

The initial conditions on the unknown function are now $q_2(0) = 0$, $\dot{q}_2(0) = 0$. Here again, the technique for obtaining the solution is standard. However, one notices that in the second member there appears a term with the same period as the proper period. In this case, the solution involves a special term linear in time. One obtains finally the solution:

$$q_2(t) = -\frac{1}{3} + \frac{29}{144}\cos t + \frac{5}{12}t\sin t + \frac{1}{9}\cos(2t) + \frac{1}{48}\cos(3t).$$

This solution contains a secular term which increases linearly with time and is thus no longer periodic. This is contrary to what one expects physically. The perturbation theory with this set of coordinates clearly exhibits limitations.

4. Now, we will perform a perturbative treatment based on angle-action coordinates. However, before proceeding it is necessary to find the canonical transformation which permits the passage from the old set (q, p) to the new one (α, I), namely $q = \sqrt{2I}\sin\alpha$, $p = \sqrt{2I}\cos\alpha$. With these new coordinates, the unperturbed Hamiltonian is $H_0(I) = I$ whereas the perturbation is $V(\alpha, J) = \frac{1}{3}q^3 = \frac{1}{3}(2J)^{3/2}\sin^3\alpha$, since at first order $J = I$.

It is easy to calculate the average values of the potential:

$$\overline{V}(J) = \frac{1}{2\pi}\int_0^{2\pi} V(\alpha, J)\,d\alpha = 0$$

$$\overline{V^2}(J) = \frac{1}{2\pi}\int_0^{2\pi} V^2(\alpha, J)\,d\alpha = \frac{5}{18}J^3.$$

Problem Solutions

The second order perturbed Hamiltonian is

$$K(J) = H_0(J) + \varepsilon K^{(1)}(J) + \varepsilon^2 K^{(2)}(J).$$

The first order correction is simply (see the summary) $K^{(1)}(J) = \overline{V}(J) = 0$, and consequently, to first order, the angular frequency $\omega(J) = \partial_J K(J)$ reduces to the unperturbed angular frequency $\omega_0(J) = \partial_J H_0(J) = 1$.

Thus, to first order

$$\omega(J) = \omega_0(J) = 1.$$

As was the case in the second question, to first order, there is no modification of the period.

To obtain the second order correction (see Problem 7.4), one must calculate the quantity

$$U(J) = \frac{\overline{V(J)^2} - \overline{V^2}(J)}{2\omega_0(J)} = -\frac{5}{12}J^2.$$

One deduces the perturbed angular frequency

$$\omega(J) = \partial_J K(J) = 1 - \frac{5}{6}\varepsilon^2 J.$$

Thus at second order

$$\omega(J) = 1 - \frac{5}{6}\varepsilon^2 J. \tag{7.11}$$

At second order, the period is modified, but, with this set of coordinates, in contrast to the old one, the system remains periodic, as it should be. It is thus necessary to employ the canonical perturbation theory.

7.3. First Canonical Correction for the Pendulum [Statement p. 348]

1. From the original Hamiltonian $H(q,p) = p^2/(2ml^2) + mgl(1 - \cos q)$, we obtain the harmonic approximation by truncating the expansion of the cosine at second order, namely $H_0(q,p) = p^2/(2ml^2) + mglq^2/2$. The natural angular frequency $\omega = \sqrt{g/l}$ is introduced in order to rewrite $H_0(q,p) = p^2/(2ml^2) + ml^2\omega^2 q^2/2$. Now, let us switch to the new variables $Q = q\sqrt{ml^2}$, $P = p/\sqrt{ml^2}$ (in order to ensure a canonical transformation $\{P,Q\} = \{p,q\}$); this substitution in H_0 allows us to write it in the standard form

$$H_0(Q,P) = \frac{1}{2}\omega(P^2 + Q^2).$$

2. The angle-action variables (α, I) are obtained using the canonical transformation $Q = \sqrt{2I} \sin \alpha$, $P = \sqrt{2I} \cos \alpha$. Expressed in terms of these new variables, the unperturbed Hamiltonian is simply $H_0(I) = \omega I$. Further developing to the quartic approximation in the expansion of the cosine adds to H_0 the perturbation $V = -\frac{1}{24} mgl q^4$, which, in terms of the new variables, is expressed as $V(\alpha, I) = -(I^2 \sin^4 \alpha)/(6ml^2)$. The perturbed Hamiltonian then reads

$$K(\alpha, I) = H_0(I) + V(\alpha, I) = \omega I - \frac{I^2 \sin^4 \alpha}{6ml^2}.$$

3. The first order correction to the Hamiltonian results from the general canonical theory of perturbations

$$K^{(1)}(J) = \overline{V}(J) = \frac{1}{2\pi} \int_0^{2\pi} V(\alpha, J)\, d\alpha.$$

With the previous value of the perturbation, the corresponding value is $-J^2/(16ml^2)$. Thus to first order, the Hamiltonian is written as

$$K(J) = \omega J - \frac{J^2}{16ml^2}.$$

The corresponding angular frequency $\omega(J) = \partial_J K(J)$ is easily deduced

$$\omega(J) = \omega - \frac{J}{8ml^2}.$$

4. The first order correction to the action is given generally by (see Formula (7.5)):

$$I^{(1)}(\alpha, J) = \frac{\left(K^{(1)}(J) - V(\alpha, J)\right)}{\omega_0(J)}.$$

In this special case, an elementary calculation provides the expression

$$I^{(1)}(\alpha, J) = \frac{J^2 \left(\cos(4\alpha) - 4\cos(2\alpha)\right)}{48ml^2 \omega}.$$

As for the correction to the angle, it can be obtained from the general relation (see Formula (7.5)) $\partial_\alpha \phi^{(1)} = \partial_J I^{(1)}$, which, in this case, gives

$$\phi^{(1)}(\alpha, J) = \frac{J}{24ml^2 \omega} \left[\frac{1}{4} \sin(4\alpha) - 2\sin(2\alpha)\right].$$

Lastly, the canonical transformation is completed by inverting the perturbative expansion while remaining consistent at this order of perturbation.

Problem Solutions

One finds $\alpha(\phi, J) = \phi - \phi^{(1)}(\phi, J)$ and $I(\phi, J) = J + I^{(1)}(\phi, J)$. With the expressions of $I^{(1)}$ and $\phi^{(1)}$ given above, we achieve the determination of the canonical transformation:

$$\alpha(\phi, J) = \phi + \frac{J}{96ml^2\omega} [8\sin(2\phi) - \sin(4\phi)]$$

$$I(\phi, J) = J + \frac{J^2}{48ml^2\omega} [\cos(4\phi) - 4\cos(2\phi)].$$

5. The Hamiltonian is conservative; its value remains constant and identified with the energy E. If we restrict ourselves to the quartic expansion only, this implies

$$E = \frac{1}{2}ml^2\dot{q}^2 + \frac{1}{2}mglq^2 - \frac{1}{24}mglq^4.$$

It will be convenient to introduce the maximum elongation (amplitude of the pendulum) q_m; one can rewrite $E = \frac{1}{2}mglq_m^2 - \frac{1}{24}mglq_m^4$. The conservation of energy can then be expressed as

$$\dot{q}^2 = \omega^2(q_m^2 - q^2)\left(1 - (q^2 + q_m^2)/12\right).$$

This is a separable differential equation which can be integrated to give the temporal evolution and the period. The latter quantity is simply 4 times the time needed to pass from angle 0 to angle q_m, that is

$$T = \frac{4}{\omega} \int_0^{q_m} \frac{dq}{\sqrt{(q_m^2 - q^2)(1 - (q^2 + q_m^2)/12)}}.$$

Performing the change of variable $u = q/q_m$ and setting $\mu = 1/(12/q_m^2 - 1)$, one can recast the expression for the period in the form

$$T = \frac{4}{\omega\sqrt{1 - q_m^2/12}} \int_0^1 \frac{du}{\sqrt{(1 - u^2)(1 - \mu u^2)}}.$$

One recognizes the form of the complete elliptic integral $K(\mu)$. Finally, at the quartic approximation, the period of the pendulum is

$$T = \frac{4}{\omega\sqrt{1 - q_m^2/12}} K\left(\frac{1}{12/q_m^2 - 1}\right). \tag{7.12}$$

The period becomes infinite as the elliptic integral becomes infinite. This happens when its argument equals 1, that is when $q_m = \sqrt{6}$ radians (about 140 deg). The other possibility $q_m = \sqrt{12}$ radians must be excluded because it is too large.

6. Let us now come to the exact case, keeping the cosine expression in the Hamiltonian. A reasoning quite similar to the one proposed above allows us to write the expression of the period in the form

$$T = \frac{4}{\omega\sqrt{2}} \int_0^{q_m} \frac{dq}{\sqrt{\cos q - \cos q_m}}.$$

Instead of the q variable, let us choose the variable u defined as $\sin(q/2) = \sin(q_m/2)u$. An elementary calculation provides the value of the period:

$$T = \frac{4}{\omega} \int_0^1 \frac{du}{\sqrt{(1-u^2)(1-\sin^2(q_m/2)u^2)}}.$$

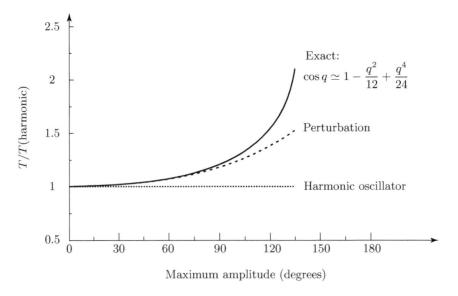

Fig. 7.4 Ratio between the period of a simple pendulum and the period corresponding to the harmonic approximation, as a function of the maximum elongation q_m calculated with the quartic approximation $\cos q \approx 1 - \frac{q^2}{2} + \frac{q^4}{24}$. The value obtained using this approximation (full line) is given by the elliptic integral of Question 5. The period tends to infinity for $q_m = \sqrt{6}$ radians or $q_m \approx 140 \deg$. The dashed line represents the result of first order perturbation theory for the quartic approximation, as given in Question 3. Practically superimposed on this latter curve is the variation for the exact period as given in Question 6. The period tends to infinity for $q_m = \pi$ radians or $q_m = 180 \deg$

Here too, we find the complete elliptic integral $K(\sin^2(q_m/2))$ and the exact expression for the period of the pendulum is

$$T = \frac{4}{\omega} K\left(\sin^2(q_m/2)\right).$$

In the particular case of a pendulum, the exact expression is finally no more cumbersome than the expression resulting from the perturbation theory. If we set $q_m \to 180\,\mathrm{deg}$, we see that, from the properties of the elliptic function, $T \to \infty$. A pendulum released from its highest position with zero velocity needs an infinite time to come back!

In Fig. 7.4, we compare the ratio of the period for various approximations to the period for the harmonic approximation, as a function of the maximum amplitude. For small amplitudes, the various approximations give similar results but for large amplitudes, the deviations may be important.

7.4. Beyond the First Order Correction
[Statement p. 349]

1. Let us pursue to second order the expansion suggested in the summary

$$\begin{aligned}
\phi(\alpha, J) &= \alpha + \varepsilon \phi^{(1)}(\alpha, J) + \varepsilon^2 \phi^{(2)}(\alpha, J) \\
I(\alpha, J) &= J + \varepsilon I^{(1)}(\alpha, J) + \varepsilon^2 I^{(2)}(\alpha, J)
\end{aligned}$$

The new Hamiltonian is $K(\alpha, I) = H_0(I) + \varepsilon V(\alpha, I)$. One substitutes in this expression the preceding expansion of I. One then performs a Taylor expansion to second order identifying the result with the final expression of the Hamiltonian which depends on J only: $K(J) = H_0(J) + \varepsilon K^{(1)}(J) + \varepsilon^2 K^{(2)}(J)$. This identification leads to the following equalities (as usual $\omega_0(J) = \partial_J H_0(J)$).

$$K^{(1)}(J) = I^{(1)}(\alpha, J)\omega_0(J) + V(\alpha, J)$$
$$K^{(2)}(J) = I^{(2)}(\alpha, J)\omega_0(J) + I^{(1)}(\alpha, J)\left(\partial_J V(\alpha, J)\right)$$
$$+ \frac{1}{2}\left(I^{(1)}(\alpha, J)\right)^2 (\partial_J \omega_0(J)).$$

2. It is easy to obtain the value of $I^{(2)}$ from the latter equality:

$$I^{(2)}(\alpha, J) = \frac{K^{(2)}(J) - I^{(1)}(\alpha, J)\left(\partial_J V(\alpha, J)\right) - \frac{1}{2}\left(I^{(1)}(\alpha, J)\right)^2 \omega_0'(J)}{\omega_0(J)}$$

which employs the simple notation $\omega_0'(J) = \partial_J \omega_0(J)$.

3. Switching from the old coordinates (α, I) to the new ones (ϕ, J) results from a canonical transformation; consequently, there is conservation of the area in phase space, a property that is expressed by

$$\oint J \, d\phi = \oint I \, d\alpha,$$

or (since for the new Hamiltonian $J = \text{const}$)

$$\varepsilon \int_0^{2\pi} I^{(1)} \, d\alpha + \varepsilon^2 \int_0^{2\pi} I^{(2)} \, d\alpha = 0.$$

This equality must be satisfied whatever the value of ε so that

$$\int_0^{2\pi} I^{(1)}(\alpha, J) \, d\alpha = 0 = \int_0^{2\pi} I^{(2)}(\alpha, J) \, d\alpha.$$

4. Let us define

$$\overline{V}(J) = \frac{1}{2\pi} \int_0^{2\pi} V(\alpha, J) \, d\alpha, \qquad \overline{V^2}(J) = \frac{1}{2\pi} \int_0^{2\pi} V^2(\alpha, J) \, d\alpha.$$

We derive $I^{(1)}$ as a function of $K^{(1)}$ and V and substitute the result in the constraint of conservation of the area. We thus obtain the first order correction to the Hamiltonian $K^{(1)}(J) = \overline{V}(J)$ and the first order correction to the action $I^{(1)}(\alpha, J) = \left(\overline{V}(J) - V(\alpha, J)\right)/\omega_0(J)$. These results are presented in the summary.

We proceed in the same way with the integral on $I^{(2)}$ (a quantity that was obtained in Question 2) which must give a vanishing contribution. It is composed of three parts. The first contribution is

$$\int_0^{2\pi} K^{(2)}(J) \, d\alpha = 2\pi K^{(2)}(J).$$

The second contribution reads

$$\int_0^{2\pi} I^{(1)}(\alpha, J) \left(\partial_J V(\alpha, J)\right) d\alpha$$

which can be transformed, after some algebra and using the expression of $I^{(1)}$ already found, into

$$\frac{2\pi}{2\omega_0(J)} \left[2\overline{V}(J) \left(\partial_J \overline{V}(J)\right) - \left(\partial_J \overline{V^2}(J)\right)\right] = \frac{2\pi}{2\omega_0(J)} \partial_J \left(\overline{V}(J)^2 - \overline{V^2}(J)\right).$$

Problem Solutions

As for the third contribution

$$-\frac{\omega_0'(J)}{2\omega_0(J)}\int_0^{2\pi}\left(I^{(1)}(\alpha,J)\right)^2,$$

a similar treatment gives the expression

$$-\frac{2\pi\omega_0'(J)}{2\omega_0(J)}\left(\overline{V}(J)^2-\overline{V^2}(J)\right).$$

Gathering all the contributions, we obtain finally

$$K^{(2)}(J)=\frac{1}{2\omega_0(J)}\left[\partial_J\left(\overline{V}(J)^2-\overline{V^2}(J)\right)+\frac{\omega_0'(J)}{\omega_0(J)}\left(\overline{V}(J)^2-\overline{V^2}(J)\right)\right],$$

which can be recast in the more convenient form

$$K^{(2)}(J)=\partial_J\left(\frac{\overline{V}(J)^2-\overline{V^2}(J)}{2\omega_0(J)}\right).$$

5. The conservation of area in phase space is equivalent to a unit Jacobian for the transformation from the old set of coordinates to the new one: $D(\phi,J)/D(\alpha,I)=1$. A well known property of the Jacobians allows us to write

$$\frac{D(\phi,J)}{D(\alpha,I)}=\frac{D(\phi,J)}{D(\phi,I)}\times\frac{D(\phi,I)}{D(\alpha,I)}.$$

Another well known formula is the following,

$$\frac{D(\phi,J)}{D(\phi,I)}=\left(\frac{D(\phi,I)}{D(\phi,J)}\right)^{-1}.$$

Taking into account these properties shows that the conservation of area is equivalent to the relation $D(\phi,I)/D(\phi,J)=D(\phi,I)/D(\alpha,I)$. But

$$\frac{D(\phi,I)}{D(\phi,J)}=[(\partial_J I)(\partial_\phi\phi)-(\partial_\phi I)(\partial_J\phi)]$$

which, owing to the properties $(\partial_\phi\phi)=1$ and $(\partial_\phi I)=0$, reduces to $D(\phi,I)/D(\phi,J)=\partial_J I$.

A similar argument gives $D(\phi,I)/D(\alpha,I)=\partial_\alpha\phi$. Thus, the conservation of area is translated into the simpler equation $\partial_J I=\partial_\alpha\phi$. Restricting to second order in ε, this equality implies $1+\varepsilon\partial_J I^{(1)}+\varepsilon^2\partial_J I^{(2)}=1+\varepsilon\partial_\alpha\phi^{(1)}+\varepsilon^2\partial_\alpha\phi^{(2)}$. The identification for the first order term gives

$$\frac{\partial\phi^{(1)}(\alpha,J)}{\partial\alpha}=\frac{\partial I^{(1)}(\alpha,J)}{\partial J}.$$

The identification for the second order term leads to the conclusion:
$$\frac{\partial \phi^{(2)}(\alpha, J)}{\partial \alpha} = \frac{\partial I^{(2)}(\alpha, J)}{\partial J}.$$
As we see from the preceding reasoning, these kinds of equations are in fact valid to all orders.

6. The expression $I^{(1)}(\alpha, J) = \left(\overline{V}(J) - V(\alpha, J)\right)/\omega_0(J)$ was already obtained. Differentiating with respect to J and integrating with respect to α, we obtain the expression of $\phi^{(1)}$:
$$\phi^{(1)}(\alpha, J) = \int d\alpha\, \partial_J \left(\overline{V}(J) - V(\alpha, J)\right)/\omega_0(J).$$

The quantity
$$I^{(2)}(\alpha, J) = \frac{K^{(2)}(J) - I^{(1)}(\alpha, J)\left(\partial_J V(\alpha, J)\right) - \frac{1}{2}\left(I^{(1)}(\alpha, J)\right)^2 \omega_0'(J)}{\omega_0(J)}$$
was derived in Question 2. All the terms appearing in this expression have been determined previously. It is enough to substitute them in the value of $I^{(2)}$. Differentiating this expression with respect to J and integrating with respect to α, we obtain the value $\phi^{(2)}(\alpha, J)$. We have now in hand all the elements necessary for the calculation of the canonical transformation at second order.

Starting from the transformation equations, and performing the necessary truncated expansions, being nevertheless consistent to all orders of perturbation, we arrive at the desired expressions:
$$\alpha(\phi, J) = \phi - \varepsilon \phi^{(1)}(\phi, J) - \varepsilon^2 \left[\phi^{(2)}(\phi, J) - \phi^{(1)}(\phi, J)\left(\partial_\phi \phi^{(1)}(\phi, J)\right)\right]$$
$$I(\phi, J) = J + \varepsilon I^{(1)}(\phi, J) + \varepsilon^2 \left[I^{(2)}(\phi, J) - \phi^{(1)}(\phi, J)\left(\partial_\phi I^{(1)}(\phi, J)\right)\right].$$

7.5. Adiabatic Invariant in an Elevator
[Statement p. 350]

1. If we denote by q the height of the ball above the ground and by Q its height above the floor of the elevator, we have the obvious relation $Q = q - h(t)$. The original Lagrangian is simply $L(q, \dot{q}) = \frac{1}{2}m\dot{q}^2 - mgq$. Owing to the preceding relation, it can be expressed in terms of the new coordinates as:
$$L(q, \dot{q}, t) = \frac{1}{2}m(\dot{Q} + \dot{h})^2 - mg(Q + h)$$
$$= \frac{1}{2}m\dot{Q}^2 - mgQ + m\dot{Q}\dot{h} + \frac{1}{2}m\dot{h}^2 - mgh.$$

Without affecting the equations of motion, it is possible to add to the Lagrangian a total derivative with respect to time of a function of coordinate and time only (see Chapter 1.11): $dF(Q,t)/dt$. Let us choose the function

$$F(Q,t) = -mQ\dot{h} + \int \left[mgh(t) - \frac{1}{2}m\dot{h}(t)^2\right] dt;$$

its derivative cancels the three last terms of the Lagrangian and adds the new term $-mQ\ddot{h}$. Finally, the equivalent Lagrangian can be written in the simpler form:

$$L(Q,\dot{Q},t) = \frac{1}{2}m\dot{Q}^2 - m(g + \ddot{h}(t))Q.$$

The momentum is easily deduced $P = \partial_{\dot{Q}}L = m\dot{Q}$ and then the Hamiltonian, $H = P\dot{Q} - L$, that is:

$$H(Q,P,t) = \frac{P^2}{2m} + (g + \ddot{h}(t))Q.$$

This property can be found directly by working in the frame of the elevator; the total acceleration is equal to the acceleration due to gravity plus the driving acceleration.

2. The Hamiltonian depends on time, but the action must be calculated by freezing the value of H at the value of the energy E at a given time so that $\ddot{h}(t) = a = \text{const}$. Let $Q_{\max}$ be the maximal height reached by the ball under these conditions. One has $E = m(g+a)Q_{\max}$. Moreover, one has $E = P^2/(2m) + m(g+a)Q$. Therefore, one can write

$$P = \pm m\sqrt{2}(g+a)^{1/2}\sqrt{Q_{\max} - Q}.$$

The action is calculated using the usual formula

$$I = \frac{1}{2\pi}\oint P\,dQ$$

or, more explicitly

$$I = \frac{\sqrt{2m}}{\pi}(g+a)^{1/2}\int_0^{Q_{\max}}\sqrt{Q_{\max} - Q}\,dQ = \frac{2\sqrt{2m}}{3\pi}(g+a)^{1/2}Q_{\max}^{3/2}.$$

Introducing the value of the energy, the action can be written in the form:

$$I(E,a) = \frac{2\sqrt{2}}{3\pi}\frac{E^{3/2}}{(g+a)\sqrt{m}}.$$

3. One assumes that the elevator accelerates slowly (as compared to the bouncing period), then decelerates slowly before stopping. For an adiabatic motion which leads it up to an acceleration $\ddot{h} = a$, the action is conserved (see the summary). Thus we have $I(a) = I(0)$, that is $E(a)^{3/2}/(g+a) = E(0)^{3/2}/g$. In other words $(E(a)/E(0))^{3/2} = (g+a)/g$. Introducing once more the maximal height, this last equality can be transformed into the desired relation:

$$\frac{Q_{\max}(a)}{Q_{\max}(0)} = \left(\frac{g}{g+a}\right)^{1/3}.$$

4. When the elevator stops, one has again $\ddot{h} = 0$, hence $(g+\ddot{h})/g = 1$ and therefore $Q_{\max}(\ddot{h}) = Q_{\max}(0)$. Subsequently the ball bounces at the same height as before the start.

7.6. Adiabatic Invariant and Adiabatic Relaxation [Statement and Figure p. 351]

1. As we saw in Problem 6.2, the phase portrait is a rectangle. Over a semi-period, from $q = -a$ to $q = a$, the momentum is positive and takes the value $p = \sqrt{2mE}$; over the other semi-period, from $q = a$ to $q = -a$, the momentum is negative and takes the value $p = -\sqrt{2mE}$. At the walls, the momentum changes sign instantaneously. The action is defined by the usual formula

$$I(E) = \frac{1}{2\pi} \oint p(E, q)\, dq;$$

in this particular case, the action is written simply as

$$I = \frac{1}{\pi} \sqrt{2mE} \int_{-a}^{a} dq,$$

or $I(E) = \dfrac{2a}{\pi} \sqrt{2mE}$ in which we set $L = 2a$; thus:

$$I(E) = \frac{L}{\pi} \sqrt{2mE}.$$

2. Between the two walls, the particle is free and its velocity remains constant. After the impact n on the fixed wall, the velocity is $v_n > 0$; the particle attains the moving wall with this velocity. In the frame attached to the wall (velocities denoted with a prime), there is an elastic rebound which implies $v'_{n+1} = -v'_n$, after which the particle bounces back with the constant velocity $v_{n+1} < 0$.

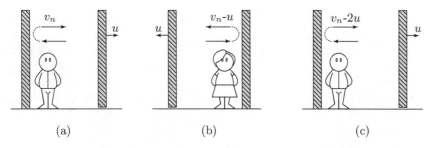

Fig. 7.5 Illustration of the velocities after the various impacts on the wall. The left wall is fixed; the right wall moves with velocity u. (a) after the impact n on the fixed wall, the velocity is v_n – (b) after the impact n on the moving wall, the velocity relative to the fixed wall is $v_n - u$ – (c) the latter velocity is $v_n - 2u$ with respect to the fixed wall

The relation between the velocities in both frames is $v' = v - u$ (it is assumed that the wall moves with a velocity $u > 0$). Consequently, one has $v'_n = v_n - u$. The condition of elastic rebound implies $v'_{n+1} = v_{n+1} - u = -v'_n = -(v_n - u)$, that is $v_{n+1} = -v_n + 2u$. Striking the fixed wall, the particle instantaneously changes the sign of the velocity (see Fig. 7.5). After the impact $n+1$, the relation is simply

$$v_{n+1} = v_n - 2u.$$

This is a recursion relation which allows us to easily express the velocity v_n after the impact n in terms of the initial velocity v_0. Explicitly, the desired relation reads:

$$v_n = v_0 - 2nu.$$

3. Instead of ascribing to the fixed wall and to the moving wall the respective abscissae $-a$ and a, let us denote them rather as 0 and X. With these new definitions, the action, calculated in the first question, becomes $I(X) = (X/\pi)\sqrt{2mE} = mXv/\pi$.

The relationship between the actions depends on the definition taken for these actions. In this question, the action is defined at the moment when the particle touches the **moving wall** at impact n. The action at this time is $I_n = mX_n v_n/\pi$ and the action at the next impact is $I_{n+1} = mX_{n+1}v_{n+1}/\pi$.

Let us write $z_n = u/v_n$. Owing to the previous question, one has $v_{n+1} = v_n(1 - 2z_n)$. It remains to calculate X_{n+1}. Let τ be the time separating the impact n and the impact $n+1$ on the moving wall. Obviously $X_{n+1} = X_n + u\tau$.

Moreover, between these two impacts, the speed remains constant at the value v_{n+1} and the distance covered is first X_n (going) $+X_{n+1}$ (coming back), so that $\tau = (X_n + X_{n+1})/v_{n+1}$. Substituting this value in the preceding equation, one finds $X_{n+1}(1 - u/v_{n+1}) = X_n(1 + u/v_{n+1})$. Moreover, $z_{n+1} = u/[v_n(1 - 2z_n)] = z_n/(1 - 2z_n)$. These equations allow us to write $X_{n+1} = X_n(1 - z_n)/(1 - 3z_n)$. Using the recursion relation between X_{n+1}, v_{n+1} and X_n, v_n in the expression of the action, we finally obtain

$$\frac{I_{n+1}}{I_n} = 1 + \frac{2z_n^2}{1 - 3z_n}.$$

The adiabaticity in the motion of the wall implies the inequalities $u \ll v_n < v_0$ and consequently $z_n \ll 1$. It follows that the actions satisfy the relation $I_{n+1}/I_n \approx 1$, or

$$I_{n+1} \approx I_n.$$

If the adiabatic condition is satisfied, there is conservation of the action between two impacts. This type of argument is often presented in textbooks.

4. The action can also be defined from the impacts on the **fixed wall**. We still have $v_{n+1} = v_n(1 - 2z_n)$. Now the quantity X_{n+1} must be calculated correctly. Let τ be the delay which separates the impact n and the impact $n+1$ on the moving wall. Of course, one has $X_{n+1} = X_n + u\tau$. At time $t = 0$, the particle sets off from the fixed wall with velocity v_n. At time τ_1, it touches the moving wall, which is placed at position X'. First we have $X' = X_n + u\tau_1$, and then $X' = v_n\tau_1$, from which we derive $X' = X_n/(1 - z_n)$.

The time needed to perform the return journey is $\tau_2 = X'/v_{n+1}$ and the wall is then positioned at $X_{n+1} = X' + u\tau_2$. Substituting τ_2 by its value and X' by its value in terms of X_n, we arrive finally at the expression $X_{n+1} = X_n/(1 - 2z_n)$. Gathering all these conclusions in the action value, we find

$$I_{n+1} = \frac{m}{\pi}\frac{X_n}{1 - 2z_n}v_n(1 - 2z_n) = \frac{m}{\pi}X_nv_n = I_n.$$

Thus the property

$$I_{n+1} = I_n$$

does not depend on the adiabaticity condition.

5. We assume a parallelepipedic box, the walls of which move with time. As we just saw, the actions for the motion of each of the axes remain constant. To simplify we set $A = 2m/\pi^2$.

Problem Solutions

First one has const $= I_x^2 = AL_x^2 \frac{1}{2}mv_x^2$ and two analogous conditions concerning the other axes. One assumes now that the dilation is homothetic in time, which means $L_x = u_x t$, etc. Thus $I_x^2 = Au_x^2 t^2 \frac{1}{2} mv_x^2$. One deduces

$$\frac{I_x^2}{Au_x^2} + \frac{I_y^2}{Au_y^2} + \frac{I_z^2}{Au_z^2} = \text{const} = t^2 \frac{1}{2} mv^2.$$

Furthermore, the volume of the box is $V = L_x L_y L_z = u_x u_y u_z t^3$, hence $t^2 = (V/(u_x u_y u_z))^{2/3}$. Substituting this expression in the previous constant relation, one obtains $V^{2/3} \frac{1}{2} mv^2 = \text{const}$. Lastly, the equipartition theorem (the particles have no mutual interaction) implies a relationship between the average kinetic energy of the particles and the temperature T, namely: $\frac{1}{2}mv^2 = \frac{3}{2}kT$, where k is the Boltzman constant. The adiabaticity condition finally leads to the law

$$TV^{2/3} = \text{const}.$$

This expression is precisely that of an adiabatic process $TV^{\gamma-1} = \text{const}$ for a monatomic ideal gaz with adiabatic coefficient $\gamma = 5/3$. We have recovered a thermodynamical law simply from pure mechanical arguments.

7.7. Charge in a Slowly Varying Magnetic Field [Statement p. 352]

1. We choose the gauge as $\mathbf{A} = (-yB, 0, 0)$. The electromagnetic potential reads $V = q_e(U - \dot{\mathbf{r}} \cdot \mathbf{A})$. With $U = 0$ and the preceding expression for $\mathbf{A}$, the potential simply becomes $V = q_e B \dot{x} y$. Let us restrict ourselves to the study of the motion in the plane Oxy.

The Lagrangian of the system is

$$L = \frac{1}{2} m(\dot{x}^2 + \dot{y}^2) - q_e B \dot{x} y.$$

One deduces the momenta $p_x = m\dot{x} - q_e B y$, $p_y = m\dot{y}$ and then the Hamiltonian $H = p_x \dot{x} + p_y \dot{y} - L$ which, after replacing the velocities by corresponding momenta, is written as

$$H(x, y, p_x, p_y) = \frac{1}{2m} \left[(p_x + q_e B y)^2 + p_y^2 \right].$$

2. We define a canonical transformation through the following generating function of the first type:

$$G_1(\phi, Y, x, y) = C \left[\frac{1}{2}(y - Y)^2 \cot(\phi) + xY \right].$$

From the general formulae concerning the transformation (see (6.9)), one deduces the corresponding momenta

$$p_x = \partial_x G_1 = CY, \qquad p_y = \partial_y G_1 = C(y - Y)\cot(\phi),$$
$$p_\phi = -\partial_\phi G_1 = \tfrac{1}{2}C(y - Y)^2(1 + \cot^2(\phi)),$$
$$p_Y = -\partial_Y G_1 = C\left[(y - Y)\cot(\phi) - x\right].$$

From the first equation, we derive $Y = p_x/C$. The substitution of this value in the second equation leads to $\cot(\phi) = p_y/(Cy - p_x)$. Using the second equation in the fourth one provides $p_Y = p_y - Cx$. Lastly, rearranging the third equation gives $p_\phi = (p_y^2 + (Cy - p_x)^2/(2C))$. To summarize, we have

$$\cot(\phi) = -\frac{p_y}{p_x - Cy}; \qquad Y = \frac{p_x}{C};$$
$$p_\phi = \frac{p_y^2 + (p_x - Cy)^2}{2C}; \qquad p_Y = p_y - Cx.$$

3. We see that the numerator of p_ϕ looks quite similar to the Hamiltonian if we choose $C = -q_e B$, a value that we henceforth adopt. We obtain the very simple relation ($\omega = q_e B/m$ is the cyclotron frequency):

$$H(p_\phi) = -\frac{q_e B}{m} p_\phi = -\omega p_\phi.$$

The ϕ and Y variables are cyclic so that p_ϕ and p_Y are constants of the motion. The ϕ coordinate appears only through trigonometric functions and hence is manifestly an angle; the associated quantity p_ϕ is therefore an action. The Y variable is a length, but it may be associated with a coordinate which could be an angle so that p_Y could be considered to be connected to an action. In any case, one remarks that the Hamiltonian is expressed in terms of actions only, without recourse to coordinates which are cyclic. This condition is sufficient to prove that the system is integrable. The canonical transformation makes these things clear.

4. Let us start from the relations given in the first question. We have first $p_x = CY = -q_e BY$. Now starting from $p_\phi = \left(p_y^2 + (Cy - p_x)^2\right)/(2C)$ and substituting $Cy - p_x$ by $p_y/\cot(\phi)$, we obtain $p_\phi = p_y^2/(2C\cos^2\phi)$. With the value for C derived previously, we deduce from this equation the value of p_y which is put in the form $p_y = -q_e B R(p_\phi)\cos\phi$, with the definition $R(p_\phi) = \sqrt{-2p_\phi/(q_e B)}$. The relation $Cx = p_y - p_Y$ allows us to obtain, with the preceding value of p_y, the expression $x = p_Y/(q_e B) + R(p_\phi)\cos\phi$. Lastly, the relation $Cy - p_x = C(y - Y)$ provides the value of y, namely $y = Y + R(p_\phi)\sin\phi$.

We have finished our study, which can be summarized as:

$$x = \frac{p_Y}{q_e B} + R(p_\phi) \cos \phi; \qquad p_x = -q_e BY$$
$$y = Y + R(p_\phi) \sin \phi; \qquad p_y = -q_e B R(p_\phi) \cos \phi,$$

with
$$R(p_\phi) = \sqrt{\frac{-2p_\phi}{q_e B}}.$$

The Hamilton equation $\dot{Y} = \partial_{p_Y} H = 0$ shows that Y is a constant of the motion; similarly $\dot{p}_Y = -\partial_Y H = 0$ shows that p_Y is constant. Lastly $\dot{p}_\phi = -\partial_\phi H = 0$ shows that p_ϕ is also constant. Obviously the same property holds for $R(p_\phi)$ and $H(p_\phi)$. The ϕ variable is the only variable which depends on time; we have in fact $\dot{\phi} = \partial_{p_\phi} H = -\omega$.

The interpretation of the new variables is now perfectly clear: from the expressions for x and y given above, we see that the trajectory of the particle is a circle with its center at abscissa $p_Y/(q_e B)$ and ordinate at Y and with radius $R(p_\phi)$. This circle is covered with the constant angular velocity ω; it is the cyclotron circle. The ϕ variable is simply the polar angle of the particle on the circle. The situation is summarized in Fig. 7.6.

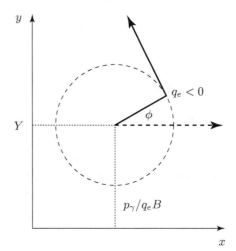

Fig. 7.6 Interpretation of the new coordinates following the canonical transformation for a particle embedded in a magnetic field

From $p_y = m\dot{y} = -q_e BR \cos \phi$, one deduces $\dot{y} = -(q_e B/m)R \cos \phi$, and from $p_x = -q_e BY = m\dot{x} - q_e By$, one deduces $\dot{x} = q_e B(y - Y)/m = (q_e B/m)R \sin \phi$. The modulus of the velocity is thus obtained as $v^2 = \dot{x}^2 + \dot{y}^2 = (q_e B/m)^2 R^2 = -2q_e B p_\phi/m^2$, a relation that could also have been obtained from $v^2 = 2H/m$.

The flux of the magnetic field across the cyclotron circle is $\Phi = BS = \pi R^2 B = -2\pi p_\phi/q_e$; in other words

$$p_\phi = -\frac{q_e \Phi}{2\pi}$$

The variable p_ϕ represents the magnetic flux across the orbit of the particle. The particle rotating on the circle generates a current $i = q_e/T = q_e \omega/(2\pi)$. The magnetic moment has a modulus $\mu = iS = i\pi R^2$. Replacing i and R by their respective values, one finds finally: $\mu = -\omega p_\phi/B = -q_e p_\phi/m$. Thus the value of the magnetic moment is

$$\mu = -\frac{q_e}{m} p_\phi.$$

5. Let us focus on the direction of the vectors. One assumes that $\boldsymbol{B}$ is directed along the Oz axis. It is easy to see (invoking the Lorentz force or the components of the velocity) that the velocity $\boldsymbol{v}$ is directed anticlockwise if $q_e < 0$ and clockwise if $q_e > 0$. In both cases, the current $\boldsymbol{i}$ is directed clockwise and the corkscrew rule shows that the magnetic moment $\boldsymbol{\mu}$ is directed along the Oz axis in the negative sense, hence collinear with $\boldsymbol{B}$ but in the opposite direction. From the proven relation $\mu = -\omega p_\phi/B$, one deduces $\mu B = -\omega p_\phi = H$. Owing to the previous discussion concerning the signs, one can write generally:

$$H = -\boldsymbol{\mu} \cdot \boldsymbol{B}.$$

6. If the magnetic field varies slowly, one learns from the theory of adiabatic invariants that the actions remain practically constant in time. Moreover, we showed that p_ϕ is an action. Thus, one can claim that p_ϕ or H remains constant in time.

7. The total Hamiltonian H is the sum of the original Hamiltonian H_0 and the perturbed potential $V(y)$: $H = H_0 + V(y)$. Switching to the new coordinates, the perturbed Hamiltonian reads

$$H(\phi, Y, p_\phi, p_Y) = -\frac{q_e B}{m} p_\phi + V(Y + R(p_\phi)\sin\phi).$$

The first Hamilton equation gives: $\dot{Y} = \partial_{p_Y} H = 0$. The relation $Y = $ const follows; thus the ordinate of the cyclotron center remains fixed. In other words, the particle drifts along the Ox axis, that is along the equipotential lines. The second Hamilton equation gives:

$$\dot{p}_Y = -\partial_Y H = -V'(y) = -V'(Y) - R(p_\phi)\sin\phi V''(Y) \approx -V'(Y).$$

The abscissa of the cyclotron center is $p_Y/(q_e B)$. Its drift velocity is thus $V_d = \dot{p}_Y/(q_e B) = -V'(Y)/(q_e B)$.

Moreover, $-V'(Y) = F$ is the force at the origin of the drift. Therefore $V_d = F/(q_e B)$. This relation is compatible with the proposed general formula:

$$V_d = \frac{F}{q_e B} \times \frac{B}{B}.$$

7.8. Illuminations Concerning the Aurora Borealis [Statement p. 354]

1. We work in a non-relativistic regime, so that the velocity of the electron is a function of the energy given by $|v| = v = \sqrt{2E/m}$. Its component on the equatorial plane,

$$v_e = v \sin \alpha = \sqrt{2E \sin^2 \alpha / m},$$

is responsible for the cyclotron motion. The corresponding cyclotron angular frequency is given by the traditional formula $\omega = q_e B_e / m$; it is independent of the velocity. In contrast, the cyclotron radius R_c depends on the velocity, since we have $R_c = v_e / \omega$ which can be calculated as:

$$R_c = \frac{\sqrt{2mE \sin^2 \alpha}}{q_e B_e}.$$

Numerical application: $B_e = 9.185\ 10^{-6}$ T, $1/\omega = 6.199\ 10^{-7}$ s, $v_e = 1.027\ 10^8$ m/s, then $R_c = 63.63$ m or $R_c/R_T = 10^{-5}$. The cyclotron period is $\tau = 2\pi/\omega = 3.89$ μs.

2. The rotational kinetic energy is $E_c = \frac{1}{2} m v_e^2 = E \sin^2 \alpha$. This energy is related to the action p_ϕ by (see Problem 7.7): $E_c = -\omega p_\phi$. We deduce the value for the action:

$$p_\phi = -\frac{mE \sin^2 \alpha}{q_e B_e}.$$

3. The variation of the magnetic field along the field lines is very slow as compared to the cyclotron radius. The action is an adiabatic invariant, which remains constant all along the electron trajectory. At abscissa s, the rotational energy is $E_c(s) = -\omega(s) p_\phi = -p_\phi q_e B(s)/m$, or, using the value of p_ϕ obtained in the preceding question:

$$E_c(s) = E \sin^2 \alpha \frac{B(s)}{B_e}.$$

The charge is subjected only to the Lorentz force which is always perpendicular to the velocity; consequently there is no work performed along the trajectory and there is conservation of the total energy. As the particle moves towards the pole, the magnetic field increases and thus the rotational energy increases to the detriment of the translational energy along the field lines. There exists an abscissa s_m for which the rotational energy is equal to the total energy $E_c(s_m) = E$. At that point, the translational velocity vanishes and the electron turns back. The previous condition is equivalent to the following condition concerning the field

$$B(s_m) = \frac{B_e}{\sin^2 \alpha}.$$

4. The translational, or longitudinal, energy is the difference between the total energy and the rotational energy

$$E_l(s) = E - E_c(s) = E - E \sin^2 \alpha \frac{B(s)}{B_e} = E\left(1 - \frac{B(s)}{B(s_m)}\right).$$

The value of the translational velocity is thus

$$v_l(s) = \sqrt{\frac{2E_l(s)}{m}} = \sqrt{\frac{2E}{m}} \sqrt{1 - \frac{B(s)}{B(s_m)}}.$$

By definition, one also has $v_l(s) = ds/dt$ so that the time delay between the two cancellations of the longitudinal velocity, between $s = -s_m$ and $s = s_m$, is given by

$$\int dt = \int_{-s_m}^{s_m} \frac{ds}{v_l(s)}.$$

The period for a forward and back journey is obviously twice this time, which leads to the value:

$$T = \sqrt{\frac{2m}{E}} \int_{-s_m}^{s_m} \frac{ds}{\sqrt{1 - B(s)/B(s_m)}}.$$

5. The force (normal to the Earth) is given by the gradient of the energy $F = |\partial_R(\mu B(R))| = 0.31 \; 10^{-4} \mu R_T^3 \times 3/R^4 = 3\mu B(R)/R$. Furthemore, at the equator, $\mu B(R) = E_c = E \sin^2 \alpha$. Thus the absolute value of the force is

$$F(R) = \frac{3E \sin^2 \alpha}{R}$$

As we saw in Problem 7.7, this force leads to a drift of the cyclotron motion with a velocity

$$\boldsymbol{V}_d = \frac{\boldsymbol{F}}{q_e B_e} \times \boldsymbol{N},$$

where $\boldsymbol{N} = \boldsymbol{B}/B$ is the unit vector along the field line. $\boldsymbol{N}$ is oriented in the south-north direction; $\boldsymbol{F}$ being radial, the drift velocity $\boldsymbol{V}_d$ is directed in the east-west or west-east direction, depending upon the sign of q_e. Finally the modulus of this drift velocity is simply $F/(q_e B_e)$ that is

$$V_d(R) = \frac{3E \sin^2 \alpha}{q_e B_e R}.$$

We show in Fig. 7.7 a number of trajectories for an electron in the Earth's magnetic field.

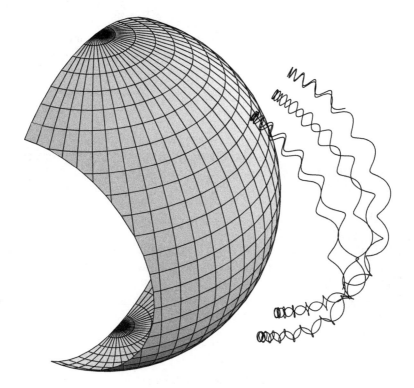

Fig. 7.7 Trajectories, in the Earth's magnetic field, calculated numerically using Hamilton's equations, for a 100 keV electron, passing at $1.6 R_T$ at the magnetic equator, and making an angle 45 deg with the magnetic field. To clarify the various motions, we have chosen a magnetism much weaker than the real one

7.9. Bead on a Rigid Wire: Hannay's Phase
[Statement and Figure p. 356]

1. The axis of rotation is perpendicular to the plane passing through O, along the Oz axis. The instantaneous rotation vector is thus $\boldsymbol{\omega} = \dot{\phi}\boldsymbol{k}$. The position M of the bead is specified by its curvilinear abscissa s referred to some arbitrary origin chosen on the wire.

   We denote by $\boldsymbol{OM} = \boldsymbol{r}(s)$ the radius vector of the bead. $\boldsymbol{t}(s)$ is the unit vector tangent to the wire at point M and $\boldsymbol{u}(s)$ the unit vector in the plane, perpendicular to $\boldsymbol{r}(s)$ at point M. The relative velocity of the bead is simply $\boldsymbol{v}_r = \dot{s}\boldsymbol{t}$ and the driving velocity is $\boldsymbol{v}_e = \boldsymbol{\omega} \times \boldsymbol{r}$. Consequently, the absolute velocity of the bead is: $\boldsymbol{v} = \boldsymbol{v}_e + \boldsymbol{v}_r = \dot{s}\boldsymbol{t} + \boldsymbol{\omega} \times \boldsymbol{r}$. Squaring, we find $v^2 = \dot{s}^2 + (\boldsymbol{\omega} \times \boldsymbol{r})^2 + 2\dot{s}\boldsymbol{t} \cdot (\boldsymbol{\omega} \times \boldsymbol{r})$. Moreover, one has $(\boldsymbol{\omega} \times \boldsymbol{r})^2 = \omega^2 r^2 - (\boldsymbol{\omega} \cdot \boldsymbol{r})^2 = \omega^2 r^2$, owing to the fact that $\boldsymbol{\omega}$ is perpendicular to the plane of the motion so that $\boldsymbol{\omega} \cdot \boldsymbol{r} = 0$. On the other hand $\boldsymbol{t} \cdot (\boldsymbol{\omega} \times \boldsymbol{r}) = \boldsymbol{\omega} \cdot (\boldsymbol{r} \times \boldsymbol{t}) = \omega|\boldsymbol{r} \times \boldsymbol{t}|$ since $\boldsymbol{\omega}$ and $\boldsymbol{r} \times \boldsymbol{t}$ are vectors which are parallel and with the same sense. With the relation $\omega = \dot{\phi}$, one deduces the kinetic energy $T = \frac{1}{2}mv^2$ in the final form:

$$T(s, \dot{s}) = \frac{1}{2}m\left[\dot{s}^2 + r(s)^2\dot{\phi}^2 + 2|\boldsymbol{r}(s) \times \boldsymbol{t}(s)|\dot{s}\dot{\phi}\right].$$

2. We neglect the square of the driving velocity $r(s)^2\dot{\phi}^2$, since it is assumed that the wire turns much more slowly than the bead on the wire $\dot{\phi} \ll \dot{s}$. Consequently, the kinetic energy can be approximated by

$$T(s, \dot{s}) = \frac{1}{2}m\left[\dot{s}^2 + 2|\boldsymbol{r}(s) \times \boldsymbol{t}(s)|\dot{s}\dot{\phi}\right].$$

   When the curvilinear abscissa is s, the radius vector is $\boldsymbol{r}(s)$ and when it is $s + ds$, the radius vector is $\boldsymbol{r}(s + ds) = \boldsymbol{r}(s) + d\boldsymbol{r}(s)$. The elementary area enclosed between the two vectors (see Fig. 7.8) is given by $dA = \frac{1}{2}|\boldsymbol{r}(s) \times \boldsymbol{r}(s+ds)| = \frac{1}{2}|\boldsymbol{r}(s) \times d\boldsymbol{r}(s)|$. Thus $2dA/ds = 2A'(s) = |\boldsymbol{r}(s) \times (d\boldsymbol{r}(s)/ds)|$. Moreover $d\boldsymbol{r}(s)/ds = \boldsymbol{t}(s)$ so that the kinetic energy can be recast in the simple form:

$$T(s, \dot{s}) = \frac{1}{2}m\left[\dot{s}^2 + 4A'(s)\dot{s}\dot{\phi}\right].$$

3. The bead slides without friction, so that the reaction force is normal to the wire and does not perform any work: the corresponding generalized force vanishes and Lagrange's equation reads: $d(\partial_{\dot{s}}T)/dt = \partial_s T$. Furthermore, $\partial_{\dot{s}}T = p = m(\dot{s} + 2A'(s)\dot{\phi})$ and $\partial_s T = 2mA''(s)\dot{s}\dot{\phi}$. In this case, Lagrange's equation is written as:

$$\ddot{s} = -2A'(s)\ddot{\phi}.$$

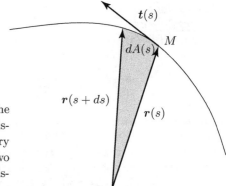

Fig. 7.8 – Radius vectors of the bead M on the wire for the abscissae s and $s+ds$. The elementary area swept out between these two points, denoted by $dA(s)$, is displayed in grey

If the wire rotates with constant angular velocity, then $\ddot\phi = 0$ so that $\ddot s = 0$, which means $\dot s = $ const. The bead retains a constant speed during its motion on the wire. On the other hand $\dot s = (p - 2mA'(s)\dot\phi)/m$ and the Hamiltonian $H = p\dot s - L$ is expressed as

$$H(s,p) = \frac{1}{2m}\left[p - 2mA(s)\dot\phi\right]^2.$$

4. In the case of a Hamiltonian which depends on time due to the presence of the function $\dot\phi$, the action must be calculated for a particular value of the time, which must be considered as frozen.

By definition, the action is $I = \frac{1}{2\pi}\oint p(s)\,ds$. Owing to the relation $p = m(\dot s + 2A'(s)\dot\phi)$ and, because, as a consequence of the previous remark, $\dot s$ and $\dot\phi$ must be frozen at values assumed at the given time, the action can be simplified to

$$I = \frac{m}{2\pi}\left[\dot s \oint ds + 2\dot\phi \oint A'(s)\,ds\right].$$

Moreover, $\oint ds = L$, the length of the wire, and $\oint A'(s)\,ds = S$, the area generated by the wire. Finally, the action is written in the form

$$I = \frac{m}{2\pi}\left[L\dot s + 2S\dot\phi\right].$$

In this expression, the velocity $\dot s$ can be deduced as a function of the action I; substituting this value in the Hamiltonian, this latter quantity can be expressed in terms of the action. Simple algebra leads to

$$H(I) = \frac{1}{2mL^2}\left[2\pi I - 2mS\dot\phi\right]^2.$$

The angular frequency follows as $\omega = \partial_I H = 2\pi[2\pi I - 2mS\dot\phi]/(mL^2)$. Lastly, the angle variable associated with the action is defined generally as $\alpha = \omega \int ds/\dot s$. An elementary calculation leads to:

$$\alpha(s) = \frac{2\pi}{L}\dot\phi.$$

5. One assumes that $\dot\phi$ varies adiabatically, which means that performing one complete revolution for the wire needs a much larger time than for one revolution of the bead. Under these conditions, we can use the fact that the action remains constant, that is $I(t) = I(0)$. Using the expression of the action given above and owing to the property $\dot\phi(0) = 0$, we finally obtain the velocity in the form:

$$\dot s = \dot s_0 - \frac{2S}{L}\dot\phi(t).$$

6. For a fixed wire; we saw that the bead moves at constant speed $\dot s = \dot s_0$ and the distance covered by the bead during the time T, the period of the wire, is $D_0 = \dot s_0 T$.

In the case of a wire rotating at angular velocity $\dot\phi(t)$, the distance covered in the same time is:

$$D = \int_0^T \dot s(t)\, dt = \int_0^T \left[\dot s_0 - \frac{2S}{L}\dot\phi(t)\right] dt = \dot s_0 T - \frac{2S}{L}(\phi(T) - \phi(0)).$$

Moreover, T is the time necessary for the wire to perform one revolution, so that $\phi(T) - \phi(0) = 2\pi$. We deduce the desired relation concerning the respective distances:

$$D - D_0 = -\frac{4\pi S}{L}.$$

7. The phase for the bead β can be defined from the property $\beta/(2\pi) = D/L$. Let us multiply the previous relation by $2\pi/L$ to find the Hannay phase $\beta_H = \beta - \beta_0$. Explicitly $\beta_H = -8\pi^2 S/L^2$.

It is remarkable that this value is independent of the velocity of the bead and the position of the rotation axis. It depends only on the geometrical form of the wire. The $-$ sign means that, compared to the motion on the fixed wire, if the bead and the wire rotate in the same sense, the bead is delayed in the case of the moving wire.

In the case of a circle $S = L^2/(4\pi)$, hence $\beta_H = -2\pi$.

Chapter 8
From Order to Chaos

Summary

In this chapter, we consider only two-dimensional autonomous systems acting in a bounded phase space

8.1. Introduction

In Chapter 6, we studied integrable systems. The description of their motion is very simple. With a correct choice of the coordinates in phase space (angle-action variables), each trajectory is a helix which rolls, at constant angular velocity, around a torus with constant actions. In Chapter 7, we emphasized the catastrophe of small divisors: the perturbation theory is singular if the ratio of certain frequencies is close to a rational number. In this chapter, we investigate this situation and show that, depending upon the initial conditions, the motion can be regular and predictable or, in complete contrast, chaotic and unpredictable[1].

To achieve this goal, it is not necessary to appeal to complicated theories; there exists a mechanical system which is very simple to solve. It leads to a numerical experiment within reach of any pocket calculator. It is the model of the **kicked rotor**: a pendulum for which the action of gravity is replaced by periodic kicks. We begin by giving its description and show that, by changing our point of view, it can be reduced to a two-dimensional autonomous system, which is integrable for a null perturbation (no kick), and quasi-integrable otherwise.

---

[1] In the short or long term.

8.2. The Model of the Kicked Rotor

The simplest mechanical system which exhibits a chaotic behaviour is the periodically kicked rotor. It is specified by an angle θ. When free, its Hamiltonian is given by $H_0(p) = p^2/(2I)$, in which I is the moment of inertia and the momentum $p = I\dot\theta$ is also the angular momentum. This Hamiltonian is a first integral. The phase space is a cylinder (θ is an angle which varies between 0 and 2π, whereas p can be any real number), and the trajectories are circles (transformed into straight lines if the cylinder is developed), since $p = \text{const}$.

Now we submit the rotor to a periodic impulse (period T and angular frequency $\omega = 2\pi/T$) which, without modification of the angle, instantaneously changes its momentum by a quantity proportional to $\sin\theta$. Without loss of generality, the period can be chosen as the unit time and the moment of inertia can be chosen as unity ($I = 1$, $T = 1$, $\omega = 2\pi$). The periodic impulse leads to an instantaneous variation of momentum $\Delta p = K \sin \theta_n$ for the nth kick (see Fig. 8.1.).

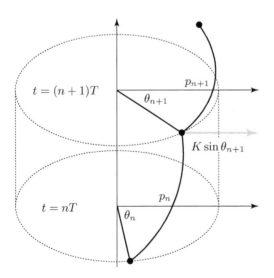

Fig. 8.1 Phase space for the kicked rotor. The cylinder represents the angle along the basic circle with the time along the generatrix. The grey arrow represents the periodic impulse which changes the angular velocity by a quantity proportional to the sine of the angle

This Hamiltonian system is one-dimensional but it is non-autonomous and non-integrable; however it is simple to solve because, between each of the kicks, the angular velocity remains constant and between the impulses n and

Summary

$n+1$ the momentum remains equal to p_n. It is convenient to investigate the angle-momentum series (θ_n, p_n) immediately after the nth kick.

It is easy to show the relationships between the points in phase space:

$$\begin{aligned} \theta_{n+1} &= \theta_n + p_n \\ p_{n+1} &= p_n + K \sin \theta_{n+1}. \end{aligned} \qquad (8.1)$$

It is difficult to imagine a simpler formulation for the equations of mechanics! They can be reduced to a mapping depending on a single parameter K which measures the intensity of the perturbation applied to the integrable system.

The mapping (8.1) is nevertheless very interesting pedagogically because, despite its simplicity, it contains intrinsically all the wealth of complicated dynamical systems. It is known as the **standard mapping**; it will be considered as a prototype for all further discussions.

Non-autonomous system/two-dimensional system

We already saw, in Chapter 4, that a non-autonomous system can be transformed into an autonomous one if we consider the time as an additional degree of freedom (see also Problem 4.9). Thus one considers the time t, no longer as the flow parameter for the Hamiltonian, but rather as a full coordinate with a corresponding conjugate momentum p_t. The system is governed by a generalized Hamiltonian $\tilde{H}(\theta, t, p, p_t) = H(\theta, p, t) + p_t$ which is independent of the new flow parameter and therefore autonomous[2].

Without the perturbation, the Hamiltonian is $\tilde{H} = p^2/2 + p_t$. It is independent of the two coordinates θ, t; in consequence, it is integrable, with the two first integrals p and p_t. The associated angle variables are respectively θ and $\alpha_t = 2\pi t/T = 2\pi t$. Actually, the Hamiltonian is periodic for both variables. The actions, the Hamiltonian and the angular frequencies are

$$\begin{aligned} I_\theta &= \frac{1}{2\pi} \oint p \, d\theta = p; \\ I_t &= \frac{1}{2\pi} \oint p_t \, d\theta = p_t \frac{T}{2\pi} = \frac{p_t}{2\pi}; \\ \tilde{H}(I_\theta, I_t) &= \frac{I_\theta^2}{2} + \omega I_t = \frac{I_\theta^2}{2} + 2\pi I_t; \\ \omega_\theta &= \dot{\theta} = I_\theta = p; \\ \omega_\alpha &= \dot{\alpha} = \omega = 2\pi. \end{aligned}$$

---

[2] We studied this point in Problem 4.9 (page 178). The Hamilton equation $t' = dt/d\tau = \partial_{p_t} \tilde{H} = 1$ shows that the time is indeed the flow parameter.

The trajectory rolls around a torus with a ratio for the frequencies equal to

$$\frac{p}{\omega} = \frac{p}{2\pi}.$$

8.3. Poincaré's Sections

For two degrees of freedom, the phase space possesses four dimensions. It is difficult to give a graphical representation of the trajectories[3]. Nevertheless, one can use the following trick: given a hyperplane (with 3 dimensions for the moment), one marks the intersection points between each trajectory and the hyperplane which is always crossed in the same sense. In general, one chooses a hyperplane which corresponds to a fixed value for one of the coordinates, for instance the second one $q_2 = const$. For the rotor, we will take $\alpha_t = 0$ (modulo 2π).

A trajectory is thus represented by a discrete set of points on the hyperplane. However, the system being autonomous, the second momentum p_2 is fixed and it is sufficient to specify, in phase space, the intersections concerning the first degree of freedom q_1 and p_1 ((θ, p) in our example). We are thus concerned with a two-dimensional plane. The intersections of the trajectory with this plane are referred to as a **Poincaré section**. The study of the system is reduced to the study of the mapping which transforms one intersection to the following one.

For a Hamiltonian system, Liouville's theorem stipulates that the Hamiltonian flow preserves volumes in phase space. In the case of a Poincaré section, a consequence of the conservation of the energy and Liouville's theorem is the conservation of the area in the Poincaré section for successive mappings. More precisely, let us imagine a set of trajectories lying within a closed contour in Poincaré's section. Each of them repeatedly intersects the section, probably at different times, but the contour of each set of these new intersections possesses the same area.

8.4. The Rotor for a Null Perturbation

In the particular case of a null perturbation $K = 0$, the Poincaré section is the representation of the simplest mapping $p_n = const = p$ and $\theta_{n+1} = \theta_n + p$.

---

[3] For a conservative system, the iso-energy surfaces are three-dimensional and one could plot the projection of these trajectories on a plane.

Summary

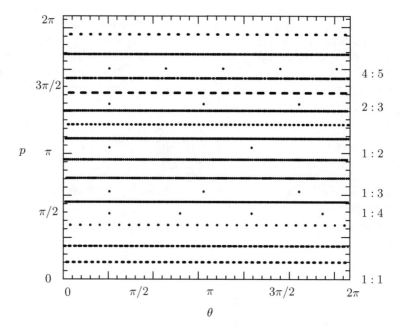

Fig. 8.2 Poincaré's section for the standard mapping with a null perturbation. For each initial condition, we represented 200 intersections. The initial conditions correspond to $\theta_0 = 1$ and different values of p_0. The values leading to a resonant torus $s:r$ are indicated at the right edge of the figure. Other values lead to non-resonant tori

As described in Chapter 6, the intersections of a trajectory (helix on a torus) with the hyperplane $\alpha_t = 0 \pmod{2\pi}$ take place on a circle (straight line $p = \text{const}$ on the developed cylinder) in phase space θ, p.

It is convenient to restrict Poincaré's section to a square for which $0 < \theta < 2\pi$ and $0 < p < 2\pi$ and to identify with a single point of this square all the points whose coordinates differ by a multiple of 2π. In other words, we work modulo 2π. The Poincaré section for a null perturbation is represented in Fig. 8.2. In this figure, one can observe the following features:

- If p is not a fraction of 2π, that is if the two frequencies are incommensurable, the intersections for a single trajectory fill in the straight line[4] $p = \text{const}$ in a dense way. We say that we are faced with **non-resonant tori**.

---

[4] Not necessarily in a uniform way. Let us imagine p very close to $2\pi/r$; we first see r segments filled with intersections and as many empty intervals. We must be patient to fill in completely the straight line.

- If $p = 2\pi/r$ (where r is an integer), which corresponds to commensurable frequencies, then, between each observation, the rotor turns by a fraction of a complete revolution. After r iterations of the mapping, we return to the original point. In this case, we speak of a **fixed point of order** r. A single trajectory appears in Poincaré's section as r aligned but distinct points.

- The same behaviour occurs if $p/(2\pi) = s/r$. In this case, the rotor performs s revolutions after r iterations and returns to the same point in phase space. In this case, we speak of a fixed point of order $s : r$. We speak also of **resonant tori** $s : r$ (or $r : s$), in the sense that the totality of the available phase space (the straight line $p = $ const) has not been explored with these initial conditions.

Consequently, depending upon the choice of initial conditions, Poincaré's section is formed by successive aligned points or by straight lines filled in a dense way.

8.5. Poincaré's Sections for the Kicked Rotor

What happens when the perturbation is switched on? The numerical experiment shows that, for K reasonably weak, the Poincaré section exhibits features analogous to those presented in Fig. 8.3 ($K = 0.75$). We recognize undulating continuous curves (KAM curves, from Kolmogoroff, Arnold, Moser), curves with elliptic shapes forming islands (stability islets) and zones with scattered points without any structure (chaotic zones).

For the value $K = 0.9716\ldots$, the undulating continuous curves in the interval $[0, 2\pi[$ disappear. The larger the perturbation, the smaller the elliptic islets and the larger the zones without structure. For a large perturbation this latter type of behaviour becomes dominant and fills in the whole phase space: a generalized chaos is installed.

What do we observe at first glance?

As we pointed out already, we observe essentially three types of distinct structures; let us comment more deeply on these.

1. **The KAM curves**

   For special initial conditions, one observes slightly undulating curves which are filled in a dense way and which cover the whole interval $[0, 2\pi[$ for θ. They are known as **KAM curves**. The amplitude of the undulation decreases as K decreases.

Summary

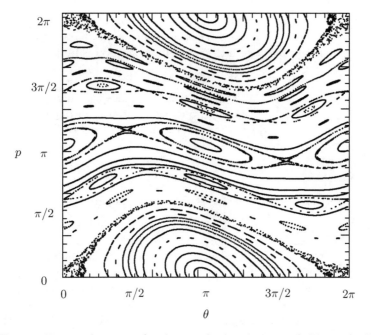

Fig. 8.3 Poincaré's section for the standard mapping with $K = 0.75$. The 36 initial conditions have been iterated 2000 times each

At the limit of a null perturbation, they become straight lines; they are the remnants of the non-resonant tori discussed previously. The most clearly visible are located on either side of the value $p = \pi$. For other initial conditions, the filling in can take more time.

We also observe between these KAM curves the presence of "islets" formed by closed curves with elliptic shapes and, on each side of these islets, zones where the points seem to be scattered at random. These structures are the remnants of the resonant tori described above.

Thus, from the straight lines of the Poincaré sections without perturbation, there remain regular curves separated by zones with more complicated structure.

All these observations are the conclusion of the famous **KAM theorem** which can be stated as follows.

Let us consider an integrable (non degenerate) system. If one adds a weak perturbation, most of the invariant non-resonant tori do not disappear. These tori, filled in a dense way by the trajectories (a single one is sufficient), form the majority in the sense that the measure of the complement of their union is small for a weak perturbation.

Another, more quantitative, point of view consists in claiming that a resonant torus $s:r$ has a destructive effect on the KAM curves in the region of phase space where the following condition is fulfilled:

$$|r\omega_1 - s\omega_2| < k(\omega_1, \omega_2)(r^2 + s^2)^{-3/4}. \tag{8.2}$$

ω_1 and ω_2 are the angular frequencies of resonance for a null perturbation and $k(\omega_1, \omega_2)$ is a finite positive quantity whose form depends on each particular case, and which diminishes as the perturbation itself weakens. Thus one understands that the tori with the smaller values of r and s are the most "devastating", since they correspond to the broadest regions of disappearance.

2. The stability islets and the stable fixed points

Consider two arbitrary KAM curves within which appear r sets of closed curves with elliptic shape whose range is less than 2π, which are called **stability islets**[5]. If one takes initial conditions located inside one of these small "ellipses", one remarks that the iterate appears inside one of the ellipses belonging to one of the other following $r - 1$ groups. For each iterate, one explores each of the other islets. With some skill, it is possible to start from a point and return practically to the same point after r iterations. The ideal point to which we exactly return after r iterations is known as a **fixed point of order** r. Obviously, any iterate of a fixed point of order r is again a fixed point of order r. In these islets, this point is stable in the sense that starting from a point in the neighborhood of the fixed point we never move far away from it after r iterations. Very often we simply refer to a fixed point as a fixed point of order 1.

In the Poincaré section represented in Fig. 8.3, one can notice the fixed point of order 1 $(\pi, 0)$ and the points $(0, \pi)$, (π, π) which are fixed points of order 2. Thus from the infinity of fixed points $p = \pi$ of the resonant torus $1:2$, the action of the perturbation preserved a single stable fixed point with its iterate.

3. Unstable fixed points and chaos

Inserted between the stability islets, there exist r regions where the curves seem to cross; indeed these are regions of concentrations of points which are not located on lines but appear to be scattered without any rule. These regions present a chaotic behaviour, for which the order of successive iterates seem completely unpredictable.

In these regions, there exist also fixed points of order r, but the behaviour of the iterates is completely different from that in the neighborhood of

---

[5] In the complement, we will show that they are still KAM curves.

stability islets. Starting from a point close to such a fixed point, the iterate appears in one of the $r-1$ other regions of the same nature, but progressively it can leave these zones and appear elsewhere in intermediate regions. The corresponding fixed point is called an **unstable fixed point of order** r. This is, in particular, the case of the fixed point of order 1 $(0,0)$, or the fixed points of order 2 $(\pi/2, \pi)$, $(3\pi/2, \pi)$ for which the clouds of chaotic points are clearly seen in Fig. 8.3.

Thus, from the infinity of fixed points $p = \pi$ of the resonant torus 1 : 2, the action of the perturbation preserves, in addition to the stable fixed point, an unstable fixed point and its iterate.

The observations discussed in the second and third remarks are the conclusions of the **Poincaré–Birkhoff theorem** which can be stated as follows:

If an autonomous system is subject to a weak perturbation (KAM theorem applicable), it possesses an infinity of fixed points. In the vicinity of a resonant torus of order r, there exists a multiple of $2r$ fixed points which are alternatively elliptic points (stable) and hyperbolic points (unstable), the iterate of a fixed point being a fixed point of the same nature.

8.6. How to Recognize Fixed Points

The point F_1 in Poincaré's section is a fixed point of order r if, after r iterations leading successively to the points F_2, F_3, ..., F_{r-1}, the next iterate F_r coincides with the original point F_1. As we pointed out already, it follows that any of the points F_2, F_3, ..., F_{r-1} is also a fixed point of same order. There exists no systematic method to determine the fixed points. Those corresponding to low orders are found intuitively; for higher order points there exist algorithms to recognize them.

Let us start from a point P_1 (q_1, p_1) and denote its nth iterate by $q_n = Q^{(n)}(q_1, p_1)$ and $p_n = P^{(n)}(q_1, p_1)$. If this point P_1 is a fixed point of order r, we have, by definition, $q_1 = Q^{(r)}(q_1, p_1)$ and $p_1 = P^{(r)}(q_1, p_1)$. What is the nature of this fixed point, stable (elliptic) or unstable (hyperbolic)?

Let us start from a point $(q_1 + \epsilon, p_1 + \nu)$ close to P_1; its rth iterate can be written to first order as $(q_1 + \epsilon', p_1 + \nu')$ with:

$$\begin{pmatrix} \epsilon' \\ \nu' \end{pmatrix} = \begin{pmatrix} \partial_q Q^{(r)} & \partial_p Q^{(r)} \\ \partial_q P^{(r)} & \partial_p P^{(r)} \end{pmatrix}\bigg|_{q_1, p_1} \begin{pmatrix} \epsilon \\ \nu \end{pmatrix}. \tag{8.3}$$

The following properties can be proved
- The matrix has a unit determinant (for a Hamiltonian system), a property we strongly advise to check.

- If the absolute value for the trace of this matrix is less than 2, the successive iterates (within this approximation) are regularly located on one of the r ellipses centered at the **stable elliptic fixed points**[6].
- If the absolute value for the trace of this matrix is greater than 2, the successive iterates are located on the branches of the r hyperbolae centered on the unstable fixed points. The iterates move away from the fixed point but draw nearer to one branch of each hyperbola, which is said to be a **stable or convergent direction**. This fixed point and its $r-1$ iterates are **unstable hyperbolic points**.
- For a stable fixed point, if one increases the intensity of the perturbation, the absolute value of the trace may increase to a value > 2. The fixed point becomes unstable and the instability is associated with the appearance of two unstable fixed points. This corresponds to the **bifurcation phenomenon** which precedes the onset of chaos.

8.7. Separatrices/Homocline Points/Chaos

For each hyperbolic[7] fixed point, there exists a curve such that the iterations of any point of this curve are again located on the curve (which is said to be invariant) and come as close as we wish to this fixed point. This curve is known as the **convergent separatrix**. There exists also another curve such that the preceding iterations come as close as we wish to this fixed point. This latter curve is known as the **divergent separatrix**. These two curves are invariant for the mapping: the iteration of any point located on the separatrix is located itself on the separatrix.

How to build these curves, at least approximatively? We start from a point close to a fixed point on the divergent branch; the iterations are close to the divergent separatrix. They are close because if we invert the mapping, one comes closer and then moves away[8] faster and faster. If we start from the convergent branch and we use the inverse mapping, we obtain points which are close to the convergent separatrix.

---

[6] They may fill in this ellipse in a dense way. They may also be located in the same regions, this property being the signature of existence of higher order fixed points. This may be the case in Fig. 8.3 in the vicinity of the fixed point $(\pi, 0)$. However, one must be very cautious: the filling in may proceed in a very inhomogeneous way.

[7] We consider a fixed point of order 1. The extension to any order does not raise any difficulty.

[8] Imagine you are going down a ridge, following it. Without visibility there is little hope that you achieve it.

Summary

It can be shown that:

- The divergent and convergent separatrices intersect at points that are images of one another. These special points are called **homocline points** (if the fixed point is of order 1) or **heterocline points** (otherwise).
- The segments of the separatrices between two intersections enclose surfaces whose areas are preserved under the mapping.
- From one iteration to the next, the intersection points come closer and closer; the conservation of area implies that the surface is stretched and deformed infinitively. This phenomenon is called a **mixture**. Two points which are initially very close move away from each other inexorably. It is close to this unstable fixed point that the **chaos phenomenon** appears.

8.8. Complements

Other KAM curves and self-similarity

The continuous curves which enclose the stable fixed points are also KAM curves: indeed, close to the fixed point, one notices that the mapping is that of a new integrable system whose ellipses are the intersections of the tori which are distorted by this approximation. One can consider this approximation as a perturbation. The same scenario can be invoked. Between these new KAM curves there appear new islets of stability and new unstable fixed points. Again, the mapping close to the fixed points inside the islets is approximately that of an integrable system but not exactly ... and everything repeats. This phenomenon is called **scale invariance** or **self-similarity**: whatever the zoom used to inspect a Poincaré section, and whatever the observed region, one finds the same miscellaneous structures.

Best rational approximation for an irrational number

A very important question arises: given two angular frequencies ω_1 and ω_2, how far are we from a destructive resonant torus, that is with the lowest values of r and s? The answer is given by the method of continuous fractions to determine the rational number r/s which is the "closest" to ω_2/ω_1. Let us assume that $\omega_2/\omega_1 = \pi$. One can write $\pi = 3 + 0.14159\ldots$ $= 3 + 1/(7 + 0.625\ldots) = 3 + 1/(7 + 1/(15 + 0.099659\ldots))= \ldots$.

The removal of the decimal part gives successive fractions 3, $\frac{22}{7}$, $\frac{333}{106}$ which are increasingly accurate rational approximations to the original irrational number. Calculating $|r\omega_1 - s\omega_2|/(r^2 + s^2)^{-3/4}$, it seems that the most dangerous torus is 333:106. However, one must remember (see Chapter 7) that the catastrophe of small denominators appears only if the Fourier analysis of the perturbation has harmonics with frequencies $333\omega_1$ and $106\omega_2$.

Problem Statements

8.1. Disappearance of Resonant Tori
[Solution p. 415] ★★

Understanding the KAM theorem using a simple example
We consider the first quadrant of integer numbers ($r > 0, s > 0$). To every rational number s/r, we associate a semi-straight line beginning at the origin and with a slope $\alpha_{s:r}$ given by $\tan(\alpha_{s:r}) = s/r$, as well as a disappearance region $2|\delta\alpha_{s:r}|$ for the resonant tori whose ratio of angular frequencies, $\omega_1/\omega_2 = \tan(\alpha_{s:r} + \delta\alpha_{s:r})$, fulfills the condition (8.2), $|r\omega_1 - s\omega_2| < K(\omega_1, \omega_2)\left(r^2 + s^2\right)^{-3/4}$, with the particular prescription $K(\omega_1, \omega_2) = k\left(\omega_1^2 + \omega_2^2\right)^{1/2}$, k being a positive constant.

1. Demonstrate the condition $|\delta\alpha_{s:r}| < kR^{-5/2}$ ($R = \left(r^2 + s^2\right)^{1/2}$) concerning the disappearance region.

   The aim of this problem is to show that the angular sum for the disappearance regions, denoted by Δ, is less than the total angular region of the first quadrant, namely $\pi/2$, when the perturbation is sufficiently small.

2. As a preliminary, show that the summation

$$\sum_{n=1}^{\infty} \frac{1}{n^{5/2}}$$

   is equal to a finite value S. Give an upper bound to S.

3. Give an upper bound to Δ.

4. Prove that, when k is less than a critical value k_c, one is sure that $\Delta < \pi/2$, that is that invariant tori still resist the perturbation. This is the essence of the KAM theorem.

8.2. Continuous Fractions or How to Play with Irrational Numbers
[Solution p. 417] ★★

A new point of view concerning an irrational number as a limit of rational numbers

Let α be an arbitrary real number. This number can be approximated by its decimal expansion, that is by the sequence of rational numbers s_0, $s_1/r_1,\ldots$, s_n/r_n where s_n is an integer and $r_n = 10^n$. An irrational number has a non-periodic infinite decimal expansion; a rational number has either a finite expansion, or a periodic infinite expansion. If one changes the numeral base, a rational number may have a finite expansion for one base and an infinite expansion for another one; for example $1/3 = 0.1$ with the base 3, but $1/3 = 0.3333\ldots$ with the base 10.

This dependency on the base implies that the above decimal approximation is not very suitable. There exists also another drawback using this type of expansion, a rather poor convergence: as $|\alpha - s_n/r_n| < 1/r_n$, one can say that the nth term of the sequence has an accuracy of order $1/r_n$. This conclusion is independent of the base.

One can thus wonder whether one can approximate the same number α by another sequence of rational numbers $\alpha_0, \alpha_1,\ldots, \alpha_n$ with an accuracy on $\alpha_n = s_n/r_n$ which would be better than $1/r_n$ and which is, in addition, independent of the base. The answer is yes and it appeals to the method of continuous fractions. In particular, a rational number always has a finite expansion.

Let $a_0, a_1,\ldots, a_n$ be a sequence of positive integers. The continuous fraction of order n, with initial term a_0 and terms $a_1,\ldots, a_n$ as successive denominators is defined as:

$$\alpha_n = a_0 + \cfrac{1}{a_1 + \cfrac{1}{a_2 + \cfrac{1}{\ldots a_{n-1} + \cfrac{1}{a_n}}}}.$$

This fraction will be referred to more simply as $\alpha_n = [a_0; a_1, a_2,\ldots, a_n]$. It is obviously a rational number written as $\alpha_n = s_n/r_n$, which is independent on the base.

1. In order to study infinite continuous fractions, it is convenient to obtain a recursion relation which allows an easy calculation of $\alpha_n = [a_0; a_1, a_2,\ldots, a_n] = s_n/r_n$ from the lower order terms $\alpha_i = [a_0; a_1, a_2,\ldots, a_i] = s_i/r_i$, with $i < n$.

   Show that the desired relation is

   $$s_n = a_n s_{n-1} + s_{n-2},$$
   $$r_n = a_n r_{n-1} + r_{n-2},$$

   with the initial conditions $s_0 = a_0$, $s_{-1} = 1$, $s_{-2} = 0$ and $r_0 = 1$, $r_{-1} = 0$, $r_{-2} = 1$.

2. Prove that
$$|\alpha_n - \alpha_{n+1}| = \frac{1}{r_n r_{n+1}}.$$
Consequently the accuracy of the continuous fraction method is $1/(r_n r_{n+1})$ instead of $1/r_n$ for an expansion on a particular base.

In the case where the sequence $a_1, \ldots, a_n$ is infinite, one speaks of an infinite continuous fraction. The limit of this fraction, from the building procedure itself, is an irrational number $\alpha = \lim(\alpha_n)$. If the sequence of denominators is periodic, Lagrange proved that α is a quadratic irrational number, that is a number of the form $\alpha = (a + b\sqrt{D})/c$, where $a, b \neq 0$, $c \neq 0$ and D are integers ($D > 1$ differing from an exact square).

3. As a very special sequence, let us consider the number $\alpha = [0; p, p, \ldots, p, \ldots]$. Calculate the value of α.

4. The irrational numbers which, in their expansion in terms of continuous fractions, converge the most rapidly toward a rational number are those which exhibit large values of a_n (after all, $a_n = \infty$ breaks the sequence and makes the number rational). What is the irrational number which is approximated the most poorly by a rational number?

8.3. Properties of the Phase Space of the Standard Mapping [Solution p. 418] ★

Simple properties of the standard mapping

1. Show that the Poincaré section for the standard mapping defined by (8.1) is symmetric with respect to the points $A = (\pi, 0)$ and $C = (\pi, \pi)$.

2. Show that the study of the standard mapping for $K < 0$ can be reduced to the case $K > 0$ simply by a translation of the angle origin.

8.4. Bifurcation of the Periodic Trajectory 1:1 for the Standard Mapping [Solution and Figure p. 419] ★ ★

Searching for fixed points of the standard mapping and appearance of a bifurcation

1. Find the fixed points of order 1 for the standard mapping defined by (8.1). Study their stability as a function of K. Give the value of K which corresponds to a transition between a stable and unstable trajectory.

Problem Statements

2. Show that the fixed points of order 2 are symmetric with respect to the point $(\pi, 0)$ and, as a consequence of the bifurcation, that they obey the equations

$$p_f = 2\pi - 2\theta_f \quad ; \quad p_f = \frac{1}{2}K\sin(\theta_f).$$

3. Study graphically the possible solutions of these equations. Check that a solution exists if the fixed point of order 1 is unstable.

4. We write the two fixed points of order 2 in the form $(\pi - \delta/2, \delta)$ and $(\pi + \delta/2, -\delta)$, δ being the root of a transcendental equation to be determined. We wish to study the stability of this trajectory. In order to do this, it is necessary to linearize the mapping in the vicinity of these two fixed points. Deduce that the stability condition is $|2 - K\cos(\delta/2)| < 2$, and that stability exists as long as $\cos(\delta/2) > 0$. What is the corresponding condition for K?

5. Let us go further. Determine the divergent and convergent directions for the fixed point of order 1, when it is unstable. Give their common direction at the limit of stability. Check that the fixed points of order 2 are placed along this direction.

8.5. Chaos–ergodicity: A Slight Difference [Solution p. 423] ⋆ ⋆

As we will see, ergodicity does not necessarily result from chaos and chaos is not always ergodic.

Imagine a configuration space for a system which is characterized by two angles α, β, i.e., exhibits the topology of a torus. Imagine now a regular motion on this torus with two angular frequencies $\omega_\alpha, \omega_\beta$, which means that the trajectory is given by $\alpha(t) = \alpha_0 + \omega_\alpha t$, $\beta(t) = \beta_0 + \omega_\beta t$. This motion is fully predictable and chaos is completely absent. The preceding properties are sufficient to completely determine the system. An illustration, which comes to mind, is that of an integrable two-dimensional system, the considered space being a torus defined by the two constant actions. The angle variables α, β and the angular frequencies $\omega_\alpha, \omega_\beta$ are determined from Hamilton's function, following the recipes given in Chapter 6. In this case, the space under consideration is a subset of phase space, for which the actions have fixed values resulting from the initial conditions.

Now, let us define the notion of ergodicity in a more mathematical way.

Let us consider an arbitrary function $F(\alpha, \beta)$, defined on the configuration space; it is obviously periodic with a period 2π both for α and β: $F(\alpha + 2i\pi, \beta + 2j\pi) = F(\alpha, \beta)$, i, j being integers.

The average value in phase space in defined by

$$\tilde{F} = \frac{1}{(2\pi)^2} \oint d\alpha \oint d\beta \, F(\alpha, \beta).$$

Now we decide to start from an arbitrary point α_0, β_0 and to follow the trajectory imposed by the dynamics (in the case of a Hamiltonian system, we would say that we follow the Hamiltonian flow), as was specified at the beginning of this problem. The F function now depends on time. One defines the time average of this function along the given trajectory by:

$$\bar{F}(\alpha_0, \beta_0) = \lim \frac{1}{T} \int_0^T F(\alpha(t), \beta(t)) \, dt, \quad T \to \infty.$$

This system is said to be ergodic in this space if, on the one hand, $\bar{F}(\alpha_0, \beta_0) = \bar{F}$ is independent of the starting point, and on the other hand the two types of average coincide: $\tilde{F} = \bar{F}$. This notion is very important in statistical physics. Ergodicity corresponds to the fact that it is equivalent to realize an average over a large number of statistical samples at a given time or over a single sample but followed all along its temporal evolution.

1. Show that the system is ergodic if the two frequencies are not commensurable.

   Hint: Perform a Fourier analysis of the function $F(\alpha, \beta)$.

2. Let us consider an ergodic system, characterized by the ergodicity property demonstrated in the preceding question. Consider an arbitrary trajectory passing initially through the point (α_0, β_0); it evolves in time exploring the configuration space. The purpose of this question is to prove that it performs the exploration in a dense way. In order to do this, let us choose a portion D of this space, wherever located, as small as we wish but with a non-vanishing measure (its area is not null: $S(D) = \int \int_D d\alpha \, d\beta \neq 0$). The principle of the demonstration is based on an ad absurdum reasoning.

   The demonstration of the previous question is valid whatever the function $F(\alpha, \beta)$, as long as it is periodic in configuration space. Let us choose a particular function $F(\alpha, \beta)$ which is equal to 1 everywhere inside the D region, and 0 everywhere outside. Assume that the trajectory does not explore this region and show that this is incompatible with the conclusions of the previous question.

Result: The trajectory of an ergodic system explores the configuration space in a dense way.

8.6. Acceleration Modes: A Curiosity of the Standard Mapping [Solution p. 425] ⋆ ⋆

The role of the momentum in Poincaré's section

Let us come back to the inexhaustible standard mapping, as defined by (8.1). We emphasized that the angle periodicity was natural; in contrast the momentum periodicity is just a practical convenience for a graphical representation. In reality, the momentum is by no mean restricted. To be convinced, we will show that there exist fixed points of order 1 in Poincaré's section, which correspond to an increase by 2π of the momentum for each impulse. We call such a trajectory an acceleration mode.

1. Such points must satisfy the condition

$$\begin{aligned} \theta_1 - \theta_0 &= 2n\pi \\ p_1 - p_0 &= 2\pi. \end{aligned}$$

It is always possible to choose initial conditions as the usual ones: $0 \leq \theta_0 < 2\pi$; $0 \leq p_0 < 2\pi$. Show that the angle (modulo 2π) for the fixed point θ_f is determined by the equation $K \sin(\theta_f) = 2\pi$. What is the condition on K for the existence of a solution? How many solutions exist? Give the expression of the momentum at each impulse. Check that it corresponds to an acceleration mode.

2. One can be interested in the stability of the solution in phase space, that is by the question of stability for the acceleration mode. Linearize the mapping in the vicinity of the fixed point and give the trace of the corresponding matrix.

3. Show that one trajectory is always unstable. Under which condition is the other one stable?

8.7. Demonstration of a Kicked Rotor?
[Solution and Figure p. 427] ⋆ ⋆ ⋆

A physical situation which simulates the standard application

To simulate a motion described by the standard mapping defined by (8.1), it is enough to impose a well chosen horizontal motion on the rotation axis of a pendulum.

A simple pendulum, of length l, of mass m, has a **vertical axis of rotation** such that **gravity does not play any role**. This axis is driven by a horizontal motion along the Ox axis, with a temporal law $a(t)$. As a generalized coordinate, we choose the angle θ between the direction of the pendulum and the Ox axis.

1. Give the traditional equation of motion for an ordinary pendulum in a constant gravitational field g and the corresponding momentum. To take into account the motion of the suspension point, it is enough to work in the accelerated frame where the pendulum is at rest. This corresponds to the change $g \to g - \ddot{a}$. Use now the hypothesis that gravity is ineffective ($g = 0$) to give the new equation of motion.

2. To recover the standard mapping (8.1), one imagines a jerky motion of the axis such that the velocity remains constant during the interval of time T and increases suddenly by v after each interval, the angle being unchanged at the moment of the jerk. The velocity $\dot{a}(t)$ thus evolves as a staircase of steps v, the jumps occurring at times $t_n = nT$ (n integer).

   Solve Lagrange's equation during the intervals at constant velocity and integrate this equation during the interval of infinitesimal time when the velocity increases by v. Deduce the mapping connecting the angle $\theta_n = \theta(t_n)$ and the momentum $p_n = p(t_n + \varepsilon)$, just after the nth kick, to the same quantities relative to the $(n+1)$th kick. Performing a suitable change of the momentum variable, show that this mapping is nothing more than the standard mapping. What is the value of the K parameter for this mapping?

3. One considers now a temporal law $a(t)$ with a sawtooth shape of period T, for which the velocity v is inverted periodically with period $T/2$ (the curve $\dot{a}(t)$ has a square wave shape). Let us define the angle $\theta_n = \theta(t_n)$ and the momentum $p_n = p(t_n + \varepsilon)$ just after the velocity changes to $-v$, and the same quantities (θ_{n+1}, p_{n+1}) obtained one period later. It is useful to introduce the intermediate quantities $\theta_i = \theta((n+1/2)T)$, $p_i = p((n+1/2)T + \varepsilon)$, obtained just after the inversion of the velocity. One can study once more the preceding question in two steps (the relation between (θ_i, p_i) and (θ_n, p_n) in a first step, the relation between (θ_{n+1}, p_{n+1}) and (θ_i, p_i) in a second step), taking great care of the sign of the velocity at each step. You should perform a change of the momentum variable in the same spirit as in the previous case, and introduce the dimensionless parameter $K = vT/l$. Give the expression of the mapping connecting (θ_{n+1}, p_{n+1}) to (θ_n, p_n) which is known as the sawtooth mapping.

4. Plot numerically the Poincaré section corresponding to the sawtooth mapping. Determine all the fixed points of order 1 (when K is not too

large, say $K < 2\pi$) and study their stability. Prove that four of them can be calculated precisely. Two others result from a transcendental equation that should be solved only in the limit of a weak value for K. Check your calculations using the picture of Poincaré's section. Notice the analogy with the fixed points of order 2 for the standard mapping.

8.8. Anosov's Mapping (or Arnold's Cat)
[Solution and Figure p. 432] ★ ★ ★

The Fibonacci sequence as an ingredient of a physical problem

A mathematical preamble

Starting from the initial values $a_1 = 1$, $a_2 = 1$, the sequence obtained by the recursion relation $a_n = a_{n-1} + a_{n-2}$ ($a_3 = 2$, $a_4 = 3$, $a_5 = 5$, ...) is known as the Fibonacci sequence. Fibonacci used it to study the evolution of rabbit populations and it is very famous. A striking property, that may be familiar, is that the ratio between two consecutive terms is equal to the golden ratio $\Phi = (1 + \sqrt{5})/2$. This relationship with the golden ratio makes it of rather common use in physical mappings. This problem is a clear illustration.

1. Demonstrate the relation $a_{n+1}a_{n-1} - a_n^2 = a_{n-1}^2 - a_n a_{n-2}$. By a recursion argument, deduce that $a_{2r+1}a_{2r-1} - a_{2r}^2 = 1$.

Physical study

We consider a very schematic Hamiltonian (it does not even have the dimension of an energy!): $H_0 = p^2/2$, where p is the momentum conjugate to the angle θ. But now the periodic kick, with a unit period, gives to the system an instantaneous variation of the momentum equal to $\Delta p = \theta$ (at the moment of the kick) modulo 2π, without changing the angle.

Be careful! in the case concerned by this mapping, the perturbation is so large that we never meet the conditions required by the KAM theorem.

2. Write down Hamilton's equations between two kicks. Deduce the mapping M connecting the angle and the momentum after two successive kicks. Check that this mapping M is linear and preserves the area in phase space.

This mapping was proposed by Anosov. It became famous after Arnold showed the successive iterations for the picture of a cat in phase space. In Fig. 8.4, we present the first two iterations of the mapping.

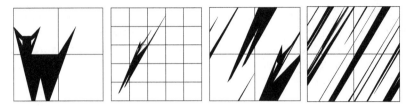

Fig. 8.4 In the first figure is the original form of the cat. The second picture shows the result of the mapping, without taking into account the topological properties of the torus. The third picture represents the real iterate of the mapping, which takes into account its properties modulo 2π. The last figure shows the result after a second iteration

Following the same arguments as those of the standard mapping, one assumes that the topology of phase space is a torus. The coordinates (θ, p) of a point in phase space are defined modulo 2π and the Poincaré section is restricted to the reference square.

3. We wish to visualize the effect of this mapping in a geometric way. Show that there exist two invariant straight lines for this mapping, that is two eigenvectors which are parallel to the vectors $(-\Phi, 1)$ and $(\Phi^{-1}, 1)$. Demonstrate that these vectors are orthogonal. After iteration, one of these eigenvectors is contracted by the ratio $\lambda_1 = 1 - \Phi^{-1}$, whereas the other is dilated by the ratio $\lambda_2 = 1 + \Phi$.

    What happens to the image of an isosceles rectangle triangle built on these two proper directions, after successive transformations? What happens to the image of a straight line? of a rectangle? of a circle?

4. Show that the matrix of the mapping can be written in the form

$$M = U^{-1} \begin{pmatrix} \lambda_1 & 0 \\ 0 & \lambda_2 \end{pmatrix} U$$

    and give the expressions of U and U^{-1} as a function of the golden ratio. Give the expression of the rth power M^r of the matrix of the mapping. Do not try to evaluate or simplify the matrix elements given in terms of the golden ratio and its inverse.

5. Relying on the first question and on recursion arguments, calculate M^r in terms of the elements of the Fibonacci sequence. Check that its determinant is unity.

6. This question is related generally to the fixed points. It will be more convenient to reason with a 2π unit and to define the reduced quantities $x = \theta/2\pi$ and $y = p/2\pi$. A fixed point of order r is defined so as to be invariant modulo 2π under the M^r mapping, so that the image after the M^r mapping of a point (x, y) is written $(x + k, y + l)$ (k and l being

positive or negative integers). Calculate these fixed points as a function of k and l.

7. In fact, all these fixed points are redundant because some of them are obtained from the others modulo 2π. To find all the non-redundant fixed points, one must restrict the values for the set of numbers (k, l) to those which lead to solutions lying in the reference square $0 \leq x < 1$; $0 \leq y < 1$. Prove that this condition restricts the region of the plane (k, l) to a parallelogram, the equation of which is to be determined.

8. Calculating the area of this parallelogram and the area of an elementary cell corresponding to one set (k, l), estimate the number N_r of fixed points of order r. You will remark anyhow that this number includes all the fixed points of the order of a divisor of r.

9. Show that all the fixed points are unstable and that none of them are compatible with a constant momentum, except the trivial fixed point of order 1: (0,0).

10. Just for fun, calculate all the fixed points of order 2. Is the estimation N_2 for this number, as determined in a previous question, reasonable?

8.9. Fermi's Accelerator
[Solution and Figure p. 438]   ★ ★ ★

A simple model explaining the acceleration of particles

To explain the presence of cosmic particles moving at very large speed, E. Fermi proposed the following mechanism. The big stellar objects are sources of very intense magnetic fields, the more intense the closer to the object. As we saw in the preceding chapter, gradients of magnetic fields act as magnetic mirrors for charged particles. If the object is itself moving at the moment of the reflection of the particle on the magnetic mirror, the speed of the particle may increase because the object carries the magnetic field along with it. At the next reflection on the mirror, the particle speed may increase again and it is not absurd to think that on average the speed increases with time. Fermi's accelerator is an elementary schematic model for such a process.

In a one-dimensional space, a particle of mass m bounces elastically between two walls; the left wall is fixed whereas the right wall is moving following the law $L(t)$. Let us denote by $L'(t)$ and $L''(t)$ respectively, the first and second derivatives of the function $L(t)$. All the velocities we speak about here are expressed as their moduli (speed), but one must be very cautious because there exists a change of sign at the moment of the reflection.

1. At time t_n, the particle strikes the right wall and bounces back from the moving wall with a speed v_n **relative to the moving wall**. It strikes the fixed left wall at time t_{n+1} and bounces back after the impact at a speed v_{n+1} **relative to the fixed wall**.

   Write down the laws which give (t_{n+1}, v_{n+1}) as function of (t_n, v_n). The phase space is defined by the plane (t, E), where E stands for the energy of the particle in the **frame attached to the wall which the particle has just struck**; show that the previous mapping preserves the area in such a phase space.

2. The definition of the motion is now changed and we denote by t_n the moment when the particle bounces on the fixed left wall and moves off with the speed v_n relative to the fixed wall, whereas t_{n+1} denotes the moment when it strikes the moving right wall and bounces back with a speed v_{n+1} relative to the moving wall. Same question as before.

3. The particle now performs a forward and backward journey bouncing alternatively between two **moving** walls with arbitrary evolution laws. Let us denote by t_n and t_{n+1} the moments of two consecutive impacts and E_n and E_{n+1} the energies of the particle, after the impact, in the frame of the wall which it has just left. Show that this mapping preserves the area in this phase space, a property which corresponds to a unit Jacobian:

$$\frac{D(t_{n+1}, E_{n+1})}{D(t_n, E_n)} = 1.$$

   It is astute to consider an intermediate situation where a fictitious fixed wall is introduced in between the two moving walls, and where the properties developed previously are applied successively.

   It may happens that $E_{n+1} > E_n$ at each iteration. This effect is known as the Fermi accelerator.

4. We consider now a situation for which the left wall remains fixed and the right wall is submitted to a periodic motion of period T. The mapping and the corresponding quantities refer to the impacts on the **fixed wall**; the speeds involved are therefore the absolute speeds. The situation is depicted in Fig. 8.5. The previous questions show that the mapping preserves the area in phase space:

$$\frac{D(t_{n+1}, E_{n+1})}{D(t_n, E_n)} = 1.$$

   We look for a fixed point of order 1, that is for a trajectory that repeats itself indefinitely. In this case, one must have $v_{n+1} = v_n$ with $t_{n+1} - t_n$, a multiple of the period T.

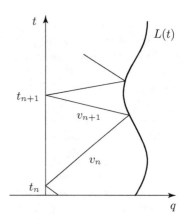

Fig. 8.5 Consecutive rebounds between a fixed wall on the left and an oscillating wall on the right. The time flows upward

One denotes by t_c the moment when the particle strikes the moving wall: $t_{n+1} > t_c > t_n$ and $L_c = L(t_c)$. Show that a necessary condition for the mapping to have a fixed point of order 1 is that the velocity of the wall at the moment of impact vanishes: $L'(t_c) = L'_c = 0$ and that $t_c = (t_{n+1} + t_n)/2$. Linearize the mapping in the vicinity of the fixed point. Show that the trajectory is stable as long as the condition $-4E/m < L_c L''_c < 0$ is satisfied.

5. Each iteration of this mapping requires the calculation of the intersecting points represented in Fig. 8.5. This calculation is not trivial and investigators usually prefer to work with a simplified version, known as Ulam's approximation. The simplification arises from the condition that the oscillation amplitude of the wall is small (with respect to the distance covered between two impacts), so that the length L remains approximately constant.

Practically without changing its position, the moving wall nevertheless induces a modification to the momentum of the particle. It is recommended to work in the spirit of the first question, that is considering the times t_n and t_{n+1} when the particle strikes the moving wall. Within this approximation, the conservation of area in phase space (t, E) no longer holds. Show that we have rather conservation of the area in the phase space (t, v). Notice the analogy with the standard mapping.

8.10. Damped Pendulum and Standard Mapping [Solution and Figure p. 443] ★ ★

Simple study for a non-Hamiltonian system

Let consider a simple pendulum with mass m, length l, whose position is specified by the angle θ with respect to the downward vertical. This pendulum is subject, in addition to a constant gravitational field g directed downward, to friction forces applied on the horizontal rotation axis and whose effects are summarized by a restoring torque $-k\dot{\theta}$.

1. Specify all the forces acting on the system. Do they arise from a potential (even a generalized one)? Is this a Lagrangian system? Write down the differential equation resulting from Lagrange's equation on the angle. Check your result with Newton's law.

2. Defining a dimensionless "time" $\tau = \omega t$, with the proper angular frequency of the pendulum $\omega = \sqrt{g/l}$, give the differential equation for the angle and show that the only parameter for this problem is the dimensionless quantity $\gamma = k/(2m\omega l^2)$. From here on, we assume $\gamma < 1$.

3. What are the equilibrium positions? Linearize the equation around each of these positions. Deduce the temporal equation for the angle $\theta(\tau)$. The momentum is defined as $p = \partial_\tau \theta$; write down the coupled equations for the angle and the momentum. Is there conservation of the area in phase space? With the help of a pocket calculator, plot the curves parametrized by the time in phase space.

4. One is interested in the angles of the pendulum for times separated by $\delta t = \delta\tau/\omega$. Lagrange's equation gives a relation between three successive angles $\theta_{n-1}, \theta_n, \theta_{n+1}$ because a reasonable approximation for the first derivative is simply $\partial_\tau \theta \approx (\theta_{n+1} - \theta_{n-1})/2\delta\tau$ and for the second derivative $\partial_\tau^2 \theta \approx (\theta_{n+1} + \theta_{n-1} - 2\theta_n)/\delta\tau^2$.

Let us set $\qquad p_n = \dfrac{\theta_{n+1} - \theta_n}{\delta\tau}; \qquad p_{n-1} = \dfrac{\theta_n - \theta_{n-1}}{\delta\tau}.$

Show that the solution of these coupled equations in the plane (θ_n, p_n) reduces to the mapping:

$$\theta_{n+1} = \theta_n + \delta\tau p_n$$
$$p_{n+1}(1 + \gamma\,\delta\tau) = p_n(1 - \gamma\,\delta\tau) - \delta\tau \sin\theta_{n+1}.$$

Calculate the Jacobian J of the mapping. Deduce the condition implying conservation of the area.

By a new definition of the momentum, show that for $\gamma = 0$, one recovers the standard mapping defined by (8.1). What is the expression of the parameter K?

5. In the case $\gamma \neq 0$, determine the eigenvalues of the matrix corresponding to the mapping linearized in the vicinity of the stable equilibrium point.

Show that, by taking two appropriate combinations of the angle and the momentum, $X(\theta,p)$ and $Y(\theta,p)$, the mapping decreases the modulus of X and Y by the same quantity to be determined. Deduce that successive mappings lead to points closer and closer to the fixed point which roll themselves up. The corresponding fixed point is known as a stable spiral point or focus.

6. Using a personal computer, plot Poincaré's sections for different initial conditions and different friction coefficients γ for the standard mapping.

8.11. Stability of Periodic Orbits on a Billiard Table [Solution p. 447] ★ ★ ★

For addicts of the game of billiards

Gaming room

Physicists working on "dynamical systems", although serious people, are very much interested in the game of billiards. The billiard cush is a plane closed curve. A billiard ball, subject to ideal reflections, is a two-dimensional autonomous system defined by its position and momentum in two directions. In contrast to standard dynamical systems, this model has the advantage of not requiring the resolution of Hamilton's equations: the motion is uniform between two rebounds. At the impact on the cush, the tangential component of the ball's velocity remains unchanged while its normal component is reversed.

Moreover, this system exhibits behaviours ranging from integrability to chaos depending upon the geometry of the table: square, ellipse, heart shape, stadium shape, triangle, ...

Concerning this latter point, the knowledge of stability for periodic orbits is very important. The aim of this problem is to establish the stability conditions for a periodic orbit of order 1, that is for a perpetual to and fro journey between two different points of the cush.

Good coordinates for the billiard game

The equivalent of the Poincaré section for the billiard table is a plane where, at each rebound, we specify the curvilinear coordinate of the nth impact, s_n, along the cush and the projection of the tangential velocity of the ball p_n at the same point. The modulus of the velocity being conserved, one can assign to it a unit value so that the projection is just the sine of the angle of incidence. The sign convention is such that $p_n > 0$ if the velocity is directed along the sense of increasing curvilinear abscissa.

The interesting problem is the study of the mapping $s_n = S(s_{n-1}, p_{n-1})$, $p_n = P(s_{n-1}, p_{n-1})$; this mapping depends of course on the shape of the billiard table. For problems concerning stability, it is sufficient to study the derivative matrix M defined as

$$\begin{pmatrix} ds_n \\ dp_n \end{pmatrix} = M \begin{pmatrix} ds_{n-1} \\ dp_{n-1} \end{pmatrix}$$

in the vicinity of a periodic trajectory.

1. Determine the derivative matrix of the mapping.

   One needs to build effectively the matrix M. The expression involves the radii of curvature for each impact R_n, R_{n-1}. A radius of curvature is taken as negative if the rebound occurs on a convex portion of the cush. Indeed the normal to the surface changes its orientation. These curvature radii are compared with the distance L covered between two impacts.

   Check that the mapping preserves the area, justifying the choice of the coordinates.

   Hints: A possible line of attack is to consider small arbitrary displacements of the impact points $t_{n-1} ds_{n-1}$ and $t_n ds_n$, where t are unit tangent vectors, oriented in the direction of increasing curvilinear abscissa (see Fig. 8.6).

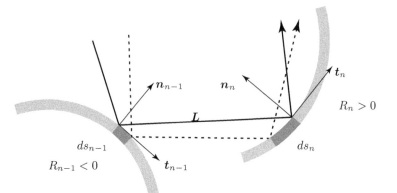

Fig. 8.6 Trajectory of the billiard ball between two consecutive impacts on the cush. The distance between these impacts is denoted L. The tangent and normal vectors to the cush are also represented for the two impacts. The dotted line represents a neighbouring trajectory

One deduces the variation of the radius vector $\boldsymbol{L}$ between the impact $n-1$ and the impact n and one calculates the variations of the coordinates $p_{n-1} = \boldsymbol{t}_{n-1} \cdot \boldsymbol{L}/L$ and $p_n = \boldsymbol{t}_n \cdot \boldsymbol{L}/L$. In the derivation, one must not forget the rotation in the direction of the tangent as we proceed along the billiard cush: $d\boldsymbol{t}_n = ds_n \boldsymbol{n}_n / R_n$ where $\boldsymbol{n}_n$ is the normal to the curve oriented inward. In addition to the tangential projections of the velocity **after** the impacts, it is interesting to introduce the related quantities q_{n-1} and q_n, the normal projections of the velocity **after** the impacts. One has the relationship $q_{n-1}^2 = 1 - p_{n-1}^2$, $q_n^2 = 1 - p_n^2$ but we must take care to note that $q_{n-1} = \boldsymbol{n}_{n-1} \cdot \boldsymbol{L}/L$ whereas $q_n = -\boldsymbol{n}_n \cdot \boldsymbol{L}/L$. The use of the Maple or Mathematica software packages may be of some help.

2. Show that the periodic trajectories of order 1 between the point A_1 (radius of curvature R_1) and the point A_2 (radius of curvature R_2), with $A_1 A_2 = L$, are stable if the following conditions are satisfied:

$$0 \leq (L/R_1 - 1)(L/R_2 - 1) \leq 1.$$

Hint: The trace of the mapping for a to and fro journey must be of modulus less than 2. A simplification results from the fact that the incidence angles vanishes. Do not forget that the linearized matrix for a forward and backward trajectory is the product of two matrices M, corresponding to the forward and backward segments.

3. Study the stability of the trajectory for the various geometries of the rebound presented in Fig. 8.7.

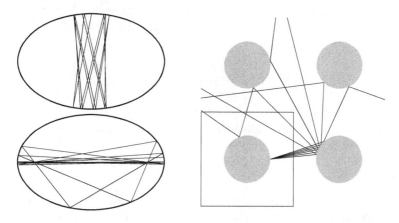

Fig. 8.7 Left upper part: trajectory on an elliptic billiard table close to the minor axis. Left lower part: trajectory on an elliptic billiard table close to the major axis. On the right part: Sinaï billiard table. A collection of impenetrable circles centered at the apexes of a squared net is equivalent to a table made of a single circle inside an impenetrable square as indicated in the figure

8.12. Lagrangian Points: Jupiter's Greeks and Trojans [solution p. 450] ★ ★ ★

This problem is a first approach to a very wide subject known as the restricted three-body problem

This is the non-integrable problem that was studied by Poincaré. A first numerical study corresponds to the famous work of the French astronomer F. Hénon.

Consider two celestial objects of masses m_2 and m_3 which, owing to their mutual attraction, rotate around each other without modifying their separation, D, which means that they move around circular orbits.

To simplify the notation, we set the total mass $M = m_2 + m_3$, and the mass asymmetry $\mu = (m_3 - m_2)/M$, so that we can write

$$m_2 = \frac{(1-\mu)M}{2} \quad \text{and} \quad m_3 = \frac{(1+\mu)M}{2}.$$

1. Give the rotational angular frequency ω of the two objects and their distances d_2 and d_3 to their center of mass denoted by O. To obtain this relation rapidly, one can equate the centripetal force and the force of attraction.

2. Consider also an asteroid with a mass m which is sufficiently weak with respect to the mass of the two objects that it does not influence their motion. The question we wish to answer is the following: what is the motion of the asteroid under the combined action of the objects, with the further condition that it remains always in the plane of the two circulating objects.

In the Galilean frame, the interaction potential of the asteroid with the two objects depends on time. It is astute to work in the frame which rotates with angular frequency ω around the Oz axis, perpendicular to the plane of the motion of the two heavy objects. Indeed, in this frame these two objects are at rest and the interaction potential is time-independent. We will seek the stationary configurations. In this particular frame, one chooses for the origin the point O and for the Ox axis the straight line which connects the two objects in the sense from object 2 to object 3.

In this frame the asteroid coordinates are x, y and their conjugate momenta are p_x, p_y. The gravitational interaction potential acting on the asteroid is denoted by $V(x, y)$.

Problem Statements

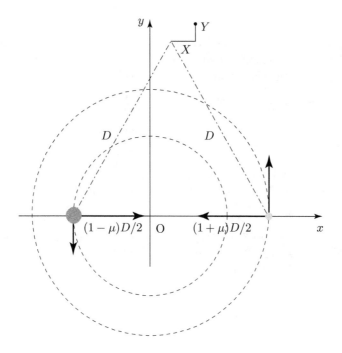

Fig. 8.8 In the frame where the two big objects are at rest, we study the motion of a third light mass. The represented distances correspond to the equilibrium configuration

Using these coordinates, give the expression of the Hamiltonian. You will employ the result of Problem 4.3, page 172: $H = H_0 - \omega L_z$ where H_0 is the Hamilton function as if the frame were Galilean and L_z the transverse component of the angular momentum. It is convenient to introduce the frequency ω instead of the gravitational constant, as was derived in the first question.

3. What is the constant of the motion E (known as the Jacobi constant)? Is the system integrable a priori? Write down the Hamilton equations.

4. We wish to determine whether or not there exist fixed configurations for the three objects. They correspond to extrema of an effective potential which contains a centrifugal term $V_e(x, y) = V(x, y) - m\omega^2(x^2 + y^2)/2$. Express this potential as a function of the distances r_2 and r_3 of the asteroid to the two other objects $V_e(r_2, r_3)$. Deduce that there exists only one extremum for $r_2 = r_3 = D$. In this equilibrium configuration, the three objects are located at the apexes of an equilateral triangle. The two possible positions for the asteroid are called Lagrangian points. Give the values of the corresponding variables $(x_L, y_L, p_{x_L}, p_{y_L})$. What is the value of the constant of the motion at these points?

5. We wish to study the stability around these Lagrangian points. To do this, one considers a small displacement X, Y of the asteroid away from the Lagrangian points (see Fig. 8.8). Show that, for small values of X, Y, we have the following expansion:

$$((X+a)^2 + (Y+b)^2)^{-1/2} = \frac{1}{L} - \frac{aX + bY}{L^3} + \frac{(2a^2 - b^2)X^2 + 6abXY + (2b^2 - a^2)Y^2}{2L^5},$$

with a and b being two arbitrary parameters and $L = \sqrt{a^2 + b^2}$. Deduce the expression of the potential to second order in X, Y.

We set $p_x = p_{x_L} + U$; $p_y = p_{y_L} + V$.

6. Give the expression of the Hamiltonian in terms of the coordinates X, Y, U, V. To be consistent, we must be sure that the transformation $(x, y, p_x, p_y) \to (X, Y, U, V)$ is canonical.

Demonstrate that, with an appropriate definition of the λ parameter, the resulting Hamilton's equations read

$$\dot{X} = \omega Y + \frac{U}{m}; \qquad \dot{Y} = -\omega X + \frac{V}{m};$$

$$\dot{U} = \omega V - \frac{m\omega^2}{8}(2X - \lambda Y); \qquad \dot{V} = -\omega U + \frac{m\omega^2}{8}(\lambda X + 10Y).$$

Notice that they are first order coupled differential equations. The solutions are combinations of proper modes[9] with angular frequency Ω.

Show that the proper frequencies result from the four solutions of the polynomial equation $r^4 - r^2 + 27/16 - \lambda^2/64 = 0$, where $r = \Omega/\omega$.

7. Deduce that there is stability in the vicinity of the Lagrangian point only if the mass asymmetry is such that $\mu^2 > 23/27$.

Check that this is indeed the case for the Sun-Jupiter couple. The groups of asteroids placed at the two symmetric Lagrangian points with respect to the Sun-Jupiter direction are called the "Greeks" and the "Trojans", with, of course, a veiled reference to Homer. Take the value $m_S = 1049\, m_J$.

Note: In fact the problem of stability is more involved. Markeev, in 1969, demonstrated the instability for two particular mass ratios, namely $\mu = (15 - \sqrt{213})/30$ and $\mu = (45 - \sqrt{1833})/90$ for which the frequencies of the two modes are commensurable, with the explicit respective values $\Omega_1/\Omega_2 = 3$ and $\Omega_1/\Omega_2 = 2$.

---

[9] See the section concerning the small amplitudes in Chapter 2.

The presentation of this three-body problem corresponds to a simplified version. There exist also Lagrangian points when the three objects are aligned. Moreover we did not study the stability of the trajectory in a direction perpendicular to the plane for the motion of the two heavy objects.

Lastly, we studied only the restricted three-body problem: the distance between the heavy objects remains always constant. More generally, for three objects with arbitrary masses moving on arbitrary orbits, there exist two configurations for which the three objects, while moving on their respective elliptic orbits, form a triangle with a variable size but always with an equilateral shape.

Problem Solutions

8.1. Disappearance of Resonant Tori
[Statement p. 396]

1. Because of the relation $s/r = \tan\alpha$, it is convenient to set $s = R\sin\alpha$, $r = R\cos\alpha$, where $R = \sqrt{s^2 + r^2}$. Similarly, one can set $\omega_1 = \Omega\sin\beta$, $\omega_2 = \Omega\cos\beta$, where $\Omega = \sqrt{\omega_1^2 + \omega_2^2}$; consequently, the function $K(\omega_1, \omega_2)$ appearing in the KAM theory reads $K(\omega_1, \omega_2) = k\Omega$. The condition for disappearance of resonating tori can be written generally:

$$|r\omega_1 - s\omega_2| < K(\omega_1, \omega_2)\left(s^2 + r^2\right)^{-3/4}.$$

With our particular prescriptions, it reads in this case $R\Omega\left|\sin(\alpha - \beta)\right| < k\Omega R^{-3/2}$. Furthermore, $\beta = \alpha + \delta\alpha$. Thus the condition for disappearance is written simply as:

$$|\sin(\delta\alpha_{s:r})| \approx |\delta\alpha_{s:r}| < kR^{-5/2}.$$

2. The series
$$\sum_{n=1}^{\infty} \frac{1}{n^\beta}$$

is convergent if $\beta > 1$; in our case $\beta = 5/2 > 1$ so that the series converges and its limit is denoted as S. The function $x^{-5/2}$ being monotonically decreasing, the series can be overestimated by an integral:

$$\sum_{n=2}^{\infty} \frac{1}{n^{5/2}} < \int_1^{\infty} x^{-5/2}\, dx = \frac{2}{3}.$$

We thus have an upper bound for the series $S_1 = 1+2/3 = 5/3$. Moreover, $1/n^{5/2} < 1/n^2$ and the series can be also overestimated by

$$S_2 = \sum_{n=2}^{\infty} \frac{1}{n^2} = \frac{\pi^2}{6}.$$

Practically, one chooses the upper bound as the smallest proposed value $\min(S_1, S_2) = S_2$. Consequently:

$$S < \frac{\pi^2}{6}.$$

3. The angular sum of all excluded regions is necessarily less than that around the irreducible rational numbers s/r:

$$\Delta < \sum_{r \text{ prime with } s} 2|\delta\alpha_{s:r}|.$$

Moreover, the latter sum is necessarily less than that concerned by all the rational numbers, irreducible or not:

$$\Delta < \sum_{r \text{ prime with } s} 2|\delta\alpha_{s:r}| < 2 \sum_{r,s=1}^{\infty} |\delta\alpha_{s:r}|.$$

Owing to these remarks and the condition found in the first question, one obtains

$$\Delta < 2k \sum_{r,s=1}^{\infty} 1/\left(s^2 + r^2\right)^{5/2}.$$

Then, the last trick relies on the inequality $s^2 + r^2 > 2rs$, which implies

$$\Delta < 2^{-3/2} k \sum_{r,s=1}^{\infty} \frac{1}{(sr)^{5/2}} \quad \text{or} \quad \Delta < 2^{-3/2} k \left(\sum_{r=1}^{\infty} \frac{1}{r^{5/2}}\right)^2 = 2^{-3/2} k S^2.$$

$$\Delta < 2^{-3/2} k S^2.$$

4. Invariant tori will resist if $\Delta < \pi/2$. This constraint is automatically fulfilled if one imposes the condition $2^{-3/2} k S^2 < \pi/2$. This will be the case of course if the perturbation, proportional to k, remains weak. Explicitly, tori resist if the perturbation is less than a critical value k_c given by:

$$k < k_c = \frac{\sqrt{2}\pi}{S^2}.$$

One can obtain a more severe constraint if one uses the upper bound for S proposed in question 2:

$$k < k'_c = \frac{36\sqrt{2}}{\pi^3} < k_c.$$

Problem Solutions 417

8.2. Continuous Fractions or How to Play with Irrational Numbers
[Statement p. 396]

1. The simplest way to demonstrate the property is to adopt a recursive method.
   - for order 0: $\alpha_0 = a_0 = s_0/r_0$. But $r_0 = a_0 r_{-1} + r_{-2} = a_0 \times 0 + 1 = 1 = r_0$, and $s_0 = a_0 s_{-1} + s_{-2} = a_0 \times 1 + 0 = a_0 = s_0$; therefore $s_0/r_0 = a_0/1 = a_0 = \alpha_0$. The property holds at this order.
   - for order 1: $\alpha_1 = a_0 + (1/a_1) = (a_0 a_1 + 1)/a_1 = s_1/r_1$. But $r_1 = a_1 r_0 + r_{-1} = a_1 \times 1 + 0 = a_1 = r_1$ and $s_1 = a_1 s_0 + s_{-1} = a_1 \times a_0 + 1$; therefore $s_1/r_1 = (a_0 a_1 + 1)/a_1 = \alpha_1$. The property holds also at this order. This is sufficient to begin the recursive process.

   Assume that the property is satisfied at order $n-1$ and let us define $\alpha'_{n-1} = [a_1; a_2, \ldots, a_n] = s'_{n-1}/r'_{n-1}$. Furthermore, the desired number is written $\alpha_n = a_0 + (1/\alpha'_{n-1}) = (a_0 s'_{n-1} + r'_{n-1})/s'_{n-1}$, whence we deduce the relation between prime and non prime quantities: $s_n = a_0 s'_{n-1} + r'_{n-1}$; $r_n = s'_{n-1}$. Owing to the assumed recursion property, we have $s'_{n-1} = a_n s'_{n-2} + s'_{n-3}$. Thus $r_n = a_n s'_{n-2} + s'_{n-3}$. Since $s'_{n-2} = r_{n-1}$ and $s'_{n-3} = r_{n-2}$, the preceding relation simply reads $r_n = a_n r_{n-1} + r_{n-2}$, which is precisely one of the two desired recursion relations.

   Lastly, $s_n = a_0 s'_{n-1} + r'_{n-1} = a_0(a_n s'_{n-2} + s'_{n-3}) + a_n r'_{n-2} + r'_{n-3} = a_n(a_0 s'_{n-2} + r'_{n-2}) + (a_0 s'_{n-3} + r'_{n-3}) = a_n s_{n-1} + s_{n-2} = s_n$ which corresponds to the second desired recursion relation. The recursion demonstration is now complete and one has:
   $$r_n = a_n r_{n-1} + r_{n-2}$$
   $$s_n = a_n s_{n-1} + s_{n-2}.$$

2. Let us first demonstrate an important property, relying on the previous recursion relations: $s_n r_{n+1} - s_{n+1} r_n = s_n(a_{n+1} r_n + r_{n-1}) - (a_{n+1} s_n + s_{n-1}) r_n = -(s_{n-1} r_n - s_n r_{n-1})$. Proceeding recursively $|s_n r_{n+1} - s_{n+1} r_n| = |s_{n-1} r_n - s_n r_{n-1}| = \ldots = |s_{-2} r_{-1} - s_{-1} r_{-2}| = |0 \times 0 - 1 \times 1| = 1$.

   Now, $$|\alpha_n - \alpha_{n+1}| = \left| \frac{s_n}{r_n} - \frac{s_{n+1}}{r_{n+1}} \right| = \frac{|s_n r_{n+1} - s_{n+1} r_n|}{|r_n r_{n+1}|}.$$

   Owing to the property just demonstrated and owing to the fact that $r_n r_{n+1}$ is a positive quantity, we obtain the desired relation:
   $$|\alpha_n - \alpha_{n+1}| = \frac{1}{r_n r_{n+1}}.$$

3. The number α is written explicitly

$$\alpha = \cfrac{1}{p + \cfrac{1}{p + \cfrac{1}{p + \cdots}}}.$$

We clearly observe the identity

$$\alpha = \frac{1}{p + \alpha},$$

which is equivalent to the second order equation $\alpha^2 + \alpha p - 1 = 0$. Only the positive root makes sense; one finds therefore:

$$\alpha = \frac{\sqrt{p^2 + 4} - p}{2}.$$

4. The irrational number which converges most poorly to a rational number corresponds to the preceding situation for which the successive denominators are the smallest possible, namely the number for which $p = 1$ in the previous question. Consequently, one obtains $\alpha = (\sqrt{5} - 1)/2$ which is manifestly related to the golden ratio $\Phi = (\sqrt{5} + 1)/2$ by $\alpha = \Phi - 1 = \Phi^{-1}$:

$$\alpha = \Phi - 1 = \Phi^{-1} = \frac{\sqrt{5} - 1}{2}.$$

8.3. Properties of the Phase Space of the Standard Mapping [Statement p. 398]

1. Let us recall the expression for the standard mapping: $\theta_{n+1} = \theta_n + p_n$; $p_{n+1} = p_n + K \sin \theta_{n+1}$. Let $M_0 = (\theta_0, p_0)$ be an arbitrary point in Poincaré's section and $M_0'(2\pi - \theta_0, -p_0)$ be the symmetric point of M_0 with respect to the point $A(\pi, 0)$. It is equivalent to the point $M_0'(2\pi - \theta_0, 2\pi - p_0)$, lying inside the standard square, and obtained from the allowed congruences.

The image of the point M_0 is $M_1 = (\theta_1, p_1) = (\theta_0 + p_0, p_0 + K \sin \theta_1)$; the image of the point M_0' is $M_1'(\theta_1', p_1') = ((2\pi - \theta_0) + (-p_0), (-p_0) + K \sin \theta_1') = (2\pi - (\theta_0 + p_0), -p_0 - K \sin(\theta_0 + p_0)) = (2\pi - \theta_1, -p_1)$. Thus M_1' is the symmetric point of M_1. Proceeding recursively, the same property holds for any further iterations. In other words, starting from two symmetric initial conditions, the full trajectories are symmetric. Since it is possible to divide Poincaré's section into two regions composed of symmetric points, one deduces that Poincaré's section is formed by symmetric trajectories.

Problem Solutions

Let us employ the same reasoning with a symmetry center chosen at $C(\pi,\pi)$. Again with the same starting point $M_0 = (\theta_0, p_0)$, one obtains the symmetric point $M'_0(2\pi - \theta_0, 2\pi - p_0)$ which is identified, up to a congruence, with the previous point M'_0. Therefore, the symmetry with respect to A is fully equivalent to the symmetry with respect to C. The conclusions resulting from the symmetry with respect to A can thus be extended to the symmetry with respect to C.

2. To an arbitrary point $M = (\theta, p)$ we associate the point $\tilde{M} = (\tilde{\theta} = \theta + \pi, \tilde{p} = p)$ obtained by a translation by π of the origin of the angles. Let $M'(\theta', p')$ be the image of M by the standard mapping; obviously $\theta' = \theta + p$, $p' = p + K \sin \theta'$. Let us associate to it the point $\tilde{M}'(\tilde{\theta}', \tilde{p}')$ obtained by the same translation. Consequently $\tilde{\theta}' = \theta' + \pi = \theta + p + \pi = \tilde{\theta} + \tilde{p}$ and $\tilde{p}' = p' = p + K \sin \theta' = \tilde{p} + K \sin(\tilde{\theta}' - \pi)$ or $\tilde{p} - K \sin \tilde{\theta}'$. In other words, the mapping which associates $\tilde{M}'$ to $\tilde{M}$ is given by:

$$\tilde{\theta}' = \tilde{\theta} + \tilde{p}$$
$$\tilde{p}' = \tilde{p} - K \sin \tilde{\theta}'.$$

One notices that it corresponds to the standard mapping but with a negative constant $K' = -K$.

8.4. Bifurcation of the Periodic Trajectory 1:1 for the Standard Mapping [Statement p. 398]

1. Let us recall the form of the standard mapping: $\theta_1 = \theta_0 + p_0$, $p_1 = p_0 + K \sin \theta_1$. The point (θ_0, p_0) is a fixed point if it corresponds to its own image; therefore one has $p_1 = p_0$ which leads to $K \sin \theta_1 = 0$, i.e., $\theta_1 = 0$ or $\theta_1 = \pi$. Moreover $\theta_1 \equiv \theta_0$. For $\theta_1 = 0$, this implies $\theta_0 = 0$, then $p_0 = 0$. For $\theta_1 = \pi$, one has similarly $\theta_0 = \pi$ and $p_0 = 0$. Therefore, we end up with two fixed points of order 1:

$$A(0,0) \quad ; \quad B(\pi, 0)$$

To study the stability of these points, it is necessary to linearize the mapping in their neighbourhood.

Close to a fixed point (θ_f, p_f), let us choose a point denoted as $(\theta_f + \varepsilon, p_f + \eta)$; the image of this point is therefore $\theta_1 = \theta_f + p_f + \varepsilon + \eta = \theta_f + \varepsilon_1$, that is $\varepsilon_1 = \varepsilon + \eta$ and $p_1 = p_f + \eta + K \sin \theta_1 = p_f + \eta + K \sin(\theta_f + \varepsilon_1) \approx p_f + K \sin \theta_f + \eta + \varepsilon_1 K \cos \theta_f = p_f + \eta_1$, so that $\eta_1 = K \cos \theta_f \varepsilon + (1 + K \cos \theta_f) \eta$. The matrix of the linearized mapping is expressed as

$$M = \begin{pmatrix} 1 & 1 \\ K \cos \theta_f & 1 + K \cos \theta_f \end{pmatrix}.$$

Its determinant is unity and its trace is just $T = 2 + K\cos\theta_f$. Close to A, $\theta_f = 0$ whence $T = 2 + K$. One always has $|T| > 2$; consequently we are dealing with an unstable hyperbolic point. Close to B, $\theta_f = \pi$ whence $T = 2 - K$. The fixed point is stable, or elliptic, as long as $|T| < 2$, that is $K < 4$. Finally:

$$
\begin{array}{lll}
A: & & \text{is always unstable; hyperbolic point} \\
B: & K < 4 & \text{is a stable or elliptic point} \\
B: & K > 4 & \text{is an unstable or hyperbolic point.}
\end{array}
$$

2. Let (θ_f, p_f) be a fixed point of order 2 for the mapping in the neighbourhood of B (the only fixed point of order 1 which is stable). From the original mapping, it is transformed into its symmetric image with respect to B, in such a way that the image (the second iterate) coincides with the starting point (the symmetric point of the image is the image of the symmetric point; see Problem 8.3). Therefore $\theta_f + p_f = 2\pi - \theta_f$ and $p_f + K\sin(\theta_f + p_f) = -p_f$. From the first equation, one obtains $p_f = 2\pi - 2\theta_f$ and from the second one $2p_f = -K\sin(\theta_f + p_f) = -K\sin(2\pi - \theta_f) = K\sin\theta_f$. The fixed points are therefore determined through the set of equations:

$$p_f = 2\pi - 2\theta_f$$
$$p_f = \tfrac{1}{2}K\sin\theta_f.$$

3. The fixed points are the intersections of the sinusoid $\tfrac{1}{2}K\sin\theta_f$ and the straight line $2\pi - 2\theta_f$. An intersection exists only if the slope of the sinusoid at point B is larger in modulus than the slope of the line at this point, that is $\tfrac{1}{2}K\cos\theta_f < -2$ or $-\tfrac{1}{2}K < -2$, therefore

$$K > 4.$$

This is the bifurcation phenomenon. When the fixed point of order 1 becomes unstable, there appears a pair of fixed points of order 2.

4. Let $p_a = \delta > 0$ for the fixed point a. From the first equation giving the fixed points, one derives $\delta = 2\pi - 2\theta_a$, that is $\theta_a = \pi - \delta/2$. The number δ is determined from the transcendental equation $\delta = \tfrac{1}{2}K\sin(\delta/2)$. This first fixed point a is thus written as $a(\pi - \delta/2, \delta)$; the second fixed point b is the symmetric point of a with respect to B, that is $b(\pi + \delta/2, -\delta)$. We remark that the momentum of this point can also be written $p_b = 2\pi - \delta$.

Let us start from a point close to a, $\theta_0 = \theta_a + \varepsilon$, $p_0 = p_a + \eta$. Its image (θ_1, p_1) is close to b, that is $\theta_1 = \theta_b + \varepsilon_1$, $p_1 = p_b + \eta_1$.

Problem Solutions

We saw in the first question that the matrix linearized in the vicinity of this point is

$$M_1 = \begin{pmatrix} 1 & 1 \\ K\cos\theta_a & 1+K\cos\theta_a \end{pmatrix}.$$

The second iterate (θ_2, p_2) is again close to a, and hence $\theta_2 = \theta_a + \varepsilon_2$, $p_2 = p_a + \eta_2$. The matrix linearized in the vicinity of this point is similarly

$$M_2 = \begin{pmatrix} 1 & 1 \\ K\cos\theta_b & 1+K\cos\theta_b \end{pmatrix}.$$

Passing from the initial point (θ_0, p_0) to its second iterate (θ_2, p_2) is achieved through the product of the two previous matrices

$$M = M_2 M_1 = \begin{pmatrix} 1 & 1 \\ K\cos\theta_b & 1+K\cos\theta_b \end{pmatrix}\begin{pmatrix} 1 & 1 \\ K\cos\theta_a & 1+K\cos\theta_a \end{pmatrix}$$

$$= \begin{pmatrix} 1+K\cos\theta_a & 2+K\cos\theta_a \\ K(\cos\theta_a + \cos\theta_b) + K^2\cos\theta_a\cos\theta_b & 1+K(\cos\theta_a + 2\cos\theta_b) + K^2\cos\theta_a\cos\theta_b \end{pmatrix}.$$

The trace of this matrix is

$$T = 2 + 2K(\cos\theta_a + \cos\theta_b) + K^2\cos\theta_a\cos\theta_b.$$

With $\cos\theta_a = \cos(\pi - \delta/2) = -\cos(\delta/2)$ and $\cos\theta_b = \cos(\pi + \delta/2) = -\cos(\delta/2)$, the trace may be expressed in a more compact form as

$$\begin{aligned} T &= 2 - 4K\cos(\delta/2) + K^2\cos^2(\delta/2) \\ &= (K\cos(\delta/2) - 2)^2 - 2. \end{aligned}$$

The fixed point a (also b) is stable if the condition $-2 < T < 2$ is satisfied. One inequality is always fulfilled; the other leads to $(K\cos(\delta/2) - 2)^2 < 4$, that is:

$$|K\cos(\delta/2) - 2| < 2.$$

If $K\cos(\delta/2) > 2$, the previous condition is equivalent to $2 < K\cos(\delta/2) < 4$. If $K\cos(\delta/2) < 2$ the same condition is equivalent to $0 < K\cos(\delta/2) < 2$. Gathering both inequalities, the condition for stability is summarized as $0 < K\cos(\delta/2) < 4$. This condition must be considered as complementary to the equation giving the value of δ, which is, as we saw, $\delta = \frac{1}{2}K\sin(\delta/2)$. An elementary calculation shows that the condition $K\cos(\delta/2) < 4$, equivalent to $\delta/2 < \tan(\delta/2)$, is always satisfied. Finally, the only interesting condition for stability is:

$$\cos(\delta/2) > 0.$$

Therefore one must have $\delta < \pi$. Moreover the relation between K and δ is $K(\delta) = 2\delta/\sin(\delta/2)$. A brief study of the curve $K(\delta)$ convinces us that, over the interval $[0, \pi]$, it is increasing and monotonic. Consequently, the condition $\delta < \pi$ is identified with the condition $K < K(\pi) = 2\pi$. Finally, this long study of stability allows us to conclude that the fixed points of order 2 obtained as the result of the bifurcation are stable as long as:

$$4 < K < 2\pi.$$

The situation is illustrated in the Fig. 8.9.

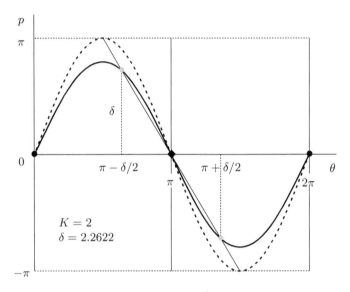

Fig. 8.9 Bifurcation around the fixed point of order 1 $(\pi, 0)$ denoted as a filled black diamond. The fixed points of order 2 are represented by two small grey circles. To better appreciate the figure, the upper part of the Poincaré section has been translated by 2π below the lower part

5. We assume that this condition is realized and we are concerned with the fixed point of order 1 $B(\pi, 0)$. The matrix linearized in the neighbourhood of this point is written explicitly

$$M = \begin{pmatrix} 1 & 1 \\ -K & 1-K \end{pmatrix}.$$

The first task is to find the eigenvalues λ, which satisfy the characteristic equation $\lambda^2 - (2-K)\lambda + 1 = 0$. Explicitly the roots are

$$\lambda_d = \frac{1}{2}\left(2 - K - \sqrt{K^2 - 4K}\right) \quad \text{and} \quad \lambda_c = \frac{1}{2}\left(2 - K + \sqrt{K^2 - 4K}\right).$$

Since $|\lambda_d| > 1$, $|\lambda_c| < 1$, the directions associated with these eigenvalues are respectively divergent and convergent. The direction of the corresponding eigenvectors is parallel to the asymptote of the hyperbole containing the iterations close to the fixed point.

It is easy to calculate the eigenvectors (up to a normalization factor) as $\boldsymbol{V_d} = (1, \lambda_d - 1)$, $\boldsymbol{V_c} = (1, \lambda_c - 1)$. When stability disappears, for $K = 4$, $\lambda_d = \lambda_c = -1$ and the directions of the asymptotes become degenerate along the vectors

$$\boldsymbol{V_d} = \boldsymbol{V_c} = (1, -2).$$

With the position of the fixed points given above, it is easily checked that $\boldsymbol{Ba} = (-\delta/2, \delta) = -(\delta/2)\boldsymbol{V_d}$ and $\boldsymbol{Bb} = (\delta/2, -\delta) = (\delta/2)\boldsymbol{V_d}$. The direction connecting the fixed points thus lies along the common asymptotes of the nascent hyperbolic point.

8.5. Chaos–ergodicity: A Slight Difference [Statement p. 399]

1. The trajectory of the system is represented by very simple equations: $\alpha(t) = \omega_\alpha t + \alpha_0$, $\beta(t) = \omega_\beta t + \beta_0$. The configuration space (α, β) possesses the topology of a torus $0 < \alpha < 2\pi$, $0 < \beta < 2\pi$. Consider a function $F(\alpha, \beta)$ defined in this space. It is obviously periodic in both variables $F(\alpha + 2\pi, \beta) = F(\alpha, \beta) = F(\alpha, \beta + 2\pi)$.

It is convenient to use a Fourier expansion and write:

$$F(\alpha, \beta) = \sum_{m,n=-\infty}^{\infty} F_{mn} e^{i(m\alpha + n\beta)}.$$

We now follow the values of the function along its trajectory; the function becomes time dependent and takes the following form:

$$F(\alpha(t), \beta(t)) = \sum_{m,n=-\infty}^{\infty} F_{mn} e^{i(m\omega_\alpha + n\omega_\beta)t} e^{i(m\alpha_0 + n\beta_0)}.$$

Let us calculate now its time average

$$\bar{F}(\alpha_0, \beta_0) = \lim_{T \to \infty} \frac{1}{T} \int_0^T F(\alpha(t), \beta(t)) \, dt$$

$$= \sum_{m,n=-\infty}^{\infty} F_{mn} e^{i(m\alpha_0 + n\beta_0)} \lim_{T \to \infty} \frac{1}{T} \int_0^T e^{i(m\omega_\alpha + n\omega_\beta)t} \, dt.$$

The value of the integral is given explicitly by

$$\frac{e^{i(m\omega_\alpha + n\omega_\beta)T} - 1}{i(m\omega_\alpha + n\omega_\beta)}$$

if $(m\omega_\alpha + n\omega_\beta) \neq 0$, otherwise it is equal to T.

We suppose that the angular frequencies are not commensurable; this means that $\omega_\alpha/\omega_\beta$ is not a rational number. This implies that $(m\omega_\alpha + n\omega_\beta) \neq 0$, except for $m = n = 0$, and that the proposed expression for the integral is always valid $\forall m \neq 0, n \neq 0$. Moreover, $0 < \left|e^{i(m\omega_\alpha + n\omega_\beta)T} - 1\right| < 2$ and hence $\lim \left(\left|e^{i(m\omega_\alpha + n\omega_\beta)T} - 1\right|\right)/T \to 0$. Consequently, in the Fourier expansion, there remains only the term with $m = n = 0$, for which the limit is 1. Finally:

$$\bar{F}(\alpha_0, \beta_0) = F_{00}.$$

This value is independent of the starting point (α_0, β_0).

Consider now the average in configuration space:

$$\tilde{F} = \frac{1}{(2\pi)^2} \oint F(\alpha, \beta)\, d\alpha\, d\beta.$$

Using the same Fourier expansion, we obtain:

$$\tilde{F} = \frac{1}{(2\pi)^2} \sum_{m,n=-\infty}^{\infty} F_{mn} \int_0^{2\pi} e^{im\alpha}\, d\alpha \int_0^{2\pi} e^{im\beta}\, d\beta$$

$$= \frac{1}{(2\pi)^2} \sum_{m,n=-\infty}^{\infty} F_{mn}(2\pi\delta_{m,0})(2\pi\delta_{n,0}) = F_{00}$$

(δ_{ij} is, as usual, the Kronecker symbol).

Thus, with the conditions of periodicity imposed on the function F, we have demonstrated the ergodicity property, namely:

$$\bar{F}(\alpha_0, \beta_0) = \tilde{F} = F_{00}.$$

2. Now we need to prove that, for an ergodic system, the trajectory explores the configuration space in a dense way. In order to do this, consider a small portion D of this space. Let us choose a particular periodic function $F(\alpha, \beta)$ defined by the equation:

$$F(\alpha, \beta) = 0 \text{ if } (\alpha, \beta) \notin D$$
$$F(\alpha, \beta) = 1 \text{ if } (\alpha, \beta) \in D.$$

Let us start from an arbitrary point (α_0, β_0) and follow the trajectory

$$\alpha(t) = \omega_\alpha t + \alpha_0, \qquad \beta(t) = \omega_\beta t + \beta_0.$$

We wish to demonstrate that, soon or later, it will explore the D region. Let us use a reasoning ad absurdum and assume that the trajectory never crosses the D domain. This means that $(\alpha(t), \beta(t)) \notin D \quad \forall t$. With the chosen function $F(\alpha, \beta)$, this condition is equivalent to $F(\alpha(t), \beta(t)) = 0$, $\forall t$. One thus deduces $\bar{F}(\alpha_0, \beta_0) = 0$.

On the other hand, the average over configuration space is equal to

$$\tilde{F} = \frac{1}{(2\pi)^2} \int\!\!\int_D d\alpha\, d\beta = \text{Area}(D)/(2\pi)^2 \neq 0.$$

Consequently, our hypothesis leads to $\tilde{F} \neq \bar{F}$ in contradiction with the ergodicity property. We thus conclude that the trajectory necessarily explores the D region, however small it may be. The space is said to be covered by the trajectory in a dense way.

8.6. Acceleration Modes: A Curiosity of the Standard Mapping [Statement p. 401]

1. We start from an arbitrary point of Poincaré's section of the standard mapping $0 < \theta_0 < 2\pi$, $0 < p_0 < 2\pi$; its image is given by $\theta_1 = \theta_0 + p_0$, $p_1 = p_0 + K \sin \theta_1$. We seek fixed points of order 1 corresponding to an increase of 2π for the momentum. We must solve the equations $\theta_1 - \theta_0 = 2n\pi$, $p_1 - p_0 = 2\pi$. The first equation can be also written as $p_0 = 2n\pi$; using the initial condition imposed on the starting point, one has $n = 0$ which leads to $p_0 = 0$ and $\theta_1 = \theta_0$. The second equation for the fixed point gives $K \sin \theta_1 = K \sin \theta_0 = 2\pi$ so that the angle for the fixed point is determined from the equation:

$$K \sin \theta_0 = 2\pi$$

or $\sin \theta_0 = 2\pi/K$. Since we have anyway $\sin \theta_0 < 1$, the constraint on the intensity of the perturbation reads:

$$K > 2\pi.$$

If this condition is satisfied, there exist two solutions for the angle θ_0. The corresponding fixed points are given by:

$$\theta_0 = \arcsin(2\pi/K); \qquad p_0 = 0$$
$$\theta_0 = \pi - \arcsin(2\pi/K); \qquad p_0 = 0.$$

With these values, we have $\theta_1 = \theta_0$ and $p_1 = 2\pi$. Let us adopt a recursive reasoning and assume $\theta_n = \theta_0$, $p_n = 2n\pi$. Using the standard mapping, the next image is characterized by $\theta_{n+1} = \theta_n + p_n = \theta_0 + 2n\pi$, which is identified with θ_0, and

$$p_{n+1} = p_n + K \sin \theta_{n+1} = 2n\pi + K \sin \theta_0 = 2n\pi + 2\pi = 2(n+1)\pi.$$

The recursion property is demonstrated. Therefore, the iteration of order n is determined by:

$$\theta_n = \theta_0$$
$$p_n = 2n\pi.$$

Passing from one iteration to the next one, the angle does not change, but the momentum increases each time by 2π. This is why we speak of an acceleration mode.

2. Let us denote as $\tilde{\theta}_0$ the angle for the fixed point and $\tilde{p}_0 = 0$ its momentum. We now choose a starting point close to the fixed point $\theta_0 = \tilde{\theta}_0 + \varepsilon_0$, $p_0 = \tilde{p}_0 + \eta_0$. Its image angle is

$$\theta_1 = \theta_0 + p_0 = \tilde{\theta}_0 + \varepsilon_0 + \eta_0 = \tilde{\theta}_1 + \varepsilon_1$$

and the image momentum is $p_1 = p_0 + K \sin \theta_1 = \tilde{p}_1 + \eta_0 + (\varepsilon_0 + \eta_0) K \cos \tilde{\theta}_0 = \tilde{p}_1 + \eta_1$. The matrix switching from (ε_0, η_0) to (ε_1, η_1) is written as:

$$M = \begin{pmatrix} 1 & 1 \\ K \cos \tilde{\theta}_0 & 1 + K \cos \tilde{\theta}_0 \end{pmatrix}.$$

The trace is equal to

$$T = 2 + K \cos \tilde{\theta}_0.$$

3. When $\tilde{\theta}_0 = \arcsin(2\pi/K)$, one calculates easily $K \cos \tilde{\theta}_0 = \sqrt{K^2 - 4\pi^2}$. Since $K > 2\pi$, one deduces $T > 2$; the corresponding fixed point is always unstable.

When $\tilde{\theta}_0 = \pi - \arcsin(2\pi/K)$, one calculates similarly

$$K \cos \tilde{\theta}_0 = -\sqrt{K^2 - 4\pi^2}.$$

One deduces $T = 2 - \sqrt{K^2 - 4\pi^2}$; one always has $T < 2$. The corresponding fixed point remains stable as long as we have the inequality: $K < 2\sqrt{\pi^2 + 4}$. To summarize, one fixed point is always unstable and the other is stable for the values of the perturbation within the interval:

$$2\pi < K < 2\sqrt{\pi^2 + 4}.$$

8.7. Demonstration of a Kicked Rotor?
[Statement p. 401]

1. The Lagrangian for the simple pendulum is given as usual by $L = \frac{1}{2}ml^2\dot{\theta}^2 + mgl\cos\theta$. The conjugate momentum is $p = ml^2\dot{\theta}$ and the equation of motion $\ddot{\theta} = -(g/l)\sin\theta$. With our axis conventions, working in the frame of the suspension point, which is governed by the temporal law $a(t)$ along the gravitational field, is equivalent to the substitution $g \to g - \ddot{a}$ in the equation of motion. Setting $g = 0$, since gravity is inefficient (at least negligible versus the acceleration of the suspension point) in this modified equation, we obtain the equation of motion for the system:

$$\ddot{\theta} = \frac{\ddot{a}}{l}\sin\theta$$

2. At time nT, the angle is $\theta(nT) = \theta_n$, the angular velocity $\dot{\theta}(nT) = \dot{\theta}_n$, the momentum $p(nT) = p_n = ml^2\dot{\theta}_n$ and the velocity of the suspension point $\dot{a} = nv$. Between the instants nT and $(n+1)T$, one has $\dot{a} = nv = $ const, hence $\ddot{a} = 0$ which leads, owing to Lagrange's equation, to $\ddot{\theta} = 0$. After integration $\dot{\theta} = $ const $= \dot{\theta}_n$, and then $p = $ const $= p_n = ml^2\dot{\theta}_n$. After a further integration, one obtains $\theta(t) = \dot{\theta}_n(t - nT) + \theta_n$. Just before the kick at time $(n+1)T$, one has $\theta((n+1)T^-) = \dot{\theta}_n T + \theta_n$. This angle does not change after the kick so that $\theta((n+1)T^+) = \theta_{n+1} = \dot{\theta}_n T + \theta_n$ that is $\theta_{n+1} = \theta_n + p_n(T/(ml^2))$.

Just before the kick $\dot{\theta} = \dot{\theta}_n$ and $p = p_n$. To determine their values just after the kick, let us integrate the Lagrange equation between $t_0 = (n+1)T - \varepsilon$ and $t_1 = (n+1)T + \varepsilon$:

$$\int_{t_0}^{t_1} \ddot{\theta}(t)\,dt = \dot{\theta}(t_1) - \dot{\theta}(t_0) = \int_{t_0}^{t_1} \frac{\ddot{a}(t)\sin\theta(t)}{l}\,dt$$

$$\approx \frac{\sin\theta_{n+1}}{l}\int_{t_0}^{t_1} \ddot{a}(t)\,dt = \frac{\sin\theta_{n+1}}{l}(\dot{a}(t_1) - \dot{a}(t_0)).$$

However, by hypothesis, $\dot{a}(t_1) - \dot{a}(t_0) = v$. Moreover

$$\dot{\theta}(t_1) = \dot{\theta}((n+1)T + \varepsilon) \quad \text{and} \quad \dot{\theta}(t_0) = \dot{\theta}((n+1)T - \varepsilon).$$

In the limit $\varepsilon \to 0$, we obtain $\dot{\theta}(t_1) = \dot{\theta}_{n+1}$ and $\dot{\theta}(t_0) = \dot{\theta}_n$. Our preceding equation reads $\dot{\theta}_{n+1} - \dot{\theta}_n = (v/l)\sin\theta_{n+1}$. Rather than angular velocities we use the momenta: $p_{n+1} - p_n = mlv\sin\theta_{n+1}$.

Therefore the desired mapping is written in the form:

$$\theta_{n+1} = \theta_n + \frac{T}{ml^2} p_n$$
$$p_{n+1} = p_n + mlv \sin \theta_{n+1}.$$

To make this mapping compatible with the standard mapping, we change momenta and define $P_n = Tp_n/(ml^2)$. The first of these equations is now identical with the corresponding one of the standard mapping: $\theta_{n+1} = \theta_n + P_n$. Multiplying the second equation by $T/(ml^2)$, it can be recast in the form of the standard mapping, if one makes the identification:

$$K = \frac{vT}{l}.$$

3. Now, we consider a velocity which takes the form of a square wave changing its sign every half-period. Let n be the label of a kick for which the velocity becomes negative. Just after, the angle is θ_n, the angular velocity $\dot\theta_n$, the momentum $p_n = ml^2 \dot\theta_n$, the velocity of the suspension point $\dot a = -v$ and its acceleration $\ddot a = 0$.

Just after the next velocity inversion, a semi-period later, the corresponding quantities are labelled θ_i, $\dot\theta_i$, p_i, $\dot a = v$. Using the results of the previous study, one can write the relation $\theta_i = \theta_n + \dot\theta_n T/2 = \theta_n + p_n T/(2ml^2)$,

$$\dot\theta_i - \dot\theta_n = \frac{\sin \theta_i}{l} (v - (-v)),$$

or $p_i = p_n + 2mlv \sin \theta_i$.

We repeat the same arguments for the other semi-period to obtain the quantities θ_{n+1}, p_{n+1} as functions of θ_i, p_i: $\theta_{n+1} = \theta_i + p_i T/(2ml^2)$, $p_{n+1} = p_i - 2mlv \sin \theta_{n+1}$. As in the previous case, let us perform the change of variable $P_n = p_n T/(2ml^2)$ and introduce the value $K = vT/l$. The preceding mappings have a simpler form: $\theta_i = \theta_n + P_n$, $P_i = P_n + K \sin \theta_i = P_n + K \sin(\theta_n + P_n)$ and $\theta_{n+1} = \theta_i + P_{n+1}$, $P_{n+1} = P_i - K \sin \theta_{n+1}$. It is sufficient to remove the intermediate quantities θ_i, p_i to obtain the desired sawtooth mapping:

$$\theta_{n+1} = \theta_n + 2P_n + K \sin(\theta_n + P_n)$$
$$P_{n+1} = P_n + K \sin(\theta_n + P_n) - K \sin \theta_{n+1}.$$

This mapping looks very similar to the squared mapping of the standard mapping; it differs only by the sign of the last term ($-K \sin \theta_{n+1}$ instead of $+K \sin \theta_{n+1}$). However this slight difference induces important modifications in Poincaré's section. This sawtooth mapping is represented in Fig. 8.10.

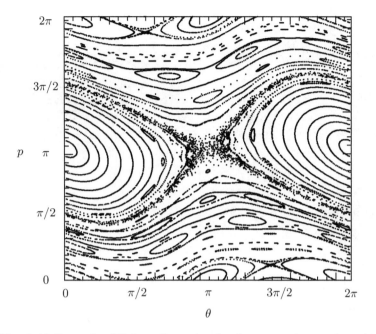

Fig. 8.10 Example of Poincaré's section for the sawtooth mapping, obtained with the value $K = 0.75$

4. Let us seek the fixed points of order 1 for this mapping. One must have $\theta_1 - \theta_0 = 2n\pi$, $P_1 - P_0 = 0$. The latter equation implies $\sin(\theta_0 + P_0) = \sin\theta_0$. There exist two families of solutions:

$$\theta_0 + P_0 = \theta_0, \qquad \theta_0 + P_0 = \pi - \theta_0.$$

First family of solutions $P_0 = 0$

From the first equation of the fixed point, one finds $2n\pi = K\sin\theta_0$ which admits roots only if $K > 2n\pi$. For the weak values of K that are of interest, one can have only $n = 0$, which implies $\sin\theta_0 = 0$. There exist two solutions $\theta_0 = 0$ and $\theta_0 = \pi$ and therefore two fixed points of order 1:

$$A(0,0); \qquad B(\pi, 0).$$

To study the stability of the fixed point (θ_f, P_f), it is necessary to linearize the mapping in its neighbourhood. Let us begin with $\theta_0 = \theta_f + \varepsilon_0$, $P_0 = P_f + \eta_0$. Substituting these relations in the first equation of the mapping and restricting to first order, one obtains $\theta_1 = \theta_f + \varepsilon_1$, where $\varepsilon_1 = \varepsilon_0(1 + K\cos\theta_f) + \eta_0(2 + K\cos\theta_f)$. Similarly, substituting in the second equation, one obtains $P_1 = P_f + \eta_1$, where $\eta_1 = \varepsilon_0(-K^2\cos^2\theta_f) + \eta_0(1 - K\cos\theta_f - K^2\cos^2\theta_f)$.

The linearized matrix therefore reads:
$$M = \begin{pmatrix} 1 + K\cos\theta_f & 2 + K\cos\theta_f \\ -K^2 \cos^2\theta_f & 1 - K\cos\theta_f - K^2 \cos^2\theta_f \end{pmatrix}.$$

It is easy to check that its determinant is unity and that the trace is equal to $T = 2 - K^2 \cos^2\theta_f$. Whatever the fixed point chosen, one has $\cos^2\theta_f = 1$ and the corresponding value of the trace is $T = 2 - K^2$. The points are stable as long as $|T| < 2$, which corresponds to $K < 2$.

$A(0,0), B(\pi,0)$ are stable if $K < 2$
$A(0,0), B(\pi,0)$ are unstable if $K > 2$.

Second family of solutions $2\theta_0 + P_0 \equiv \pi$

In this case $2P_0 + K\sin(\theta_0 + P_0) = 2n\pi = 2P_0 + K\sin\theta_0$. With the initial conditions $0 < \theta_0 < 2\pi$, one has the constraint $0 < 2\theta_0 + P_0 < 6\pi$. For the required solutions, one must restrict the choice to $2\theta_0 + P_0 = \pi, 3\pi, 5\pi$.

Finally, we must solve the following system of equations : $P_0 = (\pi, 3\pi, 5\pi) - 2\theta_0$, $P_0 = n\pi - \frac{1}{2}K\sin\theta_0$. To this end we use a graphical method, seeking the intersections of the three straight lines $\pi - 2\theta_0$, $3\pi - 2\theta_0$, $5\pi - 2\theta_0$ and the three sinusoids $n\pi - \frac{1}{2}K\sin\theta_0$ for weak values of K (it is easy to be convinced that only the values $n = 0, 1, 2$ give solutions inside the reference square). There exist four allowed solutions in the reference square. The graphical situation is depicted in Fig. 8.11.

Two fixed points are easily obtained : the central point $C(\pi,\pi)$ and the point $D(0,\pi)$. The other two, E and F, are obtained by solving a transcendental equation.

To find the position of E, we use the fact that K is small. Let us set $\theta_e = 3\pi/2 - \varepsilon$, $P_e = \eta$. We find the first relation $\eta = 2\varepsilon$, then the equation $\eta = \frac{1}{2}K\cos\varepsilon \approx K/2$. Therefore, we obtain $\eta \approx K/2$ and $\varepsilon \approx K/4$, which give the approximate position of E. The F point is obtained by symmetry arguments. In summary, the fixed points of the second family are the following:

$D(0,\pi)$; $C(\pi,\pi)$; $E(3\pi/2 - K/4, K/2)$; $F(\pi/2 + K/4, 2\pi - K/2)$.

The matrix corresponding to the linearization of the mapping in the neighbourhood of the fixed point θ_f, P_f in this family is obtained with the same technique as before. It is more involved and reads (with the definition $C_f = \cos(\theta_f + P_f)$ in order to simplify the expression)

$$\begin{pmatrix} 1 + KC_f & 2 + KC_f \\ KC_f - K\cos\theta_f - K^2\cos\theta_f C_f & 1 + KC_f - 2K\cos\theta_f - K^2\cos\theta_f C_f \end{pmatrix}.$$

Problem Solutions

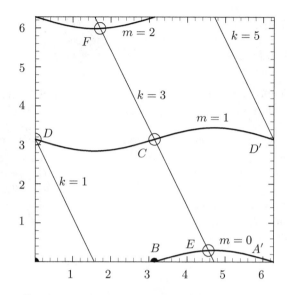

Fig. 8.11 Graphical construction to find the fixed points of the sawtooth mapping. The black dots correspond to the first family of solutions. The empty circles correspond to the second family of solutions. To calculate them, one must determine the intersections, in the reference square, of the three straight lines $k\pi - 2\theta$ ($k = 1, 3, 5$) and the three sinusoids $m\pi - \frac{1}{2}K\sin\theta$ ($m = 1, 3, 5$)

Its determinant is unity and the trace is equal to

$$T = 2 + 2K\cos(\theta_f + P_f) - 2K\cos\theta_f - K^2\cos\theta_f\cos(\theta_f + P_f).$$

For the fixed point D, the trace is explicitly equal to $T = 2 - 4K + K^2$. The stability condition is $-2 < T < 2$, which, in this case, is equivalent to $K < 4$.

For the fixed point C, the trace is $T = 2 + 4K + K^2 > 2$ and the point is always unstable.

For the point E, with the approximation of small K values, $T \approx 2 + K^2 + K^4/16 > 2$ and the point is always unstable.

The situation for the fixed point F is identical to that for point E.

To summarize,

C, E, F	are always unstable
$D, K < 4$	is stable
$D, K > 4$	is unstable

These conclusions are clearly seen in Fig. 8.10; the 3 fixed points C, E, F are embedded in a chaotic region whereas the fixed point D is the center of an islet of stability.

8.8. Anosov's Mapping (or Arnold's Cat)
[Statement and Figure p. 403]

1. Let us begin with the recursion relation for a_{n+1} and a_n in the proposed expression $a_{n+1}a_{n-1} - a_n^2 = (a_n + a_{n-1})a_{n-1} - a_n(a_{n-1} + a_{n-2}) = a_{n-1}^2 - a_n a_{n-2}$. Similarly $a_{n-1}^2 - a_n a_{n-2} = a_{n-1}a_{n-3} - a_{n-2}^2 = a_{n+1}a_{n-1} - a_n^2$. Applying the latter identity to $n = 2r$, one obtains $a_{2r+1}a_{2r-1} - a_{2r}^2 = a_{2(r-1)+1}a_{2(r-1)-1} - a_{2(r-1)}^2$. Pursuing the recursion up to the last term, this common value is equal to $a_3 a_1 - a_2^2 = 2 \times 1 - 1^2 = 1$. We thus demonstrate the required identity:

$$a_{2r+1}a_{2r-1} - a_{2r}^2 = 1.$$

2. Between two kicks, the Hamiltonian is that of a free particle $H = p^2/2$. Hamilton's equations give first $\dot{p} = -\partial_\theta H = 0$ whence $p = $ const and $\dot{\theta} = \partial_p H = p$. After the kick labelled n, the momentum is p_n and the angle θ_n. Just before the kick $(n+1)$, a time $T = 1$ later, the momentum has not changed and the angle is $\theta = p_n T + \theta_n = p_n + \theta_n$. Just after this kick, the angle is unchanged $\theta_{n+1} = p_n + \theta_n$. In contrast, the momentum exhibits an increase equal to the angle value at that time, namely θ_{n+1}; consequently $p_{n+1} = p_n + \theta_{n+1}$.

The Anosov mapping is thus defined as:

$$\theta_{n+1} = p_n + \theta_n$$
$$p_{n+1} = p_n + \theta_{n+1}.$$

It can be written in a matrix form in a totally general way (and not after linearization as we did for the study of stability in the neighbourhood of fixed points). In the second equation defined above, we substitute θ_{n+1} by its value $p_n + \theta_n$, to find $p_{n+1} = 2p_n + \theta_n$. Finally the mapping is written in matrix form as:

$$\begin{pmatrix} \theta_{n+1} \\ p_{n+1} \end{pmatrix} = \begin{pmatrix} 1 & 1 \\ 1 & 2 \end{pmatrix} \begin{pmatrix} \theta_n \\ p_n \end{pmatrix} = M \begin{pmatrix} \theta_n \\ p_n \end{pmatrix}$$

Its determinant is unity; this is also the Jacobian valid anywhere in phase space. This implies conservation of areas. Of course this mapping is linear in the whole phase space.

Problem Solutions 433

3. The eigenvalue equation reads $(1-\lambda)(2-\lambda)-1=0$, or $\lambda^2-3\lambda+1=0$. It possesses two real roots; the smaller of the two is $\lambda_1 = (3-\sqrt{5})/2 = 1 - \Phi^{-1} \approx 0.382$ and the larger is $\lambda_2 = (3+\sqrt{5})/2 = 1 + \Phi \approx 2.618$. Remember that the golden ratio $\Phi = (1+\sqrt{5})/2 \approx 1.618$ obeys the relation $\Phi^2 - \Phi = 1$, or, equivalently $\Phi = 1 + \Phi^{-1}$. It is easy to see (using the properties of the golden ratio) that the eigenvectors associated with the preceding eigenvalues are respectively (they are not normalized):

$$\boldsymbol{V}_1 = (-\Phi, 1) \quad ; \quad \boldsymbol{V}_2 = (\Phi^{-1}, 1)$$

Therefore $M\boldsymbol{V}_1 = \lambda_1 \boldsymbol{V}_1$, which corresponds to a contraction for the vector, and $M\boldsymbol{V}_2 = \lambda_2 \boldsymbol{V}_2$, which corresponds to a dilation. It can be also easily checked that $\boldsymbol{V}_1 \cdot \boldsymbol{V}_2 = -\Phi\Phi^{-1} + 1 \times 1 = 0$, which establishes the orthogonality of these vectors (the M matrix being symmetric, this is a natural property). Consequently, the mapping gives rise to a dilation by a factor 2.618 in the direction of the vector $\boldsymbol{V}_2$ and to a contraction by a factor 0.382 in the direction of the vector $\boldsymbol{V}_1$. Under the mapping, an isosceles rectangle triangle whose sides are directed along the directions of these vectors remains a rectangle triangle, since the image of the eigenvectors are still parallel to the eigenvectors and thus their directions remain orthogonal. However, for the iteration labeled i, one of the sides suffers a dilation by a factor λ_2^i while the other suffers a contraction by a factor λ_1^i. The triangle is thus stretched and becomes progressively thinner, preserving nevertheless its area. This property is illustrated in Fig. 8.12, where we represented the original triangle and its first two iterations.

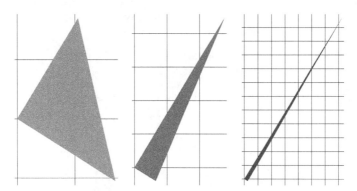

Fig. 8.12 On the left, we have represented an isosceles rectangle triangle whose sides are directed along the directions of dilation and contraction. The picture in the middle shows the first iteration by the Anosov mapping, and the right picture the second iteration. For clarity, we did not perform the congruence operations, and changed the scale

Since the mapping is linear, it transforms a straight segment into a straight segment. Moreover two parallel segments are transformed into two parallel segments. On the other hand, the mapping, while conserving areas, conserves neither angles nor lengths.

A rectangle is transformed into a parallelogram.

The most general form for a conic, in particular a circle, is a quadratic form. Because of linearity, its transforms are always quadratic forms representing conics which are of course deformed (but with the same area) as compared to the original. For a circle, which is a closed loop with a finite area, the transform is a conic with a closed loop and finite area, i.e., an ellipse.

A circle is transformed into an ellipse.

Of course, all these properties are valid before the congruence operations that bring the image back into the reference square.

4. It is well known in linear algebra that a change of basis which transforms the original matrix M to its diagonal matrix $D = \begin{pmatrix} \lambda_1 & 0 \\ 0 & \lambda_2 \end{pmatrix}$ is accomplished via a matrix built on the eigenvectors

$$S = \begin{pmatrix} -\Phi & \Phi^{-1} \\ 1 & 1 \end{pmatrix} :$$

$$D = S^{-1}MS.$$

To conform ourself to the nomenclature proposed in the statement, we write this property in the form $M = U^{-1}DU$, with the obvious change $U^{-1} = S$, therefore $U = S^{-1}$, so that the inverse matrix can be calculated with the standard technique to give:

$$U = \begin{pmatrix} -1/(\Phi + \Phi^{-1}) & \Phi^{-1}/(\Phi + \Phi^{-1}) \\ 1/(\Phi + \Phi^{-1}) & \Phi/(\Phi + \Phi^{-1}) \end{pmatrix}.$$

Finally the transformation can be summarized as:

$$M = U^{-1}DU$$

$$U^{-1} = \begin{pmatrix} -\Phi & \Phi^{-1} \\ 1 & 1 \end{pmatrix}$$

$$U = \begin{pmatrix} -1/(\Phi + \Phi^{-1}) & \Phi^{-1}/(\Phi + \Phi^{-1}) \\ 1/(\Phi + \Phi^{-1}) & \Phi/(\Phi + \Phi^{-1}) \end{pmatrix}.$$

Written in such a form, the rth power of the matrix can be straightforwardly calculated: $M^r = (U^{-1}DU)^r = U^{-1}D^rU$, in which

$$D^r = \begin{pmatrix} \lambda_1^r & 0 \\ 0 & \lambda_2^r \end{pmatrix}.$$

Problem Solutions

Using the expressions of the various matrices, one obtains finally:

$$M^r = \begin{pmatrix} \dfrac{\Phi(1-\Phi^{-1})^r + \Phi^{-1}(1+\Phi)^r}{\Phi + \Phi^{-1}} & \dfrac{(1+\Phi)^r - (1-\Phi^{-1})^r}{\Phi + \Phi^{-1}} \\ \dfrac{(1+\Phi)^r - (1-\Phi^{-1})^r}{\Phi + \Phi^{-1}} & \dfrac{\Phi^{-1}(1-\Phi^{-1})^r + \Phi(1+\Phi)^r}{\Phi + \Phi^{-1}} \end{pmatrix}.$$

This matrix is symmetric. Since all the elements of the M matrix are integers, the same property must hold for the elements of the M^r matrix given just above. The Φ and Φ^{-1} numbers being irrational, the latter property is by no means obvious for the form of M^r given previously. Fortunately we will see in the next question that this is indeed the case.

5. Using the property $\det(M^r) = (\det(M))^r$, it is immediately seen that $\det(M^r) = 1$. Nevertheless, it is instructive to build explicitly the matrix M^r. One may first remark that $M = \begin{pmatrix} a_1 & a_2 \\ a_2 & a_3 \end{pmatrix}$. Let us assume that

$$M^r = \begin{pmatrix} a_{2r-1} & a_{2r} \\ a_{2r} & a_{2r+1} \end{pmatrix}$$

and use a recursion method.

$$M^{r+1} = MM^r = \begin{pmatrix} 1 & 1 \\ 1 & 2 \end{pmatrix} \begin{pmatrix} a_{2r-1} & a_{2r} \\ a_{2r} & a_{2r+1} \end{pmatrix}$$
$$= \begin{pmatrix} a_{2r-1} + a_{2r} & a_{2r} + a_{2r+1} \\ a_{2r-1} + 2a_{2r} & a_{2r} + 2a_{2r+1} \end{pmatrix}$$

Owing to the recursion definition of the Fibonacci sequence, one has $a_{2r-1} + a_{2r} = a_{2r+1} = a_{2(r+1)-1}$, $a_{2r} + a_{2r+1} = a_{2r+2} = a_{2(r+1)}$, and $a_{2r-1} + 2a_{2r} = a_{2r} + a_{2r+1} = a_{2r+2} = a_{2(r+1)}$, $a_{2r} + 2a_{2r+1} = a_{2r+1} + a_{2r+2} = a_{2r+3} = a_{2(r+1)+1}$.

These relations allow us to write the matrix

$$M^{r+1} = \begin{pmatrix} a_{2(r+1)-1} & a_{2(r+1)} \\ a_{2(r+1)} & a_{2(r+1)+1} \end{pmatrix}$$

in the desired form. The recursion property is demonstrated. Thus

$$M^r = \begin{pmatrix} a_{2r-1} & a_{2r} \\ a_{2r} & a_{2r+1} \end{pmatrix}$$

Under this form, we can check first that all its elements are integers, and secondly that its determinant is, as expected, $a_{2r+1}a_{2r-1} - a_{2r}^2 = 1$, as shown in the first question.

6. We are seeking for fixed points of order r in the most general way. We must satisfy the equations $\theta_r - \theta_0 = 2\pi k$, $p_r - p_0 = 2\pi l$.

Let us introduce, as suggested in the statement, the reduced quantities $x = \theta/(2\pi)$, $y = p/(2\pi)$. With these conventions, we need to solve the matrix equation

$$(M^r - I)\begin{pmatrix} x_0 \\ y_0 \end{pmatrix} = \begin{pmatrix} k \\ l \end{pmatrix}.$$

We are faced with a traditional linear system; with the expression for the M^r matrix deduced in the preceding question and using the properties of Fibonacci's sequence, a short calculation leads to the solution:

$$x_0 = \frac{la_{2r} - k(a_{2r+1} - 1)}{a_{2r+1} + a_{2r-1} - 2}$$

$$y_0 = \frac{ka_{2r} - l(a_{2r-1} - 1)}{a_{2r+1} + a_{2r-1} - 2}.$$

7. To avoid redundancies, we must impose the conditions that both θ_0 and p_0 lie in the reference square; in other words we impose the restrictions $0 \leq x_0 < 1$ and $0 \leq y_0 < 1$. Translated in terms of the x_0 and y_0 expressions just obtained, these constraints are written as:

$$\frac{a_{2r+1} - 1}{a_{2r}} k \leq l < \frac{a_{2r+1} - 1}{a_{2r}} k + \frac{a_{2r+1} + a_{2r-1} - 2}{a_{2r}}$$

$$\frac{a_{2r}}{a_{2r-1} - 1} k - \frac{a_{2r+1} + a_{2r-1} - 2}{a_{2r-1} - 1} \leq l < \frac{a_{2r}}{a_{2r-1} - 1} k.$$

The first set of inequalities shows that l is comprised between two parallel straight lines with slope $(a_{2r+1} - 1)/a_{2r}$, while the second set shows that the same number is comprised between two parallel straight lines with slope $a_{2r}/(a_{2r-1} - 1)$. This means that l lies inside a parallelogram. This situation is illustrated in Fig. 8.13.

8. All the integer values (k, l) which lie inside the parallelogram are suitable. To obtain an estimation of this number, it is sufficient to divide the area of this parallelogram by the average area needed for a single pair (k, l).

This latter area is just unity since an elementary square cell with side 1 in a net whose nodes are placed on half-integer values $(1/2, 3/2, 5/2, \ldots)$ contains one and only one set of (k, l) numbers.

We must now calculate the area S of the parallelogram. Let us designate by $\boldsymbol{OA}(a_{2r}, a_{2r+1} - 1)$ the vector along one side and $\boldsymbol{OC}(a_{2r-1} - 1, a_{2r})$ the vector along the other side. The area is just given by $S = |\boldsymbol{OA} \times \boldsymbol{OC}|$. The calculation of this vector product presents no difficulty: one finds $S = a_{2r+1} + a_{2r-1} - 2$.

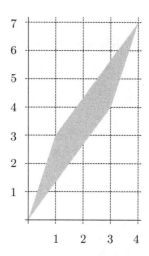

Fig. 8.13 In the plane (k, l), the grey region represents the parallelogram which contains the fixed points of Anosov's mapping. This situation corresponds to second order fixed points

This number corresponds also to the estimated number of fixed points of order r:

$$N_r = a_{2r+1} + a_{2r-1} - 2.$$

Let us emphasize nevertheless that, in this number, are counted all the fixed points of the order of a divisor of r. It is also instructive to remark that the number of fixed points tends to infinity exponentially with r, and that the parallelogram becomes more and more flat and stretched because the slopes of the parallel lines tend to a common value which is precisely the golden ratio.

9. To study the stability of fixed points of order r, one must linearize the mapping in their neighbourhood. In our particular case, the mapping is already linear so that the linearization reveals nothing new and is obviously identical with the M^r matrix itself. Its trace is equal to $T = a_{2r+1} + a_{2r-1}$. It is easy to prove that one always has

$$a_{2r+1} + a_{2r-1} > 2.$$

Consequently all the fixed points are unstable.

Let us notice also that, except $(0,0)$, there exists no fixed point with $l = 0$ (this is obvious from the definition of the parallelogram). In consequence, the momenta inevitably increase at each iteration; we are dealing with acceleration modes.

10. Let us study the fixed points of order 2. Since $a_3 = 2$, $a_4 = 3$, one calculates $a_5 = 5$. We expect to find about $5 + 2 - 2 = 5$ fixed points, among them being $(0,0)$, the fixed point of order 1. In our particular case $r = 2$ the fixed points are given by $x_0 = (3l - 4k)/5$, $y_0 = (3k - l)/5$ and the borders of the parallelogram by $\frac{4}{3}k \leq l < \frac{4}{3}k + \frac{5}{3}$, $3k - 5 \leq l < 3k$.

An exhaustive counting satisfying these conditions reveals that the fixed points of order 2 are the following:

$(0,0)$; $(1/5, 3/5)$; $(2/5, 1/5)$; $(3/5, 4/5)$; $(4/5, 2/5)$.

We count 5 such points, which is exactly the estimated number.

8.9. Fermi's Accelerator
[Statement and Figure p. 405]

1. One denotes by $L'(t) = dL(t)/dt$, $L''(t) = d^2L(t)/dt^2$ the first and second derivatives of the displacement law for the right hand wall.

   At the instant t_n, the particle bounces on the moving wall and returns towards the fixed wall with a velocity whose modulus is v_n, relative to the moving wall. At that time, the moving wall is at a distance $L(t_n)$ from the fixed wall and its velocity is equal to $L'(t_n)$. The relative velocity of the particle is $-v_n$ and, relying on the velocity addition law, the velocity is $u_n = -v_n + L'(t_n)$ in the Galilean frame of the fixed wall. The time needed to travel between the two walls is thus

   $$\frac{L(t_n)}{|u_n|} = \frac{L(t_n)}{v_n - L'(t_n)}.$$

   Therefore, the instant of the impact on the left wall is simply $t_{n+1} = t_n + L(t_n)/(v_n - L'(t_n))$. After the impact, the velocity of the particle is $u_{n+1} = -u_n = v_n - L'(t_n)$. Since the left wall is fixed, this velocity is also the velocity relative to the left wall $v_{n+1} = u_{n+1}$. So, the transformation law reads:

   $$t_{n+1} = t_n + \frac{L(t_n)}{(v_n - L'(t_n))}$$
   $$v_{n+1} = v_n - L'(t_n).$$

   We work in the space (t, E) considering, instead of v, the energy $E = \frac{1}{2}mv^2$ relative to the wall on which the particle rebounds. Let us find first the Jacobian of the transformation

   $$J = \frac{D(t_{n+1}, E_{n+1})}{D(t_n, E_n)}.$$

   Relying on the known properties of Jacobians, one can write

   $$J = \frac{D(t_{n+1}, E_{n+1})}{D(t_{n+1}, v_{n+1})} \times \frac{D(t_{n+1}, v_{n+1})}{D(t_n, v_n)} \times \frac{D(t_n, v_n)}{D(t_n, E_n)}.$$

Problem Solutions

It is easy to check that

$$\frac{D(t_{n+1}, E_{n+1})}{D(t_{n+1}, v_{n+1})} = mv_{n+1} \quad \text{and similarly} \quad \frac{D(t_n, v_n)}{D(t_n, E_n)} = \frac{1}{mv_n}.$$

Lastly $\dfrac{D(t_{n+1}, v_{n+1})}{D(t_n, v_n)} = (\partial_{t_n} t_{n+1})(\partial_{v_n} v_{n+1}) - (\partial_{v_n} t_{n+1})(\partial_{t_n} v_{n+1}).$

An elementary calculation provides

$$\partial_{t_n} t_{n+1} = 1 + \frac{L'(t_n)}{v_{n+1}} + \frac{L(t_n) L''(t_n)}{v_{n+1}^2},$$

$$\partial_{v_n} v_{n+1} = 1, \quad \partial_{v_n} t_{n+1} = -\frac{L(t_n)}{v_{n+1}^2}, \quad \text{and}$$

$$\partial_{t_n} v_{n+1} = -L''(t_n).$$

These expressions allow us to find

$$\frac{D(t_{n+1}, v_{n+1})}{D(t_n, v_n)} = 1 + \frac{L'(t_n)}{v_n - L'(t_n)} = \frac{v_n}{v_{n+1}}$$

and lead to the conclusion that:

$$\frac{D(t_{n+1}, E_{n+1})}{D(t_n, E_n)} = 1.$$

The area is preserved in this phase space.

2. Now, t_n is the instant at which the particle bounces on the left wall and returns with velocity v_n; it reaches the moving wall at the instant t_{n+1} when its distance is $L(t_{n+1})$. Since the particle travels at constant speed v_n between the two impacts, the corresponding time is $L(t_{n+1})/v_n$ and one has $t_{n+1} = t_n + L(t_{n+1})/v_n$.

At the moment of impact on the moving wall, the absolute velocity is v_n and the driving velocity of the wall $L'(t_{n+1})$. The relative velocity is $u_n = v_n - L'(t_{n+1})$. After the impact, the relative velocity changes its sign to become $u_{n+1} = -u_n$, then, from our definition, $v_{n+1} = |u_{n+1}| = v_n - L'(t_{n+1})$. To summarize, the required transformation is:

$$t_{n+1} = t_n + \frac{L(t_{n+1})}{v_n}$$

$$v_{n+1} = v_n - L'(t_{n+1}).$$

To calculate the Jacobian of the transformation, we adopt the same reasoning as before. We will nevertheless take care of the fact that the quantities $L'(t_{n+1})$ and $L(t_{n+1})$ are themselves functions of t_n and v_n through the variable t_{n+1}.

Explicitly, one finds

$$\partial_{t_n} t_{n+1} = \frac{v_n}{v_{n+1}}, \qquad \partial_{v_n} v_{n+1} = 1 + \frac{L(t_{n+1})L''(t_{n+1})}{v_n v_{n+1}},$$

and

$$\partial_{v_n} t_{n+1} = -\frac{L(t_{n+1})}{v_n v_{n+1}}, \qquad \partial_{t_n} v_{n+1} = -\frac{v_n L''(t_{n+1})}{v_{n+1}}.$$

These relations allow us to calculate $D(t_{n+1}, v_{n+1})/D(t_n, v_n) = v_n/v_{n+1}$ and thus, as in the previous question:

$$\frac{D(t_{n+1}, E_{n+1})}{D(t_n, E_n)} = 1.$$

Once again, the area in phase space is preserved.

3. One can imagine the presence of a fictitious wall, between the two moving walls, on which the particle arrives at the intermediate time T with the absolute velocity V and an absolute energy E. The situation between the right moving wall with variables (t_n, E_n) and the fictitious wall with variables (T, E) has been studied in Question 1; we proved that the Jacobian of the transformation is unity

$$\frac{D(T, E)}{D(t_n, E_n)} = 1.$$

Furthermore, the situation between the fictitious wall with variables (T, E) and the left moving wall with variables (t_{n+1}, E_{n+1}) has been examined in Question 2 where we demonstrated that the Jacobian is unity

$$\frac{D(t_{n+1}, E_{n+1})}{D(T, E)} = 1.$$

The transformation connecting the impacts between the two moving walls is described by the passage from the variables (t_n, E_n) to the variables (t_{n+1}, E_{n+1}). The Jacobian of the transformation $D(t_{n+1}, E_{n+1})/D(t_n, E_n)$ may be evaluated as

$$\frac{D(t_{n+1}, E_{n+1})}{D(T, E)} \times \frac{D(T, E)}{D(t_n, E_n)}.$$

Owing to the previous remarks, this value is unity. Thus, in the general case of two moving walls, one still has

$$\frac{D(t_{n+1}, E_{n+1})}{D(t_n, E_n)} = 1.$$

There is conservation of the area in phase phase.

An illustration of our arguments is given in Fig. 8.14.

Problem Solutions

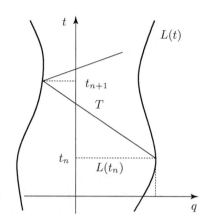

Fig. 8.14 Rebounds between two moving walls, introducing a fictitious fixed wall

4. With the conventions proposed in the statement, there exists a fixed point if after a multiple of T, the velocity v_{n+1}, after the impact $(n+1)$, is identical to that after the impact n, i.e., v_n. The relative velocity just before the impact on the moving wall is $u_n = v_n - L'(t_c)$, and after it is $u_{n+1} = -u_n = L'(t_c) - v_n$. The absolute velocity is therefore $u_{n+1} + L'(t_c) = 2L'(t_c) - v_n$. The impact on the fixed wall changes its sign $v_{n+1} = v_n - 2L'(t_c)$ so that the condition for existence of a fixed point $v_{n+1} = v_n$ implies:

$$L'(t_c) = 0.$$

Furthermore, again from the condition $v_{n+1} = v_n$, one also has $t_{n+1} = t_c + L_c/v_n$ and $t_n = t_c - L_c/v_n$. Taking the half sum of these two relations, we arrive at the desired equation:

$$t_c = \frac{t_n + t_{n+1}}{2}.$$

This particular situation is represented in Fig. 8.15.

Let us denote as t_0 the instant of impact on the fixed wall, t_c that on the moving wall, and t_1 the following instant of impact on the fixed wall; $v_0 = v_1$ are the characteristic velocities for the fixed point. One sets also

$$L_c = L(t_c), \quad L'_c = L'(t_c) = 0, \quad L''_c = L''(t_c).$$

Let us choose now a starting point on the fixed wall very close to the fixed point $t_i = t_0 + \varepsilon_0$, $v_i = v_0 + \eta_0$ which lead to the values $t'_c = t_c + \varepsilon_c$, $t_f = t_1 + \varepsilon_1$, $v_f = v_1 + \eta_1$. We first have $v_f = v_0 - 2L'(t_c + \varepsilon_c)$, which implies

$$\varepsilon_c = \frac{\eta_1 - \eta_0}{2L''_c}.$$

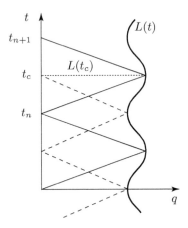

Fig. 8.15 Fixed points of the mapping between the fixed wall and the moving wall

A first equation is $t_i = t'_c - L(t'_c)/v_i$ which provides a first relation

$$\eta_1 = -2L''_c \varepsilon_0 + \left(1 + 2\frac{L_c L''_c}{v_0^2}\right)\eta_0.$$

A second equation gives $t_f = t'_c + L(t'_c)/v_f$ which provides

$$\varepsilon_1 = \frac{\eta_0}{2L''_c} - \left(\frac{1}{2L''_c} + \frac{L_c}{v_0^2}\right)\eta_1,$$

which, after replacing η_1 by its value calculated just above, is recast as

$$\varepsilon_1 = \varepsilon_0 - \left(-2\frac{L_c}{v_0^2}\right)\left(1 + \frac{L_c L''_c}{v_0^2}\right)\eta_0.$$

We have in hand all the ingredients necessary for the linearization of the mapping around the fixed point and we obtain the matrix M such that

$$\begin{pmatrix}\varepsilon_1\\ \eta_1\end{pmatrix} = M \begin{pmatrix}\varepsilon_0\\ \eta_0\end{pmatrix}.$$

The trace of this matrix is equal to $T = 2 + 2L_c L''_c/v_0^2$. There is stability of the fixed point if $-2 < T < 2$, which implies $L_c L''_c < 0$ and $L_c L''_c > -2v_0^2 = -4E/m$, so that the stability condition is summarized by the inequalities:

$$-\frac{4E}{m} < L_c L''_c < 0.$$

5. We denote as t_n the instant for impact n on the moving wall, and as t_c the instant for the next impact on the fixed wall. We already saw that $v_{n+1} = v_n - 2L'(t_{n+1})$. We have also $t_c = t_n + L(t_n)/v_n$ and $t_{n+1} = t_c + L(t_{n+1})/v_n$.

If the distance between the two walls varies slightly between the two impacts, it is legitimate to make the approximation $L(t_n) \approx L(t_{n+1}) = L$. After adding the two previous relations, we arrive at $t_{n+1} = t_n + 2L/v_n$. Thus, the Ulam mapping is written as

$$t_{n+1} = t_n + \frac{2L}{v_n}$$
$$v_{n+1} = v_n - 2L'(t_{n+1}).$$

Instead of working in phase space (t, E), we now choose the phase space (t, v). We seek the Jacobian of the transformation

$$\frac{D(t_{n+1}, v_{n+1})}{D(t_n, v_n)}.$$

With the preceding expression for the transformation, it is easy to calculate $\partial_{t_n} t_{n+1} = 1$, $\partial_{v_n} t_{n+1} = -2L/v_n^2$ and $\partial_{t_n} v_{n+1} = -2L''(t_{n+1})$, $\partial_{v_n} v_{n+1} = 1 + 4LL''(t_{n+1})/v_n^2$ and to check that the Jacobian is indeed unity:

$$\frac{D(t_{n+1}, v_{n+1})}{D(t_n, v_n)} = 1.$$

The Ulam mapping is very similar to the standard mapping. Indeed, assume that the velocity varies only slightly around a mean velocity $\bar{v}$; then $2L/v_n = (2L/\bar{v}^2)\,v_n$. Now, let us perform the replacement $t \to \theta$ and $v \to (\bar{v}^2/2L)\,p$. The first Ulam equation becomes $\theta_{n+1} = \theta_n + p_n$, which is identical with the first equation of the standard mapping. Multiply the second Ulam equation by $2L/\bar{v}^2$ and assume that the wall velocity follows a sinusoidal law $L'(t) = \sin t$; this second equation is transformed into $p_{n+1} = p_n + K\sin(\theta_{n+1})$, with $K = -4L/\bar{v}^2$. It is also identical to the second equation of the standard mapping. Under these conditions, the Ulam mapping is fully equivalent to the standard mapping.

8.10. Damped Pendulum and Standard Mapping [Statement p. 407]

1. The system is one-dimensional, the angle θ of the pendulum with the vertical being the only degree of freedom. The kinetic energy is easy to calculate, $T = \frac{1}{2}ml^2\dot{\theta}^2$. We need to calculate the generalized forces and, to do this, we perform a virtual displacement $\delta\theta$. We first consider the weight, $\boldsymbol{P} = m\boldsymbol{g}$, which produces a virtual work $\delta W = \boldsymbol{P} \cdot \delta\boldsymbol{r} = -Pl\sin\theta\,\delta\theta = Q_p\,\delta\theta$ whence the corresponding expression for the force $Q_p = -mgl\sin\theta$. This force arises from the potential $V(\theta) = -mgl\cos\theta$.

However there exists an additional generalized force due to the friction, which is precisely the moment of the friction forces $Q_f(\dot{\theta}) = -k\dot{\theta}$. This force cannot arise from a simple potential of the form $V(\theta)$ because $-\partial_\theta V(\theta)$ would be a function of θ alone, whereas Q_f is a function of $\dot{\theta}$ alone (if $k \neq 0$). Let us investigate the possibility of a generalized potential $V(\theta, \dot{\theta})$. In this case, one must satisfy the equation

$$-k\dot{\theta} = \frac{d}{dt}\left(\partial_{\dot\theta} V(\theta,\dot\theta)\right) - \partial_\theta V(\theta,\dot\theta) = \dot{\theta}\partial^2_{\theta\dot\theta}V + \ddot{\theta}\partial^2_{\dot\theta^2}V - \partial_\theta V.$$

The fact that the force does not depend on $\ddot{\theta}$ implies that $V(\theta, \dot{\theta}) = f(\theta)\dot{\theta} + g(\theta)$. The desired relation is now equivalent to $-k\dot{\theta} = g'(\theta)$, which cannot be satisfied if $k \neq 0$. The friction forces therefore do not arise from a potential, even a generalized one.

The system is not a Lagrangian system.

Nevertheless, one can safely employ the Lagrangian formulation which, in this case, is written as a Lagrange equation $d\left(\partial_{\dot\theta}T\right)/dt - \partial_\theta T = Q_p + Q_f$. It leads to the differential equation

$$ml^2\ddot\theta + k\dot\theta + mgl\sin\theta = 0.$$

It can be checked directly from Newton's equation by equating the time derivative of the angular momentum with the moment of the applied forces.

2. Let us divide the Lagrange equation by ml^2 and introduce the proper angular frequency of the pendulum $\omega = \sqrt{g/l}$; it then reads

$$\ddot\theta + \frac{k}{ml^2}\dot\theta + \omega^2\sin\theta = 0.$$

We perform the change of variable $\tau = \omega t$; the angle is now a function of τ: $\theta(\tau)$. Let us set $\theta'(\tau) = d\theta(\tau)/d\tau$, $\theta''(\tau) = d^2\theta(\tau)/d\tau^2$. One can check that $\dot\theta = \omega\theta'$, $\ddot\theta = \omega^2\theta''$. Dividing the preceding equation by ω^2 and introducing $\gamma = k/(2m\omega l^2)$, the differential equation to be solved is written in the simple form:

$$\theta'' + 2\gamma\theta' + \sin\theta = 0.$$

3. At the equilibrium point, one has $\theta = $ const, hence $\theta'' = \theta' = 0$ and thus $\sin\theta = 0$. Therefore there exist two equilibrium points:

$$\theta_0 = 0; \qquad \theta_0 = \pi.$$

Consider a point close to the solution $\theta_0 = 0$ (lower position of the pendulum) and set $\theta = \varepsilon$; then $\sin\theta \approx \varepsilon$ and the differential equation becomes $\varepsilon'' + 2\gamma\varepsilon' + \varepsilon = 0$. The solution is well known. Since $\gamma < 1$, the solution is

$$\varepsilon(\tau) = e^{-\gamma\tau}\left[a\cos\left(\sqrt{1-\gamma^2}\tau\right) + b\sin\left(\sqrt{1-\gamma^2}\tau\right)\right].$$

The system tends toward the equilibrium point but oscillates around it.

$$\theta(\tau) = e^{-\gamma\tau}\left[a\cos\left(\sqrt{1-\gamma^2}\tau\right) + b\sin\left(\sqrt{1-\gamma^2}\tau\right)\right].$$

The point $\theta_0 = 0$ is thus a stable equilibrium point.

Close to $\theta_0 = \pi$ (higher equilibrium point), one sets $\theta = \pi + \varepsilon$. The equation becomes $\varepsilon'' + 2\gamma\varepsilon' - \varepsilon = 0$. There exist two real roots, one negative $r_1 = -\gamma - \sqrt{1+\gamma^2}$, the other positive $r_2 = -\gamma + \sqrt{1+\gamma^2}$. The general solution is:

$$\theta(\tau) = ae^{r_1\tau} + be^{r_2\tau} + \pi.$$

The solution increases exponentially with time, so that this equilibrium point is unstable.

The differential equation can be recast as a set of two coupled differential equations. Setting $p = d\theta/d\tau = \theta'$, $\theta'' = dp/d\tau = p'$ and the equation is equivalent to the system:

$$\theta' = p$$
$$p' = -2\gamma p - \sin\theta.$$

4. One performs stroboscopic observations with period $\delta\tau$ and one sets $\tau_n = n\delta\tau$, $\theta(\tau_n) = \theta_n$. The derivatives are expressed as $\theta'_n = \theta'(\tau_n) = (\theta_{n+1} - \theta_{n-1})/(2\delta\tau)$, $\theta''_n = \theta''(\tau_n) = (\theta_{n+1} + \theta_{n-1} - 2\theta_n)/(\delta\tau)^2$. The differential equation is then discretized at point τ_n and written as

$$\frac{\theta_{n+1} + \theta_{n-1} - 2\theta_n}{(\delta\tau)^2} + \gamma\frac{\theta_{n+1} - \theta_{n-1}}{\delta\tau} + \sin\theta_n = 0.$$

As suggested in the statement, one defines the momenta via the identities $p_n = (\theta_{n+1} - \theta_n)/\delta\tau$. The latter equation can also be written as $\theta_{n+1} = \theta_n + \delta\tau\, p_n$. The differential equation arising from Lagrange's equation is put in the form $(p_n - p_{n-1})/\delta\tau + \gamma(p_n + p_{n-1}) + \sin\theta_n = 0$, which, after rearranging similar terms and multiplying by $\delta\tau$, gives the recursion relation $(1 + \gamma\delta\tau)p_n = (1 - \gamma\delta\tau)p_{n-1} - \delta\tau\sin\theta_n$. Finally, the transformation in phase space takes the definitive form:

$$\theta_{n+1} = \theta_n + \delta\tau\, p_n$$
$$(1+\gamma\delta\tau)p_{n+1} = (1-\gamma\delta\tau)p_n - \delta\tau\sin\theta_{n+1}.$$

The Jacobian of the transformation is explicitly equal to

$$J = (\partial_{\theta_n}\theta_{n+1})(\partial_{p_n}p_{n+1}) - (\partial_{\theta_n}p_{n+1})(\partial_{p_n}\theta_{n+1}).$$

With $\partial_{\theta_n}\theta_{n+1} = 1$, $\partial_{p_n}p_{n+1} = (1-\gamma\delta\tau)/(1+\gamma\delta\tau) - (\delta\tau)^2\cos\theta_{n+1}/(1+\gamma\delta\tau)$ and $\partial_{\theta_n}p_{n+1} = -\delta\tau\cos\theta_{n+1}/(1+\gamma\delta\tau)$, $\partial_{p_n}\theta_{n+1} = \delta\tau$, it is easy to demonstrate that the Jacobian is equal to:

$$J = \frac{1-\gamma\delta\tau}{1+\gamma\delta\tau}.$$

There is conservation of the area if $J = 1$, that is if $\gamma = 0$, or if friction is absent.

Consider this latter case and denote $P_n = \delta\tau\, p_n$. The transformation can now be written:

$$\begin{aligned}\theta_{n+1} &= \theta_n + P_n \\ P_{n+1} &= P_n - (\delta\tau)^2\sin\theta_{n+1}.\end{aligned}$$

This transformation is identical to the standard mapping if one makes the identification $K = -(\delta\tau)^2$.

5. The stable fixed point is $\theta_0 = 0, p_0 = 0$. One chooses a starting point (ε, η) close to this point and one linearizes the mapping given in the preceding question to obtain its image. The transformation matrix is written explicitly

$$M = \begin{pmatrix} 1 & \delta\tau \\ -\dfrac{\delta\tau}{1+\gamma\delta\tau} & \dfrac{1-\gamma\delta\tau-\delta\tau^2}{1+\gamma\delta\tau} \end{pmatrix}.$$

The trace is equal to $T = (2-(\delta\tau)^2)/(1+\gamma\delta\tau)$ and its determinant is $D = (1-\gamma\delta\tau)/(1+\gamma\delta\tau)$. The characteristic equation $\lambda^2 - T\lambda + D = 0$ admits two complex conjugate roots σ, σ^*. The corresponding eigenvectors for the M matrix are X, Y. Therefore $X_{n+1} = \sigma X_n$ and $Y_{n+1} = \sigma^* Y_n$. Since $|\sigma| = |\sigma^*| = \sqrt{D}$, one deduces

$$\left|\frac{X_{n+1}}{X_n}\right| = \left|\frac{Y_{n+1}}{Y_n}\right| = \sqrt{D}.$$

D being less than 1, these latter equalities prove that $X_n \to 0, Y_n \to 0$ following successive iterations. The quantities (ε, η) are related to (X, Y) by a linear transformation and it follows that $\varepsilon_n \to 0, \eta_n \to 0$ with time.

Whatever the starting point (close to the fixed point), the mapping brings it progressively to the fixed point iteration after iteration. This fixed point is known as an attractor, a focus or a spiral point.

Problem Solutions

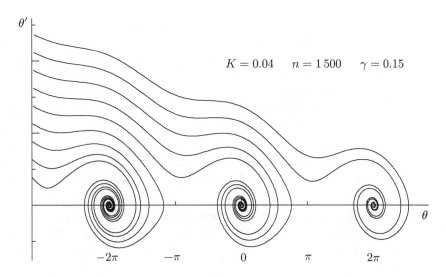

Fig. 8.16 Poincaré section corresponding to the damped pendulum, starting from 9 different initial conditions. The various parameters are indicated in the figure

6. The Poincaré section corresponding to the damped pendulum is represented in Fig. 8.16 for a number of iterations equal to 1500, with the parameters $K = 0.04$ and $\gamma = 0.15$, and for 9 different initial conditions.

8.11. Stability of Periodic Orbits on a Billiard Table [Statement and Figure p. 409]

1. We choose an origin O and a point A on the cush which is specified by its curvilinear abscissa $s = \overparen{OA}$. At point A is defined the unit vector along the tangent $t(s)$, oriented in the sense of increasing s, the unit vector along the normal $n(s)$, oriented towards the interior of the table, and the unit vector perpendicular to the plane of the table z, such that the trihedron (t, n, z) is direct.

Let B be a point close to A, such that $ds = \overparen{AB}$, where the tangent is t' and the normal n'. The directions along n and n' intersect at a point C. The circle of center C and radius $R = CA$ is called the osculatory circle, and R is the radius of curvature at the point A. The convention is such that
$$d\boldsymbol{t} = \boldsymbol{t}' - \boldsymbol{t} = \frac{ds}{R}\boldsymbol{n} = \frac{ds}{R}(\boldsymbol{z} \times \boldsymbol{t}).$$
With such a definition, the radius of curvature is positive, $R > 0$, if concavity is directed inward, and negative, $R < 0$, if it is directed outward.

At the impact $(n-1)$, the ball attains the cush at point A with abscissa s_{n-1} and, **after** the impact, the reflection angle is i_{n-1}. The second coordinate is defined as $p_{n-1} = \sin i_{n-1} = (\boldsymbol{v}_{n-1})_t/|\boldsymbol{v}_{n-1}|$ (tangential component of the velocity normalized to 1). At the next impact, the ball touches the cush at point B, with abscissa s_n, after which the reflection angle is i_n and the other coordinate p_n.

Let us set $\boldsymbol{L} = \boldsymbol{AB}$ and let $\boldsymbol{D} = \boldsymbol{L}/L$ be the unit vector along the direction of propagation between two impacts. It is easy to show that, in all cases, $p_n = \boldsymbol{t}_n \cdot \boldsymbol{D}$. It is also convenient to introduce the notation $q_{n-1} = \boldsymbol{n}_{n-1} \cdot \boldsymbol{D}$. Be careful because $q_n = -\boldsymbol{n}_n \cdot \boldsymbol{D}$. Moreover we have the relation $q_n^2 + p_n^2 = 1$.

Let A' be a point close to A, with abscissa $s'_{n-1} = s_{n-1} + ds_{n-1}$ and B', a point close to B, with abscissa $s'_n = s_n + ds_n$. One has $\boldsymbol{A'B'} = \boldsymbol{L'} = \boldsymbol{L} + d\boldsymbol{L}$; it is easy to show that $d\boldsymbol{L} = ds_n \boldsymbol{t}_n - ds_{n-1}\boldsymbol{t}_{n-1}$, then, from $\boldsymbol{L} \cdot d\boldsymbol{L} = L\,dL$, that $dL = \boldsymbol{D} \cdot d\boldsymbol{L} = ds_n p_n - ds_{n-1} p_{n-1}$. It can be shown also, for instance using the vector $\boldsymbol{N} = \boldsymbol{z} \times \boldsymbol{D}$ and with $\boldsymbol{t}_n = p_n \boldsymbol{D} + q_n \boldsymbol{N}$, that $\boldsymbol{t}_n \cdot \boldsymbol{t}_{n-1} = p_n p_{n-1} - q_n q_{n-1}$. Similarly, one can calculate

$$d\boldsymbol{D} = \frac{d\boldsymbol{L}}{L} - \frac{\boldsymbol{D}\,dL}{L} = \frac{ds_n}{L}\boldsymbol{t}_n - \frac{ds_{n-1}}{L}\boldsymbol{t}_{n-1} - \boldsymbol{D}\frac{ds_n p_n - ds_{n-1}p_{n-1}}{L}.$$

After these rather tedious preliminary calculations, one can deduce

$$dp_{n-1} = d(\boldsymbol{t}_{n-1} \cdot \boldsymbol{D}) = d\boldsymbol{t}_{n-1} \cdot \boldsymbol{D} + \boldsymbol{t}_{n-1} \cdot d\boldsymbol{D}$$
$$= ds_{n-1}\frac{q_{n-1}}{R_{n-1}} - ds_{n-1}\frac{q_{n-1}^2}{L} - ds_n \frac{q_n q_{n-1}}{L},$$

a formula that can be inverted to find:

$$ds_n = \frac{1}{q_n}\left(\frac{L}{R_{n-1}} - q_{n-1}\right)ds_{n-1} - \frac{L}{q_n q_{n-1}}dp_{n-1}.$$

A very similar calculation leads to

$$dp_n = q_n\left(\frac{q_n}{L} - \frac{1}{R_n}\right)ds_n + \frac{q_n q_{n-1}}{L}ds_{n-1}.$$

If ds_n is replaced by its value obtained previously, one arrives at

$$dp_n = \frac{q_{n-1}R_{n-1} + q_n R_n - L}{R_n R_{n-1}}ds_{n-1} + \frac{(L/R_n - q_n)}{q_{n-1}}dp_{n-1}.$$

We finally succeed in writing the transformation in the required form

$$\begin{pmatrix} ds_n \\ dp_n \end{pmatrix} = M \begin{pmatrix} ds_{n-1} \\ dp_{n-1} \end{pmatrix},$$

Problem Solutions

with the following expression[10] for the transformation matrix M:

$$M = \begin{pmatrix} \dfrac{1}{q_n}\left(\dfrac{L}{R_{n-1}} - q_{n-1}\right) & -\dfrac{L}{q_n q_{n-1}} \\ \dfrac{1}{R_n R_{n-1}}(q_{n-1} R_{n-1} + q_n R_n - L) & \dfrac{1}{q_{n-1}}\left(\dfrac{L}{R_n} - q_n\right) \end{pmatrix}.$$

The calculation of the determinant requires some care. Expanding all the terms, it can be expressed in the form $\det(M) = (q_n q_{n-1})/(q_n q_{n-1}) = 1$. Therefore, one obtains

$$\det(M) = 1.$$

The transformation preserves the area in phase space (s_n, p_n).

2. The fixed points (s_n, p_n) of the mapping do not change from one impact to the other; this means that the ball bounces perpetually between point A_1 (where the curvature radius is R_1) and point A_2 (where the curvature radius is R_2). Moreover, the angle of incidence at each impact is 0, which implies $p_1 = p_2 = 0$, $q_1 = q_2 = 1$. To study the stability of these fixed points, one must linearize the mapping corresponding to one period in the neighbourhood of these fixed points. Furthermore, $M(\text{period}) = M(\text{backward}) M(\text{forward})$. Using the convenient notation $X_i = (L/R_i) - 1$, the above considerations allow us to simplify the M matrices which are written:

$$M(\text{forward}) = \begin{pmatrix} X_1 & -L \\ \dfrac{1 - X_1 X_2}{L} & X_2 \end{pmatrix}$$

and

$$M(\text{backward}) = \begin{pmatrix} X_2 & -L \\ \dfrac{1 - X_1 X_2}{L} & X_1 \end{pmatrix}$$

In fact, we are interested only in the trace of $M(\text{period})$ and it is sufficient to calculate the diagonal elements and take their sum. One finds $T = [X_1 X_2 - 1 + X_1 X_2] + [-1 + X_1 X_2 + X_1 X_2] = 2(2X_1 X_2 - 1)$. The trajectory is stable if this trace lies in the interval $]-2, +2[$. This condition is equivalent to $0 < X_1 X_2 < 1$, or:

$$0 < \left(\dfrac{L}{R_1} - 1\right)\left(\dfrac{L}{R_2} - 1\right) < 1.$$

---

[10] One can arrive more quickly at the result using the calculation of Problem 4.13, valid for reflections on two parallel planes, and remarking that the curvature induces a change of the incidence angle by a quantity ds/R.

3. For an elliptic billiard table and close to the minor axis, one has $R_1 = R_2 > 0$ and $L < 2R_1 = 2R_2$. One has the property $0 < L/R_i < 2$ and therefore $-1 < L/R_i - 1 < 1$ whence

$$0 < \left(\frac{L}{R_1} - 1\right)\left(\frac{L}{R_2} - 1\right) = \left(\frac{L}{R_1} - 1\right)^2 < 1.$$

This trajectory fulfills the stability criterion; it is stable.

On the contrary, for a trajectory close to the major axis, one still has $R_1 = R_2 > 0$, but this time $L > 2R_1 = 2R_2$ from which one deduces $1 < (L/R_1 - 1)(L/R_2 - 1) = (L/R_1 - 1)^2$. The trajectory is unstable. The discussion concerning this type of trajectory merits a more detailed development according to the passage or not through the focus.

In the case of rebounds on two convex surfaces, one has $R_i < 0$, hence $L/R_i - 1 < -1$ and thus $(L/R_1 - 1)(L/R_2 - 1) > 1$ which implies an unstable trajectory. This is, in particular, the case of the Sinaï billiard table for which all the surfaces are convex. This type of table has been extensively studied, particularly for its interest concerning chaotic quantum spectra.

8.12. Lagrangian Points: Jupiter's Greeks and Trojans [Statement and Figure p. 412]

1. Let us consider first the two heavy objects, with masses m_2 and m_3. We know that they follow paths which are homothetic to that of a fictive mass which rotates around a circle of center O. The considered motion is circular; thus object 2 rotates around O at a distance d_2 with an angular velocity ω and object 3 rotates around O at a distance d_3 with the same angular velocity ω. In this way, the two objects remain always aligned with O and at a fixed distance D from one another.

The gravitational force experienced by an object due to the influence of the other is Gm_2m_3/D^2; this force is equal to the mass of the considered object multiplied by the centripetal acceleration. Therefore $m_2 d_2 \omega^2 = Gm_2m_3/D^2 = m_3 d_3 \omega^2$. An elementary calculation provides first $\omega^2 = MG/D^3$. Then, replacing the masses in terms of μ and ω^2 by the value just obtained, one finds the required distances $d_2 = \frac{1}{2}(1+\mu)D$ and $d_3 = \frac{1}{2}(1-\mu)D$. It can be checked that these values are compatible with the position of the center of mass at the origin. In summary, one has

$$\omega = \sqrt{\frac{MG}{D^3}}; \qquad d_2 = \frac{1}{2}(1+\mu)D; \qquad d_3 = \frac{1}{2}(1-\mu)D.$$

2. Let us work in the frame centered at the center of mass O, and which rotates with angular velocity ω (with respect to a Galilean frame). In this frame the two heavy objects are at rest and the straight line which connects them is chosen as the Ox axis.

Now, one considers an asteroid of mass m, submitted to the gravitational influence of the two heavy bodies. The corresponding potential is written $V(r_2, r_3) = -mm_2 G/r_2 - mm_3 G/r_3$. Since the distance r_i between the asteroid and the object i is a function of the coordinates x, y of the asteroid in this frame, the potential can also be expressed as $V(x, y)$. In this frame the heavy bodies are located at a fixed position and the gravitational interaction potential for the asteroid does not depend explicitly on time.

The price to pay for such a simplification, as we saw in Problem 4.3, is the addition of a supplementary contribution $-\omega L_z$ (in our particular case L_z is the component of the angular momentum of the asteroid on an axis perpendicular to the rotation plane) to the original Hamiltonian H_0. The latter Hamiltonian is just $H_0 = (p_x^2 + p_y^2)/(2m) + V(x, y)$ and the component of the angular momentum $L_z = xp_y - yp_x$. We obtain the final expression of the Hamiltonian:

$$H(x, y, p_x, p_y) = \frac{1}{2m}\left(p_x^2 + p_y^2\right) - \omega(xp_y - yp_x) + V(x, y).$$

Let us notice, as we saw in Problem 4.3, that the centrifugal term is absorbed in the angular momentum term $L_z = xp_y - yp_x$ which is different from the kinetic momentum $\sigma_z = m(x\dot{y} - y\dot{x})$.

3. The Hamiltonian is time independent $\partial H/\partial t = 0 = dH/dt$ and its value remains constant on the trajectory, this constant being known as Jacobi's constant:

$$H(x, y, p_x, p_y) = E = \text{const}.$$

The system has two degrees of freedom, but only one first integral. A priori, it is not integrable.

Hamilton's equations give respectively $\dot{x} = \partial_{p_x} H = p_x/m + \omega y$, $\dot{y} = \partial_{p_y} H = p_y/m - \omega x$, and $\dot{p}_x = -\partial_x H = \omega p_y - \partial_x V$, $\dot{p}_y = -\partial_y H = -\omega p_x - \partial_y V$. Denoting the components of the gravitational attraction as $f_x = -\partial_x V$, $f_y = -\partial_y V$, these equations can be written in the form:

$$\dot{x} = \frac{p_x}{m} + \omega y; \qquad \dot{y} = \frac{p_y}{m} - \omega x$$
$$\dot{p}_x = \omega p_y + f_x; \qquad \dot{p}_y = -\omega p_x + f_y.$$

4. At the fixed points, one has $\dot{x} = \dot{y} = 0$ that is $p_x = -m\omega y$, $p_y = m\omega x$. Differentiating, one finds further $\dot{p}_x = -m\omega \dot{y} = 0 = m\omega^2 x + f_x$ and $\dot{p}_y = m\omega \dot{x} = 0 = m\omega^2 y + f_y$. Consequently, one obtains the equations determining the fixed points

$$f_x(x,y) = -m\omega^2 x, \qquad f_y(x,y) = -m\omega^2 y.$$

Owing to the definitions $f_x = -\partial_x V$, $f_y = -\partial_y V$ and

$$-m\omega^2 x = -\frac{1}{2}m\omega^2 \partial_x(x^2+y^2), \qquad -m\omega^2 y = -\frac{1}{2}m\omega^2 \partial_y(x^2+y^2),$$

the equilibrium conditions are equivalent to the search for the extremum $\partial_x V_e(x,y) = 0$, $\partial_y V_e(x,y) = 0$ of an effective potential

$$V_e(x,y) = V(x,y) - \frac{1}{2}m\omega^2(x^2+y^2).$$

Instead of the variables x, y, it is more convenient[11] to retain the variables r_2, r_3. We already saw that

$$V(r_2, r_3) = -\frac{mMG}{2}\left(\frac{1-\mu}{r_2} + \frac{1+\mu}{r_3}\right).$$

Now, we must calculate $x^2 + y^2 = r^2$, the square of the distance of the asteroid to the origin. From the well known properties of triangles, one has first $r_2^2 = r^2 + d_2^2 - 2rd_2\cos\alpha$ where α is the angle between Ox and the direction of the asteroid. Similarly, one has $r_3^2 = r^2 + d_3^2 - 2rd_3\cos(\pi-\alpha)$. Extracting the value $\cos\alpha$ from the first equation and substituting it in the second one, one obtains after rearranging $r^2 = (d_2 r_3^2 + d_3 r_2^2)/D - d_2 d_3$, or

$$r^2 = \frac{1}{2}(1+\mu)r_3^2 + \frac{1}{2}(1-\mu)r_2^2 - \frac{1}{4}(1-\mu^2)D^2.$$

We have in hand all the ingredients required to express the effective potential in the form $V_e(r_2, r_3)$. The equilibrium condition $\partial_{r_2} V_e(r_2, r_3) = 0$ provides the relation $MG = \omega^2 r_2^3$, or, using the value of ω^2 found in the first question, $r_2 = D$. The second equilibrium condition $\partial_{r_3} V_e(r_2, r_3) = 0$ gives similarly $r_3 = D$. In the frame under consideration, there exist two equilibrium positions, such that the asteroid forms with the heavy objects an equilateral triangle. These particular points are known as Lagrangian points.

---

[11] There exist also Lagrangian points on the line that connects the heavy bodies. In this case, r_2 and r_3 are no longer independent variables and the treatment presented here is no longer valid.

Problem Solutions

The coordinates of these points are easily obtained. Consider the point with a positive ordinate. Its abscissa is just in the middle of the segment joining the heavy bodies, that is

$$x_L = \frac{1}{2}\left[\frac{(1+\mu)D}{2} - \frac{(1-\mu)D}{2}\right] = \frac{\mu D}{2}.$$

Its ordinate corresponds to the height of the equilateral triangle, namely $y_L = \sqrt{3}D/2$. The momenta follow from $p_{x_L} = -m\omega y_L$ and $p_{y_L} = m\omega x_L$. To summarize, one finds:

$$x_L = \frac{1}{2}\mu D; \qquad y_L = \frac{1}{2}\sqrt{3}D$$

$$p_{x_L} = -\frac{1}{2}m\omega\sqrt{3}D; \qquad p_{y_L} = \frac{1}{2}m\omega\mu D.$$

In contrast with usual situations, the values of the momenta at the equilibrium points do not vanish.

At these points (in the rest of this question, we drop the index L for typographical simplicity), one has $p_x = -m\omega y$, $p_y = m\omega x$ and hence

$$\frac{p_x^2 + p_y^2}{2m} = \frac{1}{2}m\omega^2(x^2+y^2) = \frac{1}{8}m\omega^2 D^2(\mu^2+3).$$

One has also $-\omega(xp_y - yp_x) = -m\omega^2(x^2+y^2) = -\frac{1}{4}m\omega^2 D^2(\mu^2+3)$ and lastly $V(x,y) = -m\omega^2 D^2$. Gathering the three contributions, one obtains the values of E, explicitly:

$$E = -\frac{1}{8}m\omega^2 D^2(\mu^2+11).$$

5. Let us start from a point $x = x_L + X$, $y = y_L + Y$ close to the equilibrium point ($X/D, Y/D \ll 1$) and seek the expression of the potential at this point, restricting ourselves to first order in $X/D, Y/D$.

One has

$$\frac{1}{\sqrt{(X+a)^2 + (Y+b)^2}} =$$

$$= L^{-1}\left(1 + \frac{2aX+2bY}{L^2} + \frac{X^2+Y^2}{L^2}\right)^{-1/2}$$

$$\approx L^{-1}\left[1 - \frac{2aX+2bY}{2L^2} - \frac{X^2+Y^2}{2L^2} + \frac{3(2aX+2bY)^2}{8L^4}\right]$$

$$= \frac{1}{L} - \frac{aX+bY}{L^3} + \frac{(2a^2-b^2)X^2 + (2b^2-a^2)Y^2 + 6abXY}{2L^5}$$

which is the desired expansion. To simplify, let us set $\tilde{X} = X/D$, $\tilde{Y} = Y/D$. One must first calculate

$$\begin{aligned} r_3 &= \sqrt{(x + (1-\mu)D/2)^2 + y^2} \\ &= \sqrt{(x_L + X + (1-\mu)D/2)^2 + (y_L + Y)^2} \\ &= \sqrt{(X + D/2)^2 + \left(Y + D\sqrt{3}/2\right)^2}. \end{aligned}$$

We apply the preceding formula with $a = D/2$, $b = D\sqrt{3}/2$; in this case $L = D$. One obtains

$$\frac{1}{r_3} = D^{-1}\left(1 - \frac{1}{2}\tilde{X} - \frac{\sqrt{3}}{2}\tilde{Y} - \frac{1}{8}\tilde{X}^2 + \frac{5}{8}\tilde{Y}^2 + \frac{3\sqrt{3}}{4}\tilde{X}\tilde{Y}\right).$$

Now, one has

$$\begin{aligned} r_2 &= \sqrt{(x - (1+\mu)D/2)^2 + y^2} \\ &= \sqrt{(x_L + X - (1+\mu)D/2)^2 + (y_L + Y)^2} \\ &= \sqrt{(X - D/2)^2 + (Y + D\sqrt{3}/2)^2}. \end{aligned}$$

We apply the expansion with now $a = -D/2$, $b = D\sqrt{3}/2$. Compared to the previous case, it is sufficient to change a into $-a$, thus changing the sign of the odd powers of $\tilde{X}$. One obtains now

$$\frac{1}{r_2} = D^{-1}\left(1 + \frac{1}{2}\tilde{X} - \frac{\sqrt{3}}{2}\tilde{Y} - \frac{1}{8}\tilde{X}^2 + \frac{5}{8}\tilde{Y}^2 - \frac{3\sqrt{3}}{4}\tilde{X}\tilde{Y}\right).$$

A last calculation leads to the expression of the potential:

$$V(X,Y) = -m\omega^2 D^2 + m\omega^2 \left[\frac{1}{2}\mu DX + \frac{\sqrt{3}}{2} DY + \frac{1}{8}\left(X^2 - \lambda XY - 5Y^2\right)\right]$$

with the value $\lambda = 6\sqrt{3}\mu$.

6. Let us now set $p_x = p_{x_L} + U$, $p_y = p_{y_L} + V$. Starting from the elementary Poisson brackets on the original variables x, y, p_x, p_y, it can be easily shown that we also have the elementary Poisson brackets for the new variables $\{X,Y\} = 0 = \{U,V\} = \{X,V\} = \{Y,U\}$, and $\{X,U\} = 1 = \{Y,V\}$. These properties are sufficient to ensure the canonicity for this change of variables.

Problem Solutions

A long but straightforward calculation allows us to write the Hamiltonian in terms of the new variables:

$$H(X,Y,U,V) = E + \frac{U^2 + V^2}{2m} - \omega(XV - YU)$$
$$+ \frac{m\omega^2}{8}(X^2 - 5Y^2 - \lambda XY).$$

Since the transformation is canonical, there is invariance of the form of Hamilton's equations. Therefore $\dot{X} = \partial_U H = U/m + \omega Y$, $\dot{Y} = \partial_V H = V/m - \omega X$, and $\dot{U} = -\partial_X H = \omega V - m\omega^2(2X - \lambda Y)/8$, $\dot{V} = -\partial_Y H = -\omega U + m\omega^2(\lambda X + 10Y)/8$. The new Hamilton's equations are:

$$\dot{X} = \frac{U}{m} + \omega Y; \qquad \dot{Y} = \frac{V}{m} - \omega X$$
$$\dot{U} = \omega V - \frac{1}{8}m\omega^2(2X - \lambda Y); \qquad \dot{V} = -\omega U + \frac{1}{8}m\omega^2(\lambda X + 10Y).$$

We are thus faced with coupled first order differential equations. The method to solve them is well known: one searches for the proper modes in the form $X = X_0 e^{i\Omega t}$, $Y = Y_0 e^{i\Omega t}$, $U = U_0 e^{i\Omega t}$, $V = V_0 e^{i\Omega t}$. Substituting these values in the differential equations, we obtain a linear system which possesses a non trivial solution only if the determinant of the corresponding matrix vanishes, a property that translates as:

$$\begin{vmatrix} -i\Omega & \omega & 1/m & 0 \\ -\omega & -i\Omega & 0 & 1/m \\ -m\omega^2/4 & \lambda m\omega^2/8 & -i\Omega & \omega \\ \lambda m\omega^2/8 & 5m\omega^2/4 & -\omega & -i\Omega \end{vmatrix} = 0.$$

Setting $r = \Omega/\omega$, this condition is equivalent to the characteristic bisquared equation: $r^4 - r^2 + (27/16 - \lambda^2/64) = 0$.

7. There is stability for the motion if the root is a real number, in particular if $r^2 > 0$. This means that r^2 is also a real number. Since the sum of the roots is equal to 1, at least one of the two roots is positive. The required condition is therefore that the discriminant is positive, that is $1 - 4(27/16 - \lambda^2/64) > 0$, or $1 - 27/4 + \lambda^2/16 > 0$ or $\lambda^2 > 92$. Using the relation between λ and μ, this inequality is equivalent to

$$\mu^2 > \frac{23}{27}.$$

For the Sun-Jupiter system $\mu = 1048/1050$ and $\mu^2 = 0.996 > 23/27 = 0.851$. The asteroids, Greeks and Trojans, at the Lagrangian points of Jupiter's orbits, follow stable trajectories.

Bibliography

In this bibliography, we deliberately restricted ourselves to the textbooks or papers that we effectively consulted and used practically. The subject of this book being very broad, it is obvious that the following list is far from exhaustive and there certainly exist many other excellent books. A more complete bibliography can be found as references in some of the textbooks mentioned below.

Textbooks that are recommended for undergraduate students which provide a good overview

L. Landau and E. Lifchitz, *Mechanics*, Pergamon Press, 1969.

   Unavoidable basic textbook.

V. Arnold, *Mathematical Methods of Classical Mechanics*, Springer, 2nd edition, 1989.

   Reference textbook which, and this is an additional quality for specialists, introduces and uses the tools of differential geometry.

H. Goldstein, *Classical Mechanics*, Addison-Wesley, Londres, 2nd edition, 1980.

   Basic textbook for point mechanics and continuous medium mechanics. It is very complete, especially for relativitic aspects. The discussion concerning chaos is absent.

H. Goldstein, C. Poole and J. Fasko, *Classical Mechanics*, Addison-Wesley, Londres, 3rd edition, 2002.

   This third edition of the famous "Goldstein" book takes advantage of the experience of two new authors for renewing some obsolete aspects of the previous editions, in particular concerning relativistic topics. This edition addresses the notion of chaos.

I. Percival and D. Richards, *Introduction to Dynamics*, Cambridge University Press, 1982.

Excellent textbook that we recommend for its clarity and its pedagogical emphasis, written by specialists in this domain.

J.V. José and E.J. Saletan, *Classical Dynamics*, Cambridge University Press, 1998.

This very pedagogical textbook takes time to explain and illustrate every new notion. Its contains many explicit diagrams and figures and is a source of many interesting and miscellaneous exercices.

S. Hildebrandt and A. Tromba, *Mathématiques et formes optimales*, Belin, Pour la Science, 1986 (in French).

A real masterpiece of clarity which explains the various aspects of the least action principle.

G.L. Kotkin and U. Serbo, *Collection of Problems in Classical Mechanics*, Pergamon Press, 1971.

An excellent source of very beautiful original problems on various aspects of classical mechanics.

I. Stewart, *Does God Play Dice?*, Penguin, new edition, 1990.

Remarkable popular work concerning chaos, understandable to all. It addresses all interesting aspects, including those which go beyond Hamiltonian systems.

Poincaré, collection "Les génies de la science", Pour la Science, n°4, November 2000 (in french).

Useful for all.

Textbooks for undergraduate and graduate students which we used as reference works concerning the summaries of this book

C. Lanczos, *The Variational Principles of Mechanics*, University of Toronto Press, 1970.

Very marginal work, which deals with subjects generally forgotten by other authors and which is, consequently, an invaluable complement.

H.G. Schuster, *Deterministic Chaos: An Introduction*, VCH Verlagsgellschaft Germany, 1987.

This book easy to read, illustrated with many figures, is an introduction to chaos; it is not restricted to Hamiltonian systems.

Bibliography

M. Tabor, *Chaos and Integrability in Nonlinear Dynamics*, John Wiley & Sons, New York, 1988.

Very concise and complete, this textbook has the advantage of avoiding sophisticated mathematical tools.

A.J. Lichtenberg and M.A. Lieberman, *Regular and Stochastic Motion*, Springer-Verlag, 1983.

This book is addressed to specialists.

E. Ott, *Chaos in Dynamical Systems*, Cambridge University Press, 1993.

This clear and complete textbook is concerned with general dynamical systems and thus goes beyond the ambition of the present work.

M.C. Gutzwiller, *Chaos in Classical and Quantum Mechanics*, Springer-Verlag, New York Inc., 1990.

A book for students interested in quantum mechanics. Nevertheless, the part devoted to classical mechanics is well presented.

E.N. Lorentz, *Essence of Chaos*, University of Washington Press, 1995.

Easily readable, this book addresses all topics concerning chaos.

F. Scheck, *Mechanics: From Newton's Law to Deterministic Chaos*, Springer-Verlag, 3rd edition, 1999.

A book close to ours in spirit, but which employs more abstract mathematical notations.

W. Yourgrau and S. Mandelstam, *Variational Principles in Dynamics and Quantum Theory*, Pitman, 3rd edition, 1968.

Original and epistemological analysis of Hamilton's theory. Many historical references and connections with other domains in physics.

N. Rasband, *Dynamics*, John Wiley & Sons, 1983.

Many aspects are addressed in this book, but the mathematical notations are abstract.

K.T. Alligood, T.D. Sauer and J.A. Yorke, *Chaos, an Introduction to Dynamical Systems*, Springer-Verlag, 1997.

This book is clear and complete, but reserved for students with a good mathematical training.

Books which were consulted occasionally to address very special points. The interested reader may thus go further by looking at them

L. Landau and E. Lifchitz, *The Classical Theory of Fields*, Pergamon Press, 1994.

L. Landau and E. Lifchitz, *Theory of Elasticity*, Pergamon Press, 1986.

J.D. Jackson, *Classical Electrodynamics*, John Wiley & Sons, New York, 1962.

A. Messiah, *Quantum Mechanics*, North Holland, 1963.

L.E. Ballentine, *Quantum Mechanics*, World Scientific Publishing Co., 1998.

B. Cagnac and J.C. Pebay- Peyroula, *Modern Atomic Physics*, MacMillan interacting publishing, 1975.

J. Bass, *Cours de mathématiques*, Masson, 1977 (in French).

J.W.S. Rayleigh, *Theory of Sound*, Dover, 1945.

R.P. Feynman, *Lectures on Physics*, Addison-Wesley Publishing Company, 1963.

S. Weinberg, *Gravitation and Cosmology*, John Wiley & Sons, 1972.

G. Bruhat, *Optique*, Masson, 1992 (in French).

W.H. Press, S.A. Teukolsky, W.T. Vetterling and B.P. Flannery, *Numerical Recipes*, Cambridge University Press, 1992.

R. Campbell, *Théorie générale de l'équation de Mathieu et de quelques autres équations*, Masson, 1955 (in French).

C. Rosensweig and J.B. Krieger, *Exact Quantization Conditions*, *J. Math. Phys.*, 9, 849, 1968.

F.L. Moore, J.C. Robinson, C.F. Barucha, Bala Sundaram and M.G. Raizen, Atom Optics Realisation of the Quantum Delta-kicked Rotor, *Phys. Rev. Lett.* 75, 4598, 1995.

H. Stapelfeldt and T. Seideman, *Aligning Molecules with Strong Laser Pulses*, *Rev. Mod. Phys.* 75, 543, 2003.

INDEX

acceleration mode, 401, 425, 437
acoustical frequency, 75, 108
action, 112
    function, 233, 239
    functional, 112
    variable, 284
adiabatic invariant, 341, 346
Aharonov–Bohm effect, 246
d'Alembert principle, 12, 14
angle–action variables, 283
Anosov's mapping, 403, 432
aphelion, 223, 229
areal velocity, 56
Arnold's cat, 403, 432
atomic chain, 75, 107
attractor, 446
aurora borealis, 354, 379
autonomous, 167
axial frequency, 95
axle, 19, 39

bead, 16, 28, 186, 224, 356, 382
bifurcation, 187, 228, 394, 398, 419
billiard, 183, 409, 447
Binet equation, 56, 133, 134, 143,
    163, 222, 228
blade, 71, 102
boost, 304
brachistochrone, 148

calculus of variations, 114
canonical
    perturbation, 342, 348
    transformation, 283, 334
caustic, 243, 264

centrifugal force, 23, 50
chaos, 390, 392, 395
chaos–ergodicity, 399, 423
Compton, 23, 47
conjugate
    point, 119, 138
    variable, 52
conservative, 167
constant of the motion, 53, 167,
    282, 288
constraint equation, 40
continuous fraction, 396, 417
convergent
    direction, 394
    separatrix, 394
Coriolis force, 12, 17, 23, 49
Coulomb problem, 302
Curie principle, 125
cyclic coordinate, 53
cycloid, 31
cyclotron
    frequency, 96, 352, 376
    motion, 379
    radius, 379

damped pendulum, 407, 443
declination angle, 69
degree of freedom, 10
divergent separatrix, 394
drift, 66, 67, 91, 354, 381

ecliptic, 68, 98
elastic
    bar, 64
    moment, 71

electromagnetic potential, 51
electrostatic lens, 243, 265
ellipsoid, 150
elliptic
    coordinate, 248, 276
    integral, 349
    point, 168
energy, 53, 167
equinox precession, 68, 97

Fermat
    path, 236
    principle, 122, 144, 181
Fermi accelerator, 405, 438
Fibonacci sequence, 403, 435
field, 112
fine structure, 293
    constant, 292, 294, 318
first integral, 281, 288
fixed point, 169, 390, 392
flexion vibration, 71, 102
flow, 167, 204
    parameter, 287, 304, 337
Foucault pendulum, 59, 79
free fall, 242, 261
friction force, 12

Galilean transformation, 304, 338
general action, 253
generalized
    acceleration, 10
    coordinate, 9
    force, 12
    momentum, 52, 165
    potential, 51
    velocity, 10
generating function, 289, 297, 298,
        301, 302, 324, 330, 332,
        353, 375
generator, 284, 287, 303, 336
geodesic, 241, 261
group
    speed, 239
    velocity, 252, 279

gyroscope, 21, 44, 46

Hamilton
    equation, 167
    function, 166
    principle, 111
Hamilton–Jacobi, 233, 235
Hamiltonian, 166
Hannay's phase, 356, 382
harmonic oscillator, 295, 298, 302
Heiles, 62
Hénon, 62
Hénon and Heiles potential, 84
heterocline point, 395
holonomic, 13, 40
    constraint, 18, 116, 123
homocline point, 395
hoop, 16, 28, 186, 224
Huygens
    construction, 238, 243
    pendulum, 17, 31
hyperbolic point, 168

index, 236
inertial force, 23, 48
integrable system, 281
integral constraint, 115, 159
involution, 282, 323
isochronous, 17

Jacobi theorem, 236
Jupiter's
    Greeks, 412, 450
    Trojans, 412, 450

KAM
    curve, 390
    theorem, 391
Kepler problem, 174, 292, 295, 314
kicked rotor, 385
kinetic energy, 10
Koenig theorem, 18, 76, 99

Lagrange

Index 463

equation, 10, 52
function, 52
multiplier, 13, 19, 41, 115, 150, 159
Lagrangian, 52
point, 412, 413, 450, 452
system, 52
Landau levels, 312
Laplace law, 127, 158
least action principle, 111, 118, 135
Legendre transform, 166
libration, 227
Liouville theorem, 167
Lorentz
force, 116, 131, 200
transformation, 304, 339

magnetic
flux, 246
forces, 12
moment, 378
magnetron frequency, 96
Mathieu, 211
equation, 180
Maupertuis principle, 121, 141, 235, 245, 268
Mercury, 128, 162
Minkowski metric, 176
mixture, 395

Noether theorem, 54, 58, 78
non-resonant torus, 389

optical
frequency, 75, 108
path, 214

Painlevé integral, 58, 77
parabolic coordinate, 247, 271
parametric resonance, 170
Paul trap, 180, 211
Penning trap, 67, 94
perihelion, 197, 223, 229
perturbation theory, 341, 342

phase
portrait, 37, 168, 190, 220, 227, 229
space, 165
speed, 239
velocity, 252, 279
Poincaré section, 183, 388
Poincaré–Birkhoff theorem, 393
Poisson bracket, 281, 327, 329
precession of perihelia, 129, 164
prolate shape, 249
propagator, 169, 208, 212
proper mode, 57

quadrupolar approximation, 249
quadrupole interaction, 97
quantum, 112
quasi-integrable system, 341

reaction force, 16, 19, 28
reduced
action, 121, 234, 239, 253
mass, 55
resonant torus, 286, 390, 396, 415
reverse pendulum, 178, 207
revisiting theorem, 168
rope, 16, 27
rotating frame, 65, 89, 172, 192
rule EBK, 285
Runge–Lenz vector, 173, 195
Rydberg constant, 294, 318

saddle point, 120
sawtooth mapping, 402, 428
scalar potential, 51
scale invariance, 395
Schwarzschild
metric, 129, 162
radius, 129, 164
secular term, 348, 362
self-similarity, 395
separation of variables, 236
separatrice, 168
shearing modulus, 64

sidereal day, 69
sine-Gordon equation, 73, 105
sling, 15, 26
small denominators, 345
Snell–Descartes law, 144, 215
soap film, 125, 154
"soft" mode, 63, 86
solitary wave, 73, 105
Sommerfeld atom, 293, 316
spiral point, 446
square well, 351
stability islet, 392
stable
    elliptic fixed point, 394
    node, 168
standard mapping, 387, 398, 401, 402, 418, 425, 428
Stark effect, 247, 271
surface tension, 127, 158

Toda net, 62
torus, 282

trajectory, 11
transversal wave, 64, 88
turn indicator, 21, 43
turning point, 168, 274, 278

Ulam
    approximation, 407
    mapping, 443
unstable
    hyperbolic point, 394
    node, 168

vector potential, 51
virial theorem, 186, 223, 230
virtual
    displacement, 12
    work, 12

wave front, 242, 243, 263, 264
wheel jack, 14, 24

Young modulus, 72

CPSIA information can be obtained
at www.ICGtesting.com
Printed in the USA
LVHW10*1055220818
587750LV00008BA/54/P